ENGINEERING LIBRARY

Planning and Design of Engineering Systems

D0083629

This comprehensive introduction to the scope and nature of engineering offers students a commonsense approach to the solution of engineering problems. Case studies and real-world examples are used to illustrate the role of the engineer, the type of work involved and the methodologies employed in engineering practice.

It focuses on civil engineering design and problem solving, but also more generally covers creativity and problem solving, social and environmental issues, management, communications, law and ethics. Its scope runs from the planning, design, modelling and analysis phases to the implementation or construction phase. It begins by outlining a conceptual framework for undertaking engineering projects then provides a range of techniques and tools for solving the sorts of problems that commonly arise.

It is an extensively revised and extended new edition which has been written for introductory courses in undergraduate engineering programs. It is also intended for non-specialist readers who seek information on the nature of engineering work and how it is carried out.

The authors are all members of the School of Civil and Environmental Engineering at the University of Adelaide.

Also available from Taylor & Francis

Risk Management in Projects

Martin Loosemore, John Raftery, Charles Reilly, David Higgon

Hb: ISBN 978-0-415-26055-8
Pb: ISBN 978-0-415-26056-5

Reliability-Based Design in Geotechnical Engineering

Kok Kwang Phoon

Hb: ISBN 978-0-415-39630-1

Examples in Structural Analysis

William M.C. McKenzie

Hb: ISBN 978-0-415-37053-0
Pb: ISBN 978-0-415-37054-7

Project Management Demystified 3rd Edition

Geoff Reiss

Pb: ISBN 978-0-415-42163-8

Introduction to Design for Civil Engineers

A.W. Beeby & R.S. Narayanan

Pb: ISBN 978-0-419-23550-7

Project Planning and Control

David G. Carmichael

Hb: ISBN 978-0-415-34722-6

Structures: From Theory to Practice

Alan Jennings

Hb: ISBN 978-0-415-26842-4
Pb: ISBN 978-0-415-26843-1

Planning and Design of Engineering Systems

Second Edition

Graeme Dandy, David Walker, Trevor Daniell and Robert Warner

LONDON AND NEW YORK

First published 1989
by Unwin Hyman Ltd.
Reprinted 2000 by E & FN Spon

Second edition published 2008
by Taylor & Francis
2 Park Square, Milton Park, Abingdon, Oxon OX14 4RN

Simultaneously published in the USA and Canada
by Taylor & Francis
270 Madison Ave, New York, NY 10016, USA

Taylor & Francis is an imprint of the Taylor & Francis Group, an informa business

© 2008 Graeme Dandy, David Walker, Trevor Daniell and Robert Warner

This book has been prepared from camera-ready copy provided by the authors
Printed and bound in Great Britain by The Cromwell Press, Trowbridge, Wiltshire

All rights reserved. No part of this book may be reprinted or reproduced or utilised in any form or by any electronic, mechanical, or other means, now known or hereafter invented, including photocopying and recording, or in any information storage or retrieval system, without permission in writing from the publishers.

The publisher makes no representation, express or implied, with regard to the accuracy of the information contained in this book and cannot accept any legal responsibility or liability for any efforts or omissions that may be made.

British Library Cataloguing in Publication Data
A catalogue record for this book is available from the British Library

Library of Congress Cataloging in Publication Data
Planning and design of engineering systems / Graeme Dandy . . .
[et al.]. -- 2nd ed.
p. cm.
Includes bibliographical references and index.
ISBN 978-0-415-40551-5 (hardback : alk. paper) --
ISBN 978-0-415-40552-2 (pbk. : alk. paper)
1. Engineering. 2. Engineering design. I. Dandy, G. C.
TA145.D235 2007
620′.0042--dc22
2007000017

ISBN10: 0–415–40551–3 Hardback
ISBN10: 0–415–40552–1 Paperback
ISBN10: 0–203–96080–7 ebook

ISBN13: 978–0–415–40551–5 Hardback
ISBN13: 978–0–415–40552–2 Paperback
ISBN13: 978–0–203–96080–6 ebook

Contents

PREFACE

Planning and design are the key activities which, together with management, allow any engineering project to be taken from the initial concept stage through to successful implementation. Each engineering project, whether in the most traditional or the newest developing field, relies for its success on the application of the basic processes of planning, design and management. Our prime aim in this book is to show how the processes of planning and design are carried out.

However, the underlying purpose of the book is somewhat broader: to explain the nature of engineering and to describe the type of work engineers undertake. The book therefore deals with the problems engineers are called on to solve, and, most importantly, the simple, common sense methodologies that are used to solve engineering problems. It also describes some quantitative tools that are used in undertaking the work of engineering planning and design.

The book has been written for students who are commencing their studies of engineering, and for lecturers who are presenting classes to students in the early semesters of an engineering course program. It is also intended for non-specialist readers who seek information on the nature of engineering work and how it is carried out. Some of the more advanced material in the later chapters may be presented in the later years of an engineering degree program.

The purpose of the book has not changed substantially from that of the first edition, which appeared in the late 1980s. Since then, however, the need for introductory courses in undergraduate engineering programs has become more generally recognised by engineering academics. Courses which provide an introduction to engineering and to engineering planning and design are now an integral part of engineering undergraduate programs.

This book retains much of the material from the first edition, but it has been extensively extended, revised and updated. In particular, the chapters on creativity, problem solving, and on social and environmental aspects have been extensively rewritten and expanded. There are new chapters on management, communications, law and ethics. In our treatment of management we deal not only with project management but also with team work, team building and with the management of people. We have included additional case studies and examples to highlight the application of the underlying concepts. By including a number of extension exercises that require wide reading, sometimes from original sources, we have also attempted to challenge our students and to help them prepare for life-long learning.

The text emphasises the fact that engineering is an essential part of the framework of our society, and is therefore closely linked with various social, political, legal and environmental questions and problems.

The first edition of the book has served as the text for an introductory subject on engineering planning and design. This has been included in the first-year syllabus of all engineering courses at the University of Adelaide for many years, as well as at other institutions worldwide. Some of the later chapters are treated in subsequent years of our civil engineering course. The extensive rewriting we have undertaken for the second edition reflects our ongoing experience in presenting the material, as well as significant changes that have occurred in engineering itself.

The majority of the examples presented in the book come from the field of civil and environmental engineering, which is the area of engineering expertise of the authors. Nevertheless, we believe that the ideas and content are generally applicable and relevant to all fields of engineering, and beyond.

The material contained in this new edition can be used in various ways. We have divided the major portion of the book into two parts: the first ten chapters set out a conceptual framework for undertaking engineering projects, while Chapters 11 to 13 provide a range of techniques and tools for solving the sorts of problems that commonly arise.

The conceptual framework provided by Chapters 1–10 is designed to introduce engineering as a profession, to set out what engineering work involves and how it is undertaken, and to describe an approach to engineering problem-solving that is both flexible and formal. We see all ten chapters as being important for any first year engineering course. At the University of Adelaide, material selected from these chapters is also applied in a first year design course for civil and environmental engineering students.

The techniques and tools presented in Chapters 11–13 are designed to allow quantitative assessment of design and planning problems. Lecturers presenting first year courses can select from the range of topics that are covered here. These chapters contain material that is beyond what would normally be covered in a first year introductory course and is ideal for higher-level courses in an engineering program.

Graeme Dandy,
David Walker,
Trevor Daniell,
Robert Warner,

Adelaide, 6th June, 2007

CHAPTER ONE

Engineers in Society

In this introductory chapter we discuss the nature, history, and scope of engineering work and the role of the engineer in society. In broad terms, engineering is concerned with improving the quality of human life by providing and maintaining the complex physical infrastructure which is necessary for the functioning of modern society. Engineers make extensive use of scientific and mathematical knowledge. However, their work is distinguished from the work of scientists and mathematicians by an emphasis on the practical use of knowledge to solve problems related to the physical infrastructure as economically and efficiently as possible, in an environmentally and socially responsible way.

1.1 INTRODUCTION

In his account of engineering in the ancient world, Sprague de Camp (1963) argues that engineering has played a pivotal role in the development of modern society and suggests that the story of engineering is as important as the history of kings, generals, philosophers and politicians in explaining the state of our world today. Unfortunately, the nature and purpose of engineering work, both in past civilisations and in our own modern society, is not widely appreciated in the community, despite the fact that modern life would be impossible, almost unthinkable, without the physical infrastructure that is provided through engineering work. In this introductory chapter our aim is to provide an overview of the nature, the history, and the scope of engineering work and the essential role it plays in society.

The English word engineer seems to date from the seventeenth century and to be based on the much earlier Latin word *ingenium*, which has been translated as talent or mental power (Lienhard, 2000). Tracing back to the Latin roots we find that an *ingenium* was an ingenious device or siege engine used by the Roman army to attack fortifications. But long before the word engineer came into common use, people were carrying out works that are now recognised as engineering.

It has been argued that the first people to undertake true engineering works were the Sumerians, who lived in what today is southern Iraq from 4500 BC to 1750 BC. The land they inhabited was "hot and dry, with soil that is arid and wind-swept and unproductive [but by 3000 BC] the Sumerians ... turned [it] into a veritable Garden of Eden and developed what was probably the first high civilization in the history of man" (Kramer, 1963). This transformation was the result of a deliberate and highly organised effort to modify their surroundings using a combination of technological skill and central organisation. As Kramer points out, "the construction of an intricate system of canals, dykes, weirs, and reservoirs demanded ... engineering skill and knowledge. Surveys and plans had to be prepared which involved the use of levelling instruments, measuring rods, drawing, and mapping."

It is possible to delve further back in time from 3000 BC, and still find evidence of collaborative efforts in public construction that should be considered as examples of early engineering. According to Mithen (2003), between 9600 and 8500 BC the town of Jericho, part of modern Palestine, was partly surrounded by a wall 3.6 metres high and 1.8 metres wide at the base. Initially it was assumed that the wall was for security but given the fact that the wall only ran around part of the city it has been suggested that it may have been built to prevent flooding. Inside the wall there was a tower 8 metres high and 9 metres in diameter complete with an internal staircase. Mithen suggests that "such architecture was completely unprecedented in human history, ... at least a hundred men working for a hundred days would have been necessary to build the wall and tower". As was also the case with the Sumerians, the evidence here of technological skill, combined with planning, design, and organisational ability is, in our opinion the very essence of engineering.

1.2 SOCIETY AND ITS ENGINEERING INFRASTRUCTURE

The world's first engineering works, associated with early civilisations such as those in Jericho and Iraq, resulted in public buildings, fortifications, roads and irrigation works. Today we use the term physical infrastructure for such works. The purpose of the physical infrastructure in those societies was to enhance the quality of life of the people through the provision of shelter, clean water, sewerage, defence and protection, transport and communications. The early development of better metallic tools and weapons, from copper, bronze and iron, also contributed to an improved physical infrastructure and hence to improvements in the quality of life of the people.

In today's world, society's need for a physical infrastructure to enhance the quality of life has not diminished. Indeed, these needs have increased enormously over the centuries and continue to expand. As in the past, we rely on a physical infrastructure that provides buildings, clean water, effective systems for transport and communications, and various other services such as those for the removal and treatment of refuse and sewage. It is the role of engineers to provide, maintain and extend the physical infrastructure that society relies on to function effectively and to prosper. Society also relies on the organisational ability of engineers to manage the construction and operation of such infrastructure.

The infrastructure we use today is immeasurably more sophisticated and more efficient than it was in past civilisations. For example, the ancient Romans relied on their roads to maintain communications throughout their extensive empire. It could thus have taken months for orders from Rome to be dispatched to the outermost frontiers of the empire. Today, through modern technology, we can make visual and audio contact with others almost instantaneously around the planet and into space via various forms of telecommunication. These communication channels also allow us to send complex documents and data. In addition to cheap and rapid communication, people today can make use of facilities such as the information sources provided by computers and the Internet, mass entertainment through radio and television, and rapid ground and air transport, both local and global.

Engineering and civilisation

When Juan de Grijalva set sail from Havana on April 8, 1518 he was in search of an ancient civilisation; one that Spanish sailors had discovered quite by accident the previous year. Three sailing ships, blown off course by a severe storm, had happened upon ancient ruins of a splendid city. There, according to Gallenkamp (1960) "an astonishing sight was visible in the distance: rising up as though an outgrowth of the native limestone were a high wall enclosing a series of terraced pyramids and palace-like buildings constructed of carefully fitted stones".

The constructions were more than just the work of skilled stonemasons: they showed a high level of planning, design and organisation and the Spaniards were convinced they had found a lost civilisation. This assessment came well before they saw any of its population, before they had studied its language, its art, its legal or political systems. The evidence of technological expertise, engineering skill and organisation on a significant scale was all that was required.

The discovery of the Maya civilisation has many parallels, for example the Incas in South America and the Aztecs in central and southern Mexico. In each case an advanced civilisation had been recognised, initially at least, on the basis of its architecture and engineering infrastructure. On further study, the characteristics of each civilisation have been discovered. Sometimes there were surprises: the Incas had no written language, nor had they developed the wheel, yet the fact that they had achieved a state of civilisation was beyond question on the basis of their engineering prowess alone.

Maya ruins at Palenque, in Chiapas, Mexico. © J. Nield.

The engineering infrastructure has developed progressively over thousands of years in response to the expanding needs of society. At the same time society itself has evolved within the context of its available engineering infrastructure. The engineering infrastructure is thus interwoven into the fabric of the society it supports. The continuing history of science, engineering and technology is so intimately associated with the history of civilisation that it is hardly meaningful to separate one from another.

Engineering developments thus have a feedback influence on society and can bring about profound changes in our way of life. Engineering-induced changes in society can be as significant and far-reaching as those brought about by political, social and commercial developments. For example, the transportation system in a city tends to grow in extent and capacity to meet the needs of the inhabitants. In early cities, and up to the time of the Industrial Revolution, patterns of urbanisation were dictated by the need for workers to be within walking distance of their place of work (Richards, 1969; White, 1978). This therefore limited the size of cities and dictated, to a large extent, population distributions. Following the development of mechanised transport, the scope for expansion increased considerably with a symbiotic relationship between population and transport routes. This comes about because the construction of a new road, for example, tends to encourage people, businesses and other developments to it. This then leads on through reducing travel times and costs to certain parts of the city to a positive feedback mechanism spurring further growth. Gordon and Richardson (1997), in a paper discussing the relative merits of compact cities, state that in 1890 the effective radius of US cities was approximately 2 miles (3.2km) and this was based on pedestrian access. By 1920 it had grown to 8 miles (12.9km) as a result of the development of public transport and by 1950 was at 11 miles (17.7km) with the widespread use of the private car. In the 1970s the construction of urban freeways had allowed the radius to increase to 20 to 24 miles (32.2–38.6km).

But the relationship between engineering and society goes well beyond the physical. Consider for a moment one of the major cities of the world, such as Sydney, Paris, London, New York, or Rome. It is very likely that in thinking about these places a mental image forms: for Sydney the Opera House or the Harbour bridge (Figure 1.1); for Paris the Eiffel tower; for London Big Ben or the Houses of Parliament; New York brings visions of the Empire State building, for Rome perhaps the Colosseum. These are all significant constructions, and all form part of each city's infrastructure. But in many ways they are more than just physical structures; they are part of the city's very identity and if lost, a part of the city would be lost forever. When the World Trade Center buildings in New York were destroyed in 2001, it was more than just the loss of an iconic pair of buildings and the lives within. The strike was aimed at the very essence of New York City and its existence.

1.3 ENGINEERING IN HISTORY

The Middle East and South America are not the only locations where evidence of early engineering works can be found. Records of engineering works of great complexity and large magnitude show that these were also undertaken in (for example) ancient Egypt, the Roman Empire, India, and China. Many of the early engineering works concerned irrigation for agriculture, water supplies for cities, the construction of roads, and the fortification of cities. Everywhere there was evidence of the level of organisation required. As Rivers (2005) has pointed out, the Great Wall of China, which was started around the 6th or 7th Century BC, "could not have been constructed solely with the technology of cutting stone from quarries and then transporting those stones to the designated sites. Unless an elaborate organisation existed in Chinese society, the Great Wall would never have been built, then or now."

Figure 1.1 Sydney Harbour bridge.

The construction of the pyramids (Figure 1.2) was one of the major engineering feats of the ancient world. It is not widely appreciated that the construction technique evolved from earlier structures built from sun-dried bricks where the inward sloping walls overcame the problems of rain-induced slumping and deterioration. When the same method was used for stone it gave the structures almost unlimited life, although there was thought to be at least one failure at Meidum during modifications to an earlier step structure (Lehner, 1997). The pyramids surviving to this day also show some damage that occurred due to tomb robbers and later populations removing the outer stones for their own purposes. In addition to the structural aspects of the designs, much ingenuity went into the construction in an attempt to foil tomb raiders. This included the use of false passages and the construction of chambers that were sealed off by heavy slabs that could be lowered into position when required. This was achieved using a cleverly designed system of props, supported on chambers of sand, where the sand could be removed by spilling (Lehner, 1997).

While many of the examples described so far have been in the application of what would now be called civil engineering, there is also ample evidence of early developments in mechanical, mining and chemical engineering. Around 9800 BC there is evidence of the production of what is today known as Plaster of Paris from the heating of gypsum in wood-fired kilns in Mesopotamia. At times between 8000 and 7000 BC the people inhabiting modern day Turkey were mining copper ore and beating it into beads, hooks and metal sheets, and at a somewhat earlier time the people of Mesopotamia were forming copper plates into small tubes as a form of jewellery (Mithen, 2003).

Figure 1.2 The sphinx and pyramid, near modern Cairo. © R Seracino.

An ancient engineer's life and its rewards

Sprague de Camp (1963) gives an account of an engineer who made good in Egypt, a land of rigid class lines not easily crossed. However, engineers succeeded in crossing them, since engineering ability is not a common gift. The architect Nekhebu (from 24th Century BC) told on his tomb the story of his rise from humble beginnings:

His Majesty found me a common builder; and His Majesty conferred upon me the offices of Inspector of Builders, then Overseer of Builders, and Superintendent of a Guild. And His Majesty conferred upon me the offices of King's Architect and Builder, then Royal Architect and Builder under the King's Supervision. And His Majesty conferred upon me the offices of Sole Companion, King's Architect and Builder in the Two Houses.

Nekhebu was also awarded other titles and rewarded with gold, bread and beer. How little has changed!

Life and engineering in ancient Rome

Much has been written about the engineering accomplishments of the Romans, most of it positive. But life in ancient Rome must have been a mixture of experiences for many of its citizens. According to Carcopino (1941), Rome in the 2nd Century had a population of 50,000 citizens living in approximately 1,000 private mansions (domus) and a further 1.1 to 1.6 million in 46,602 apartment blocks (insulae). While the private mansions were, by all accounts, magnificent, life in the apartment blocks must have been less comfortable with inadequate heating and cooling, no running water or provisions to handle sewage, and fire being a constant concern. As early as the 3rd Century BC three story masonry apartment blocks were quite common and around this time an edict limited the height of buildings to 20 metres, although some buildings rose to 5 or 6 storeys. However, they were often poorly built and in danger of collapse. There was a law that limited the thickness of the outside walls to approximately 450 mm and this led to inadequate construction.

According to Carcopino (1941), who may have been using a little artistic licence, *"the city was constantly filled with the noise of buildings collapsing or being torn down to prevent it; and the tenants of their insula lived in constant expectation of it coming down on their heads."*

Moving on in time we find techniques developed by the Greeks in the fields of military and civilian engineering taken over by the Romans, who developed and applied them to an unprecedented degree. The scale of achievement of the Roman engineers in the construction of roads, bridges, aqueducts, baths, and public buildings throughout the ancient world was in many cases not equalled again until the 20th Century. Their works remain impressive even by today's standards of accomplishment. Remains of Roman civilian constructions and roads are still to be found (and in use) in Europe and throughout the limits of their empire. Water was supplied to Rome through eleven aqueducts, the largest of which was over 100 km in length, with the water directed not only over the familiar multi-level arched aqueducts which traversed the plains (examples of which survive in Segovia, Spain and Nimes, France), but also through extensive systems of tunnels. Water was reticulated in the city by means of pipes. Although engineers are often attributed with improving the health of populations by the provision of clean potable (drinkable) water, it has been speculated that the use of lead water pipes may have been responsible for extensive lead poisoning among the population of Rome before its fall in the 5th Century.

Early engineering was based largely on experience and empirical rules. According to Straub (1952) "In spite of the remarkable scientific standard, especially of the Greeks, in the spheres of mechanics and statics there was hardly any connection between theory and practice, and hardly any attempt to apply the scientific knowledge to practical purposes, in the sense of modern engineering." By the 18th Century the situation was changing, although the changes were taken up at different rates in different countries. In France, for example, the École des Ponts et Chaussées was established to train engineers with mathematics, geometry, and mechanics all on the syllabus. Florman (1976) suggests a later shift, placing it around 1850. Prior to this time, he argues, although there were many fine engineers and engineering works the engineering itself had been practised more as a craft than a profession, relying more on commonsense and time-honoured experience than the application of scientific principles. According to Kirby et al. (1990) the rapid development of this style of formal training came when it was found that designs based on scientific principles were more economical than those based purely on experience.

By the late 1700s engineers were sufficiently well established to be able to form professional societies. A group in England, with John Smeaton who is thought to be the first person to refer to himself as a civil engineer, formed a society of engineers in 1771. In 1828 that society became the Institution of Civil Engineers which continues to this day. Their definition of engineering highlights both the wide range of their skills and the emphasis on working for the benefit of humanity:

> *[Engineering is] ... the art of directing the great sources of power in nature for the use and convenience of man, as the means of production and of traffic in states, both for external and internal trade as applied in the construction of roads, bridges, aqueducts, canals, river navigation, and docks for international intercourse and exchange, and in the construction of ports, harbours, moles, breakwaters and light-houses, ... navigation by artificial power ... construction and application of machinery, and in the drainage of cities and towns.*

Specialisation within the ranks of the non-military engineers had already begun to occur in the early 19th Century, with the development of steam power for factories and locomotion. This then led to the formation of societies for mechanical, mining, electrical and chemical engineering over the following 80 years. Since 1928 these societies have also had a number of offshoots that have formed once specialisation had increased to a level where it was possible to differentiate the members and their skills. A useful way to demonstrate the progress that has been made is to consider a recent definition of engineering from the US Accreditation Board for Engineering and Technology (quoted in Voland, 2004):

> *[Engineering is] ... the profession in which a knowledge of the mathematical and natural sciences, gained by study, experience, and practice, is applied with judgement to develop ways to utilize, economically, the materials and forces of nature for the benefit of mankind.*

Note that the long list of particular skills has been replaced with the general theme of working for the benefit of mankind. Notice also the introduction of the idea that the work must be carried out economically. For students in their early years of engineering study at university the emphasis is often on developing a knowledge of mathematics and natural sciences, but it is how this knowledge is applied with judgement for the benefit of mankind that forms a central theme of this book.

In the next section we move on from the past and consider engineering work and how it is carried out nowadays. It will be seen that it is much more than the design of structural elements, chemical processes, or electrical circuits and that, as the title of this book suggests, there is planning before design.

1.4 THE NATURE AND SCOPE OF ENGINEERING WORK

In the middle of the 20th Century, civil, electrical, mechanical, chemical and mining engineering were thought of as the main branches of engineering. These have been further sub-divided to make way for more specialisation with newer branches including aeronautical, aerospace, agricultural, automotive, biomedical, coastal, computer systems, electronic, environmental, mechatronic, medical, optical, rehabilitation, and transport engineering, to name but a few. Although engineering work has tended to become progressively more specialised, even the most diverse fields of engineering have various common characteristics which can be used to identify the basic nature of engineering work. For example, a common characteristic which has applied since the 19th Century is the use of scientific knowledge, technology and mathematics. Also, there are various activities which are common to the different fields and branches of engineering. These include planning and design, analysis, implementation, research and development, project management, and sales and marketing of engineering products and services. Taken together, these activities provide a good indication of the nature and scope of engineering work. We therefore consider some of them briefly in relation to the conduct of engineering.

Engineering projects

In broad terms, engineers develop, maintain, and improve the physical infrastructure, and thereby improve the conditions of life in the community. Because of the unlimited needs and demands of the community on the one hand, and the limited availability of resources such as people, energy, materials, and money, on the other, it is essential that engineering work be carried out efficiently, with minimum use of scarce resources, and in limited time. For this reason, modern engineering work is usually undertaken in the form of goal-directed projects.

Typical examples of engineering projects include the design, construction, and operation of a chemical plant to produce fertilizer, the design and manufacture of a clutch system for a new model car, the adaptation and programming of a microprocessor for the operation of a programmed washing machine, and the widening and strengthening of a bridge to allow increased traffic. A large part of modern engineering work is also concerned with the maintenance and upkeep of existing facilities. Although minor maintenance work may be undertaken by paraprofessional staff on a routine basis, any substantial maintenance and repair work needs to be executed as a carefully planned engineering project. Engineering projects and their management will be discussed further in Chapter 5.

Planning and design

Planning and design are activities in which the details of an engineering system or a process are determined to the extent necessary to allow implementation to be undertaken. For example, before work can start on the manufacture of a device or the construction of a building, a design has to be carried out, and then the processes of manufacture or building have to be carefully planned. The initial planning and design phase in an engineering project is crucially important, and usually lengthy.

The terms planning and design are not mutually exclusive and there is no consistency in the technical usage of these words. Nevertheless, the tendency in this book will be to refer to the design of physical systems and components, and the planning of operations and processes, i.e. to the design of hardware and the planning of software. Chapter 3 focuses on various aspects of planning and design and the way the activities are undertaken. According to this usage, one refers to the design of a bridge, and to the planning of the process of bridge construction. Design work is thus undertaken with the purpose of determining precise details of any devices, machines, or physical systems which are to be constructed or manufactured as part of the project, while the planning work is undertaken to determine the details of the procedures and processes which form part of the project.

Planning and design are crucial activities in the conduct of any engineering project. The quality of this work inevitably has a decisive effect on the success of the project. Planning and design are required in almost every type of engineering project, irrespective of the branches which may be involved. A large part of this book is devoted to the question of how the basic engineering activities of planning and design are undertaken. Planning and design are also complex activities; so complex that they can only be carried out in an iterative manner, with much trial and error.

Engineering Management

While the trend to increasing specialisation assists in some respects by providing engineers with highly focussed skills, it does have a side effect: there is now a much greater need for a large number of specialists to work together on a project and this creates increased need for teamwork and the management of projects by further specialists. If one looks at the typical billboard outside any major construction it is hard not to be impressed by the large number of specialists required to bring the project to fruition.

Engineering management is concerned with the planning and organisation of an engineering project in order to ensure that it can be undertaken efficiently and carried out successfully through all stages to completion. As the scale and complexity of a project increase, so too does the need for careful and effective planning and management. For example, a small project such as the installation of a roundabout in an existing street may require a team of 4 to 5 professionals including a project supervisor, project secretary, design engineer and a draftsperson. By the time the project size is measured in millions of dollars the team is more like 10 to 12 people with additional engineers with specialised knowledge, and an increase in support staff also. Engineering project management will be discussed further in Chapters 3 and 5. The skills required for teams will be outlined in Chapter 6.

1.5 ENGINEERING, SCIENCE AND MATHEMATICS

It has already been pointed out that, because of the unlimited demands which society makes on its limited resources, engineering projects usually have to be carried out at minimum cost, and in limited time. To achieve maximum efficiency, use is made of all available relevant knowledge. In particular, engineers draw wherever possible on scientific principles and mathematics as well as current technology. However, such knowledge is in itself insufficient, and usually has to be supplemented, for example by engineering research and development (R&D) work, by experiment, and by knowledge gained from past experience.

Although engineers are users of scientific and mathematical knowledge, there are important features which distinguish the work of the engineer from that of the scientist and mathematician. The engineer has an active role to play in using resources to satisfy community needs. In contrast, the scientist is more concerned with understanding and modelling the physical and biological world, and is thus engaged in acquiring knowledge. The active work of the engineer is also distinct from that of the mathematician, who is concerned with logical structures which may or may not be observable in the physical world. While the prime engineering concern is with achievable community goals, and not with the discovery of new scientific or mathematical knowledge as such, good engineering work often leads to new scientific, mathematical and engineering knowledge and new technologies.

1.6 ENGINEERING AS CREATIVE PROBLEM SOLVING

We have seen in this chapter how engineering is concerned with providing, maintaining and improving the physical infrastructure that human society needs in order to exist and to flourish. Engineering work is largely concerned with solving the problems that arise from community needs, as these relate to the physical infrastructure.

One of the characteristics of the modern engineered infrastructure is its complexity. To appreciate this complexity it is only necessary to examine closely a typical piece of the physical infrastructure, say a road system that allows traffic to flow into, out of, and around a modern city. Such a road system typically includes hundreds of kilometres of multi-lane access highways and even more kilometres in a grid of smaller urban and suburban roads and streets. The access highways include one or more circumferential ring roads that allow traffic to bypass the city as well as the radial roads that bring the traffic into and out of the grid of local roads and streets. While intersections in the main access and ring roads will consist typically of cross-overs and fly-overs to allow the traffic to separate and flow smoothly in different directions, most of the smaller city streets and roads will intersect at grade and have traffic lights to control and improve the flow. There will be many bridges, embankments and retaining walls to provide grade separation where needed and to traverse geographic obstacles such as rivers. Another important component of the road system is required to provide traffic control and management. This in itself is a complex system, consisting of traffic sensors at key locations to measure traffic volumes, a sophisticated plan for optimising the traffic flows in the various directions, associated computer programs and computing facilities, and control devices, including lights and warning signs, which communicate instructions and information to drivers.

The complexity of such a road system makes it physically impossible for any one human to appreciate all the details and keep them in mind, let alone to understand the detailed functioning of each component from second to second in time. Furthermore, it would be physically impossible for a single engineer working alone to carry out the design and supervise the construction of such a system in a reasonable amount of time. Nevertheless, such pieces of engineering infrastructure have to be designed and constructed economically and in a short time span. They also have to operate safely, efficiently, and near faultlessly.

Engineers thus have to be able to deal with complexity. To do this they use a simple decomposition process, whereby each complex entity is progressively broken down into simpler and simpler components until a stage is reached where it is possible to understand how each individual part operates and how the various parts interact with each other. This decomposition approach will be discussed in some detail in Chapter 2, where some basic concepts will be introduced to formalise this approach.

In addition to dealing with the problem of complexity, engineers also have to be able to solve ill-defined and open-ended problems (those without a unique solution). Community needs, when first expressed, tend to be vaguely formulated. Indeed, one of the first steps to be taken in engineering work is to identify the real needs of the community that are to be satisfied, as distinct from initial community perceptions. In this regard engineers cannot simply be problem-solvers but must act as problem-framers (Donnelly and Boyle, 2006), applying their skills and experience and recasting the problem to ensure optimal outcomes. This, and other aspects of engineering, will be developed in the following chapters.

Even when a problem has been clearly formulated there will usually be a number of alternative engineering approaches that can be legitimately taken. Engineering problems rarely have simple, unique solutions and for this reason engineering work is often described as open-ended problem solving. In Chapter 3 of this book we look at a methodology for solving open-ended engineering problems.

Dealing with complexity and solving ill-defined, open-ended problems are two of the prime characteristics of engineering work. Before we look at these aspects of engineering, we need to emphasise that two quite different and contrasting ways of thinking need to be used for these different activities. A logical, deductive, analytic way of thinking is required in order to apply the decomposition approach to complexity. On the other hand, it is necessary to be able to think laterally and innovatively when dealing with open-ended, design-type problems. Indeed, a purely analytic approach can be positively counter-productive. The contrasting ways of thinking may be described as convergent thinking, which is needed in the case of analytic work, and divergent thinking, for ill-defined design-type work. Engineers have to be skilled in both these ways of thinking and these will be dealt with in Chapter 4.

1.7 ENGINEERING IN THE 21ST CENTURY

The National Academy of Engineering of the United States (2006) published a list of the 20 greatest engineering achievements of the 20th Century. The full list is given as:

- Electrification
- Aeroplane
- Electronics
- Agricultural Mechanization
- Telephone
- Highways
- Internet
- Household Appliances
- Nuclear Technologies
- Petroleum and Petrochemical Technology
- Automobile
- Water Supply and Distribution
- Radio and TV
- Computers
- High performance materials
- Spacecraft
- Imaging
- Health Technology
- Laser and Fibre Optics
- Air Conditioning and Refrigeration.

The list of achievements is of interest for a number of reasons. Firstly, it is a graphic illustration of the relevance of engineering work to everyday life. Engineers are working on key aspects of relevance to society and are directly responsible for the current material quality of life that much of the planet enjoys. Secondly, the list throws down the challenge to current engineers (and engineers in training) to work creatively. A comparable list in 100 years time is unlikely to include any of the entries in the current list. Thirdly, the list is interesting for what it hides rather than what it shows. The list is a series of successful outcomes; clearly defined solutions to what were presumably clearly defined problems. But is that really true? One of the key arguments in this book is that much of engineering is not clear cut, and the problems are not clearly defined, but complex and open-ended. This means that many traditional problem solving methods simply will not work for engineering problems.

As an example, consider the development of radio. Radio today, with its AM (amplitude modulation) and FM (frequency modulation) means of transmission, is an integral part of everyday life. It was based on the discovery by Heinrich Hertz of electromagnetic waves in 1887. By 1901 Guglielmo Marconi had developed the technology to the stage where he was able to transmit radio signals across the Atlantic Ocean (discussed in more detail in Chapter 4). But if we look back to when it was being developed we find that what exists now, and the way it works, is quite different from what might have been expected from the simple problem that was believed to exist. Put simply, the original problem was to allow communication between two distant points. Early solutions to this problem included the telegraph and telephone. In each case a single sender transmitted a signal to a single receiver and this was viewed as the problem: how a sender and receiver could communicate between two points some distance apart. Radio was seen as an attractive alternative since it did not require wires for the signal and therefore had advantages. Yet radio today works on a broadcast philosophy where a single sender transmits to the environment in general and the signal can be received by anyone with an appropriate receiving apparatus. And this is the strength of radio: it has become a medium of mass communication.

Following on from the list of engineering achievements in the 20th Century, it is instructive to ask ourselves what are likely to be the next important technologies to be developed. An article in Technology Review (2005), designed to give readers a sample of things to come, lists 10 emerging technologies. These are:

- Airborne networks
- Quantum wires
- Silicon photonics
- Metabolomics
- Magnetic-resonance force microscopy
- Universal memory
- Bacterial factories
- Enviromatics
- Cell phone viruses
- Biomechatronics.

While not wanting to put too much weight on any of the predictions, nor to describe them all in detail, looking through the list does bring out some interesting conclusions regarding the near future of engineering and technology. First and foremost is the fact that with existing societal needs the current stock of engineering infrastructure will survive and will need to be maintained and developed. Also, while the basic needs of society do not change markedly, the means by which engineers satisfy those needs change enormously. At the same time the frontiers for engineering are expanding, and expanding rapidly in ways that were science fiction only a few decades ago.

Physical infrastructure is getting smarter

An emerging trend in engineering is the implantation of small sensors in engineering systems so that performance can be monitored. For example, sensors in buildings and structures are used to monitor stress and vibrations, each of which can point to impending trouble long before it manifests itself. In some cases the sensor is built from sophisticated electronics but recent developments have shown that much can be gained from quite simple arrangements. For example, carbon fibres embedded in concrete during construction as a strengthening measure have been found to show variable resistance based on whether the material is under compression or tension (Hansen, 2005). This means that performance can be monitored at any chosen location without the need to have embedded anything beforehand. Research is also proceeding investigating the potential for these devices to power themselves using vibrational energy from the structure, thus overcoming the need for external power supplies.

Engineering systems are becoming more complex

As already mentioned, a key feature of modern infrastructure is the interconnectivity of the elements leading to highly complex systems. This in turn leads to some far-reaching and perhaps unexpected side effects when one component fails or behaves in an unexpected manner. A good example of a complex system is modern electricity power generation which involves massive networks and, because of this, include the potential for significant unexpected side effects. The power failure on August 14th, 2003 which affected a large proportion of North America and Canada including the cities of New York, Detroit, Cleveland and Toronto, and which left up to 60 million people without power is a good example. According to reports, the speed at which the blackout spread was unexpected, as was its extent. In a matter of seconds, circuit breakers installed to protect electrical installations from sudden, potentially damaging power surges tripped taking out entire regions. In total, more than 20 power plants were temporarily shut down, including nine nuclear reactors in four American states. A task force report suggested that the interconnectivity of the nation's power grid, which is the reason for its success most of the time, leaves it vulnerable to this kind of massive failure.

The potential for further instances of this sort of failure means that engineers will increasingly be involved in modelling complex systems as part of their designs. It is likely that much can be learned from nature, where the web of life shows the resilience that is possible in complex systems.

Engineering is finding increasing application in medicine

Biomedical engineering (or bioengineering) is becoming increasingly important and this trend is likely to continue. Three of the items on the list of emerging technologies are in this area. Metabolomics concerns diagnostic testing for diseases by analysing sugar and fat molecules that are the products of metabolism. Bacterial factories make use of engineered bacteria to produce drugs to combat diseases such as malaria. Biomechatronics is the area of engineering that mates robotics with the human nervous system leading to a new generation of artificial limbs that behave like the real thing and that can be controlled directly by the person's brain.

According to Citron and Nerem (2004) bioengineering has been recognised as an established branch of engineering since the late 1970s. They identify two key applications for the profession: diagnostic imaging (ultrasound, computer assisted tomography, CAT and functional magnetic resonance imaging, fMRI) and implanted therapeutic devices (e.g. cardiac rhythm stimulators, prosthetic heart valves, vascular stents, implanted drug pumps and neurological stimulators). Together these have had a significant effect on the health of large sections of the population using ideas and methods that were unimaginable a few decades ago. For example, Citron and Nerem (2004) estimated over half a million people had heart pace-maker implants and a further 300,000 had implantable defibrillators to deal with life-threatening heart arrhythmias.

The application of nanotechnology in medicine is another area where engineers are currently making significant progress. Dowling (2004) reports on current and future developments including better artificial implants, sensors to monitor human health, and improved artificial cochleae and retinas. She warns that some of these will not be realised for at least ten years – not a particularly long lead time for many. Duncan (2005) gives further examples that include nanomedicines that aim to diagnose problems and then deliver appropriate drugs, to promote tissue regeneration and repair. She states that "these ideas may seem like science fiction but to dismiss them would be foolish".

Engineering devices are getting smaller

Engineers over a number of fields are developing and using technology on a minute scale. Computer engineers are currently constructing and planning computer circuits where wires are getting down to sizes measured in nanometres (billionths of a metre). In the list of emerging technologies, quantum wires are extremely low resistance wires using carbon nanotubes, where the diameter of single walled tubes is in the order of nanometres. To put that in perspective, the DNA double helix has a diameter of 2 nm while an average human hair has a diameter around 80,000 nm. This continues the push in an area generally called nanotechnology which has been developing since the first tentative steps in 1959 (Gribbin and Gribbin, 1997).

One of the early promoters in the area of nanotechnology was physicist and Nobel prize winner Richard Feynman who, in 1959, instigated two prizes; one for the first person to build an electric motor that would fit inside a 0.4mm cube, and a second for anyone who could reduce printed text by a factor of 25,000 which could then be read by an electron microscope. He was surprised a year later when someone arrived at his office carrying a fairly large box. It contained a microscope (required to see the motor) and at that stage he knew he was going to have to pay up (which he did).

The second prize took 26 years before it was claimed by a graduate student who had written the first page of A Tale of Two Cities by Charles Dickens at the required scale. At that scale, as a matter of interest, it has been said that all 24 volumes of the Encyclopaedia Britannica could be written onto the head of a pin (Feynman, 1999).

In the last few years the size of motors has been reduced even further with a recent one consisting of a gold rotor on a nanotube shaft measuring 500 nanometres across (Sanders, 2003). One consequence of such miniaturisation is the reduction in power required to run these devices and this then points to the next trend in engineering: new sources of power.

Power is being harvested rather than mined

The industrial revolution of the 18th Century was driven by the advent of machines that replaced human and animal powered devices with those that could provide significantly increased output through the use of concentrated energy sources. Wood was quickly replaced by coal which in turn was replaced by petroleum products. In some situations oil has been replaced by fissionable (radioactive) material. At the turn of the 21st Century oil is the dominant source of energy for much of the world's transport, heating and industry. The advantages of oil are its relative abundance, its concentration of energy and its low cost in relation to other energy sources. As the world considers alternative energy sources such as solar, tidal, wind (see Figure 1.3) and wave it encounters problems based on energy density. The energy in the wind or in the waves is much less dense and requires significantly more physical infrastructure to harvest it. It is likely that this may lead to more distributed power systems, and to a change in the mindset of engineers planning and designing systems to make use of the renewable energy systems.

Engineers will have to achieve sustainability

The world population has reached a stage where it is having a noticeable effect on the earth and its ecosystems. When Svante Arrhenius first suggested in 1896 that carbon dioxide in the atmosphere could cause global warming he was thinking of natural sources of the gas and did not believe that man-made releases could have any effect given the very small amounts involved at the time (Bengtsson, 1994). Now, one hundred years on, the situation is quite different and the general consensus of the scientific community is that humans do have the potential to make an impact on the climate of the whole earth and may have already started to do so. The phenomenon has been called global warming but this rather innocuous term downplays its significance. It is thought likely that the world is on the threshold of large scale global climate change and that this in turn will lead to significant weather pattern changes and significant ocean current changes.

At this time much is being asked and expected of engineers and their ability to develop a technological fix. In the past engineers have always managed to satisfy society's needs for increased energy, faster communication, safer transport and more reliable water sources. It may be that, once again, society is asking the wrong questions and engineers must be creative and responsible in their response.

Engineering societies around the world now include in their code of conduct the requirement that engineers promote and work in a sustainable manner and this will become more important over the next few decades.

Figure 1.3 Starfish Hill wind farm, South Australia.

1.8 SUMMARY

Society cannot and does not develop without infrastructure. It requires its roads, buildings, communication networks, transportation systems, and all the other components that allow society to function, and in turn promote and foster the development of the arts and sciences. But the physical infrastructure is more than just roads and buildings that society requires for its functioning; it becomes an integral part of the society and it develops along with society and society in turn develops along with it.

The history of engineering has shown up a number of trends and these continue to this day. A key trend is an increasing specialisation and this is demonstrated by the development of the many new strands of engineering. This brings many advantages but is not without its consequences. A key outcome is the need for engineers to manage themselves in teams involving specialists from a wide range of disciplines. This throws extra emphasis on the need for proper and detailed planning of engineering works and the need for engineers in general to be familiar with the general principles of management.

The engineering profession is being shaped by trends that have been developing for some time. Engineers are working with technologies that are on a scale previously unimagined: small and large, and each has thrown up challenges. The small (nanotechnology) is taking engineers into areas such as biomedical engineering that have developed only recently. The large scale projects are demonstrating the need to understand massive complex systems and to be able to deal with unexpected side effects that may be very difficult to predict during the design phase.

Finally, perhaps the greatest challenge for the engineer in the next one hundred years is to develop the means of satisfying the needs of an increasing world population with high expectations, while doing so in a way that is sustainable and that ensures that the planet continues to be a safe, hospitable and predictable place on which to live.

PROBLEMS

1.1 Find the employment section of a wide-circulation newspaper and look for engineering positions. For job descriptions that fall outside the traditional forms of civil, electrical, mechanical, chemical and mining engineering make some attempt to classify them as specialisations under the relevant traditional form.

1.2 Computers are listed as one of the 20th Century's most important achievements. In light of the famous statement rightly or wrongly attributed to Thomas J. Watson (Maney, 2003) who was Chairman of IBM in 1943 (*I think there is a world market for about five computers.*) discuss what problem the Chairman might have anticipated computers would solve.

1.3 Consider the problem of establishing a small scientific station based in Antarctica, to be permanently occupied by 20 people. The base is to have shipping access, and will receive supplies once a year in December. Make a complete list of the components of the engineering infrastructure which will be needed to run the base successfully.

1.4 Develop a list of five engineering achievements that you think may be on the list for the greatest in the 21st century. Which branches of engineering will be responsible for these, and do these branches exist yet?

1.5 Make a list of the tasks involved in setting up a raceway on public roads in or near a city centre. Consider the time that each task would involve and develop a timeline. Assume that traffic disruption should be minimised.

REFERENCES

Bengtsson, L., 1994, Climate Change, Climate of the 21st Century. *Agricultural and Forest Management*, 72, pp. 3–29.

Carcopino, J., 1941, *Daily Life in Ancient Rome*. (Peregrine), 365pp.

Citron, P. and Nerem, R.M., 2004, Bioengineering: 25 Years of Progress – But Still Only a Beginning. *Technology in Society*, 26, pp. 415–431.

Donnelly, R. and Boyle, C., 2006, The Catch-22 of Engineering Sustainable Development. Forum, in *Journal of Environmental Engineering*, ASCE, 132(2), pp. 149–155.

Dowling, A.P., 2004, Development of Nanotechnologies. *Nanotoday*, December issue, pp. 30–35.

Duncan, R., 2005, Nanomedicine Gets Clinical. *Nanotoday*, August issue, pp. 16–17.

Feynman, R., 1999, *The Pleasure of Finding Things Out*. (Penguin), 270pp.

Florman, S.C., 1976, *The Existential Pleasure of Engineering*. (St. Martin's Press), 160pp.

Gallenkamp, C., 1960, *Maya. The Riddle and Rediscovery of a Lost Civilization*. (Frederick Muller Limited), 219pp.

Gordon, P. and Richardson, H.W., 1997, Are Compact Cities a Desirable Planning Goal? *Journal of the American Planning Association*, 63(1).

Gribbin, J. and Gribbin, M., 1997, *Richard Feynman. A Life in Science*. (Viking), 301pp.

Hansen, B., 2005, New Concrete Can Monitor Itself. *Civil Engineering, The Magazine of the American Society of Civil Engineers*, 75(11), p. 35.

Kirby, R.S.; Withington, S.; Darling, A.B. and Kilgour, F.G., 1990, *Engineering in History*. (Dover Publications), 530pp.

Kramer, S.N., 1963, *The Sumerians. Their History, Culture and Character*. The University of Chicago Press, 355pp.

Lehner, M., 1997, *The Complete Pyramids*. (Thames and Johnson), 256pp.

Lienhard, J., 2000, *The Engines of Our Ingenuity*. (Oxford University Press), 262pp.

Maney, K., 2003,. *The Maverick and His Machine: Thomas Watson, Sr. and the Making of IBM.* (John Wiley & Sons). ISBN 0471414638.

Mithen, S., 2003, *After the Ice. A Global Human History 20,000-5000BC*. (Phoenix), 622pp.

National Academy of Engineering, 2006, Greatest Engineering Achievements of the 20th Century. http://www.greatachievements.org/ (Downloaded 12/4/06)

Richards, B., 1969, Urban Transportation and City Form. *Futures*, pp. 239–251.

Rivers, T.J., 2005, An Introduction to the Metaphysics of Technology. *Technology in Society*, 27, pp. 551–574.

Sanders, R., 2003, Physicists Build World's Smallest Motor Using Nanotubes and Etched Silicon. UC Berkeley News. http://www.berkeley.edu/news/media/releases/2003/07/23_motor.shtml. Downloaded 26/3/06.

Sprague de Camp, L., 1963, *The Ancient Engineers*. (Rigby Ltd.), 408pp.
Straub, H., 1952, *A History of Civil Engineering*. (translated by E. Rockwell). (C. Nicholls & Co.), 258pp.
Technology Review, 2005, 10 Emerging Technologies. *Technology Review*, 108(5), pp. 44–52.
White, R., 1978, *Transport*. In: Environment and Man, Volume 8, The Built Environment. Eds. J. Lenihan and W.W. Fletcher, (Blackie), pp. 82–119.

CHAPTER TWO

Engineering Systems Concepts

One of the characteristics of the engineering infrastructure is its complexity. To deal with complexity, engineers use a process of decomposition whereby a complex system is broken down into successively smaller and simpler component parts until it is possible to study and analyse the individual parts and the way they interact with each other. In this chapter we introduce a number of systems concepts that formalise and facilitate this decomposition process.

2.1 DEALING WITH COMPLEXITY

We have seen in Chapter 1 that a particular characteristic of the engineering infrastructure is its complexity. A special methodology is needed to deal with this complexity and to allow engineering systems to be designed, constructed and maintained. Conceptually, this methodology is simple: it involves breaking a complex entity down into progressively smaller and smaller component parts until each part is so simple that it is possible to study the behaviour of the parts and how they interact with each other. The decomposition approach can be used to deal both with complex things and with complex problems. It is essential to the successful completion of any large engineering project.

This methodology requires a form of convergent, or deductive, thinking, which is also used in engineering analysis. The approach is not used solely by engineers: it is needed in many different fields of study, ranging from biology to corporate planning and from computer science to business management. To formalise the decomposition process, we now introduce and explain a number of basic systems concepts.

2.2 SYSTEMS AND PROCESSES

The word **system** is used frequently, if rather vaguely, in everyday speech in phrases such as water supply system, communications system, political system and biological system. The word has already been used informally in this way in Chapter 1. In everyday usage, the word means some complex "thing" that is organised in some way. We shall make frequent use of the word, but with the following specific meaning:

> *A system is a collection of inter-related and interacting components that work together in an organised manner to fulfil a specific purpose or function.*

For example, a city office building is a system, the main components being the structural (load-bearing) skeleton, the cladding (the outer protective surfaces), the usable inner spaces, the mechanical services (including elevators and air conditioning), the electrical services, the information technology services, the water supply service, and the waste and sewage extraction services. The function of this system is to provide an attractive, sheltered, serviced working space in a convenient location for office workers.

The word **process** is also used frequently but rather vaguely in everyday conversation. We speak of manufacturing processes, deterioration processes, and political, biological and chemical processes. In such usage, the word suggests some sequence of events that involves a progressive change in time. The corrosion that occurs in the small metal fasteners that join two sheets of cladding metal together in a building is thus a process. The slow, progressive, complex changes that occur in a rain forest over time, either naturally or as the result of pollution and atmospheric warming, constitute a process. In sewage treatment works a number of chemical and biological processes are employed to treat the wastewater. The word process is also applied to the complex sequence of events that occur in a human brain. In this book we shall use the word process with the following meaning:

A process is a sequence of inter-related activities that proceeds in time.

Often a process occurs in, or is generated by, a system or a component of a system. The progressive cracking over time of a reinforced concrete floor slab in a building is a process that takes place in a component of a structural system. The seasonal movements and related cracking that occur in a house constructed on reactive soil, which swells in the presence of moisture and shrinks in its absence, constitute a process. The construction of a bridge and the design of a building are processes. The planning and the design phases of an engineering project are also processes.

2.3 MODELLING AND ANALYSIS OF SYSTEMS AND PROCESSES

In order to construct new items of infrastructure, and also to maintain and improve the existing infrastructure, we have to be able to predict how engineering systems and processes will behave in the future, under both normal and unusual operating conditions. We have to be able to make such predictions before the systems and processes even exist.

The term **analysis** is applied to the activity of predicting how an engineering system and its components behave when subjected to a specified set of operating conditions and environmental conditions. For example in the design of a new bridge, a quantitative analysis may be carried out to check whether a proposed design will perform adequately, without excessive deflection or vibration, when subjected to the design loads. Another analysis may be used to check that there is adequate strength to resist any severe overloads that might occur, for example as the result of flooding or earthquake. Analysis is an essential step in the processes of design and planning.

Analysis can also be used to study the behaviour of an existing system, for example to detect inadequacies and to devise improvements or modifications to overcome the inadequacies. An analysis of an existing bridge might be carried out to predict its response to progressively increasing traffic loads. If it is found that the bridge is not going to perform adequately in the future, then further analysis may be used to check whether a proposed set of strengthening procedures will be adequate. Similarly, analysis may be used to investigate the response of the bridge to daily, seasonal and annual variations in ambient temperature.

The processes involved in engineering planning and design

Complex engineering infrastructure projects are carried out in separate but interrelated phases, such as planning, design, implementation and maintenance. Each of these phases is a complex process and, following the decomposition approach, is broken down progressively into many simple, interrelated tasks that can be handled individually. In this way, a large engineering project, requiring many thousands of hours of work, can be completed in a relatively short time period by groups of people working simultaneously on different tasks. Clearly, great care needs to be exercised in the initial breakdown of the work, and also in the planning, management and continuing coordination of the work of the separate groups, in order to ensure that the desired outcome will actually be achieved.

As the scale and complexity of a project increases, so too does the need for careful and systematic planning and design. A systematic approach to project planning has evolved progressively over many years and is used in all of the major engineering disciplines. Impetus was given in the 1960s to the development of a "systems" approach to engineering planning and design by the need to coordinate extremely large national projects such as the NASA space program in the United States.

The processes of engineering planning and design are explained in Chapter 3 and various techniques, developed to assist in project planning, are discussed in Chapter 5.

In order to undertake the analysis of an engineering system, we have to deal with the problem of complexity. Real engineering systems are extremely complex when observed closely and it is not possible or even desirable to take full account of all of these complexities. The degree of complexity in any real system is such that we have to simplify and idealise it in order to understand it. The process of creating a simplified, idealised representation of the system is called **modelling**, and the simplified representation is a **model**. The purpose of the model is to mimic the behaviour of the real system to an acceptable level of accuracy. The model provides the starting point for any analysis of the behaviour of the physical system.

In current engineering work, the modelling is usually theoretical and the resulting model takes the form of mathematical equations and relations. The subsequent analysis may require complex computations that are best undertaken by computer. In such cases we are using **mathematical models**. In the past, however, before computing facilities were available, much greater reliance was placed on **physical models**. These were usually scaled down, simplified versions of the real engineering system. They were constructed in the laboratory and tested in a way that was intended to simulate the conditions under which the real system would operate. The results of the model tests were interpreted in order to predict the behaviour of the real system. The massive improvements in computing hardware and software that have occurred in recent decades have allowed mathematical modelling and analysis to be undertaken for an ever-broadening range of engineering problems.

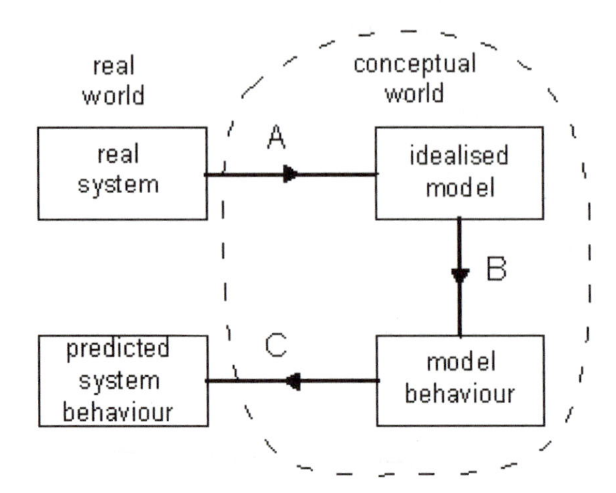

Figure 2.1 Modelling and analysis to predict system behaviour

The processes involved in the theoretical modelling and analysis of a physical system are shown in the schematic in Figure 2.1, which deals with the case where the analysis is used to predict how an existing system will respond to a specific set of operating conditions and inputs. The three processes in Figure 2.1 are represented by arrowed lines and denoted as A, B, and C. The modelling process, A, produces an idealised, conceptual model, which is a simplification and idealisation of the real system. The analysis of the system model is process B. It is carried out for a specific set of operating conditions and inputs. If the model is very accurate, then the behaviour of the model provides a good prediction of the behaviour of the real system. Nevertheless, this is not always the case and it is important to look at the simplifications and approximations that have been introduced in the modelling process in order to identify possible differences between the simplified conceptual model and the complex real system. These differences can lead to error and need to be considered when the results of the analysis are used to predict the behaviour of the real system. A process of interpretation and evaluation is therefore shown in Figure 2.1 as C.

To complete the picture, the operation of the real system, under the same operating conditions and inputs as were assumed in the analysis, is shown as D in Figure 2.2. The operation of the real system is of course independent of the analysis. If the modelling, analysis and interpretation are good there will be a close correspondence between the actual system behaviour, shown in Figure 2.2, and the predicted behaviour in Figure 2.1.

real world

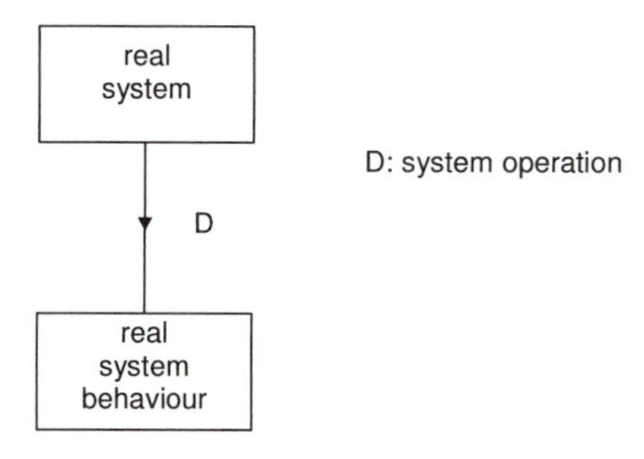

Figure 2.2 Real system behaviour

Mathematical models may be used to study the behaviour of existing systems and to identify possible improvements that can be made. They may also be used to study a proposed new system that does not yet exist. Although a slightly different sequence is then needed, the processes are similar to those already discussed and shown in Figures 2.1 and 2.2.

Some very complex systems still do not lend themselves to theoretical analysis and require physical modelling. For example, coastal engineering processes may be so complex that they have to be simulated using physical models, which are constructed in large laboratory tanks (e.g. Figure 2.3).

Figure 2.3 Physical model of a coastal breakwater

Special pieces of equipment, including wave-making machines, are used in the tank to create waves and tides. Coastal models can thus be used to study complex processes such as the progressive movement of sand along a beach front, or the alleviating effect on destructive wave action of constructing a breakwater to create a harbour. The physical model shown in Figure 2.3 was used to study protective measures for a town beach and boat launching facility. The wave paddle, to the right of the picture, generated waves that were measured at a number of positions both outside and inside the protected area.

The processes involved in the use of a physical model to predict the behaviour of a complex real system parallel those used in mathematical modelling and analysis and are shown in Figure 2.4.

The construction of the physical model requires simplification and idealisation, as was the case for mathematical modelling, but it also requires the physical construction of the model. These steps are shown as A_1 and A_2 in Figure 2.4. In some situations the conceptual process represented by A_1 might turn out to be a form of conceptual modelling.

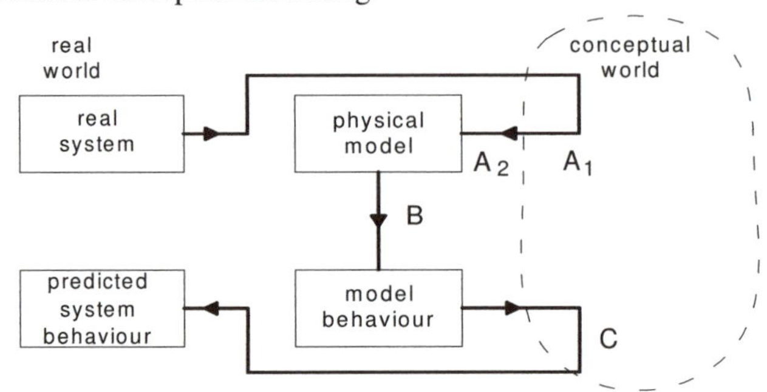

A_1 : modelling (idealising, simplifying)

A_2 : modelling (physical construction of model)

B : model testing

C : interpretation

Figure 2.4 Use of physical modelling to predict system behaviour

Although the terms analysis and modelling are used interchangeably by some authors, we distinguish between them in this book. For us, modelling is the first but distinct conceptual step that is undertaken before any analysis can be undertaken. It involves simplification and idealisation. In contrast, the second step is analytical and mathematical. The aim in modelling and analysis is to achieve a reliable prediction of the way the actual system will behave. As we have already mentioned, verification and accuracy are important issues and we will discuss them later in this chapter.

2.4 SYSTEMS CONCEPTS

There are various additional concepts that can facilitate the modelling and analysis of engineering systems and we shall now introduce them. These concepts are also applicable to processes but for reasons of brevity our discussion will focus on their application to systems. Their relevance to processes will be obvious. The concepts apply equally to real-world systems and processes and to the theoretical and physical models that we use to study real world systems.

Sub-systems, components and system hierarchy

It is characteristic of an engineering system that it consists of **components** which in turn are made up of **sub-components**. Each component of a system, when considered closely, is thus a system in its own right. For example, a water treatment plant, which is a component of a water supply system, is itself made up of various components, such as mixing and holding tanks, control valves and measuring devices. The term **subsystem** is therefore used interchangeably with component.

It is also characteristic of a system that it will itself act as a component in a larger system, in conjunction with other systems. Thus, the water supply system can be regarded as a component of the city system, whose other components include a transportation system, a communication system, and a sewerage system. The term **system hierarchy** describes this structure whereby a higher-level system is composed of components that are themselves systems and made up of components (or sub-systems) and so on. The concept of system and components will be used in Chapter 3 to assist in identifying and defining an engineering problem by treating it as part of a wider engineering problem, and, implicitly, the system as part of a wider system.

Just as a physical system can be regarded as a set of interrelated components, so can a process be broken down into a number of sub-processes. Each of these sub-processes can in turn be broken down into sub-sub-processes. The design process is thus undertaken in phases which are, in effect, sub-processes.

Of course, the decomposition process is not in any way new or novel. In another context, Jonathan Swift wrote:

A flea hath smaller fleas that on him prey,
and these have smaller fleas to bite 'em.
And so proceed ad infinitum.

System boundaries

In many engineering systems the various components can be distinguished one from another because there are clear and distinct physical boundaries. This was the case with the components of the water supply system previously discussed. However, the boundaries are not always clear-cut, and in some situations it may be necessary to introduce artificial boundaries. For example, arbitrary boundaries are used in large, massive structural systems, such as dams or building skeletons or aircraft frames, in order to carry out an analysis by the so-called finite element method. Thus, in the analysis of a massive dam wall to determine the combined effects of water pressure and earthquake motion, the wall may be conceptually broken up, by a grid of systematically located boundary surfaces, into a very large number of small brick-like pieces, or "finite elements". By first analysing the behaviour of the individual elements, and then the way they interact locally with each other, the behaviour of the complete wall can be established. The breakdown of the system into component elements in this case is not based on any physical boundaries, and, from a physical point of view, is arbitrary.

As another example, the fuselage and wings and tail structure are broadly distinguishable components of the frame of an aircraft and in the design of an airframe it is advantageous to deal separately with these components. While the components are distinguishable, the precise boundaries between them may not be clear. In such circumstances, arbitrary boundary surfaces are chosen, so that separate analyses can be carried out for each component. Of course, the local interactions across the boundaries are important, as we shall see, and depend in part on where the boundaries have been located.

The concepts of system, sub-system, hierarchy and boundary provide a useful starting point for engineers to describe, define, analyse and design complex engineering systems and processes. With these concepts the processes of analysis and design can be broken down into tasks of manageable size. The decomposition approach, coupled with the idealisations and simplifications introduced in the modelling and analysis, allows us to deal with large engineering systems that would otherwise be far too complex to handle.

Input and output

No physical system, whether engineered or natural, is self-sufficient and cut off from the rest of the world. On investigation, a system is always found to interact in some way with other neighbouring systems and hence to be a component in a wider system. It is therefore necessary to look not only at an individual system, or for that matter an individual system component, but also at the way the system (or component) interacts with its neighbours. A system interacts with other systems by means of **inputs** and **outputs**.

These usually (but not always) take the form of a flow of matter, energy or information through the boundaries of the system. The inflow, or inputs, may be thought of as the influence of the rest of the world on the system we are considering. The output, or outflow, from a system becomes input to other systems and hence represents the effect of the system on the rest of the world. In the same way, a component of a system interacts with other neighbouring components through inputs and outputs. A system can be represented schematically as a box with lines indicating inputs and outputs, as in Figure 2.5.

Figure 2.5 System with inputs and outputs

The inputs and outputs may vary in time, and are therefore shown as time functions: $y_1(t)$, $y_2(t)$, $y_3(t)$, $z_1(t)$, $z_2(t)$. A system and its components, each with inputs and outputs, can be represented schematically, as in Figure 2.6, by boxes to denote the components and by lines between the boxes to represent the inputs and outputs.

The input and output lines joining any two individual boxes represent the direct interactions between these two components. Additional indirect interactions may of course occur between two components via other intermediate components. The inputs and outputs to the overall system are inputs and outputs to individual components, as shown in Figure 2.6.

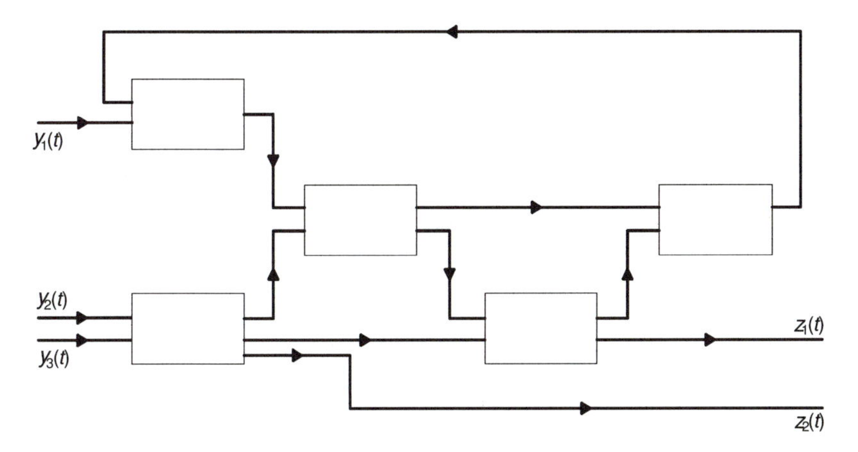

Figure 2.6 System and components

Returning to the example of the water supply system, we have as system inputs the inflow of water into the reservoirs from the catchment areas. The quantity and quality of the water flowing from reservoir to reservoir and from component to component are input and output variables for the components. A simplified schematic of the flow of water into, through and out of such a system is shown in Figure 2.7. This figure does not show information on the quality and quantity of water in each component, which are important inputs and outputs. Other inputs and outputs, such as electrical energy and the chemicals used for water treatment, are also not represented.

As an example of inputs and outputs that do not fall into the categories of matter, energy and information, we can consider a structural system whose function is to pick up the applied loads and carry them through the system and into the supports. Figure 2.8 shows a simple pin-jointed frame. The frame is the system, while the tensile and compressive members and the joints are the components. The inputs and outputs for the components are forces. The loads applied at joints A and B are the inputs to the system and the reactions at joints C, D and E are the system outputs. In a more complex structural system the inputs and outputs between neighbouring components might be stress fields distributed across the contact areas between components, together with the associated strain fields.

In the earlier discussion of system boundaries it was mentioned that a real system might, or might not, display clear-cut natural internal boundaries among components. The boundaries among systems and components tend to be clear cut if the sub-components separate themselves out from each other into clusters, with relatively few, but clear, inputs from one cluster to another. Figure 2.9 shows schematics of two systems, one with natural internal boundaries, and the other without such natural boundaries.

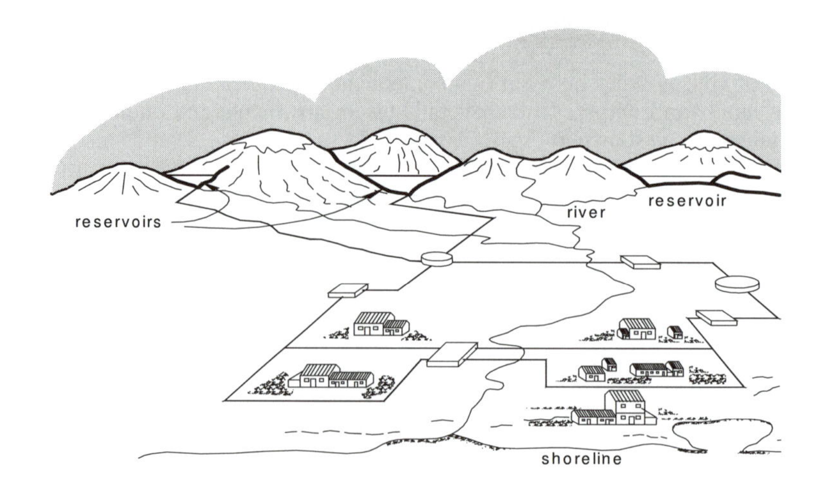

(a) simplified layout of city water supply system

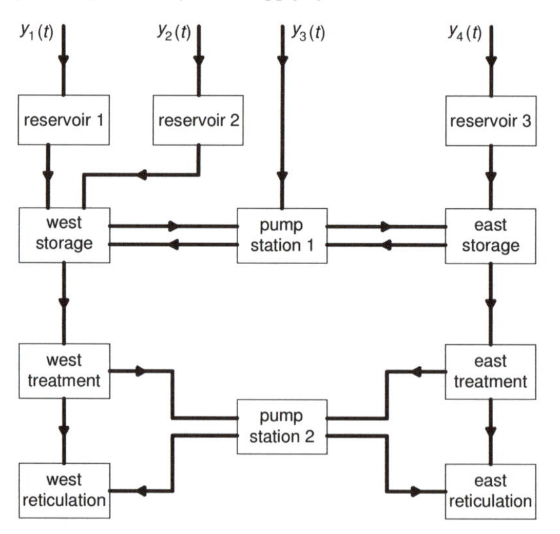

(b) system components

Figure 2.7 City water supply system

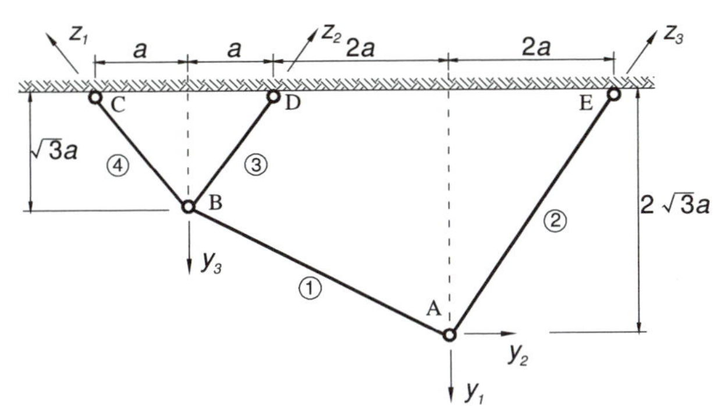

(a) tension-bar system

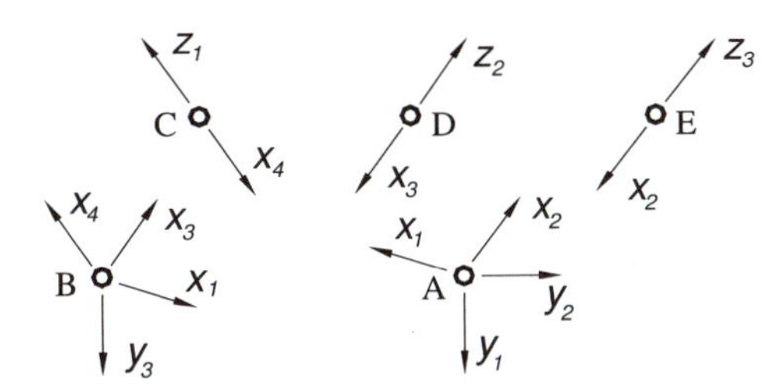

(b) forces acting at joints

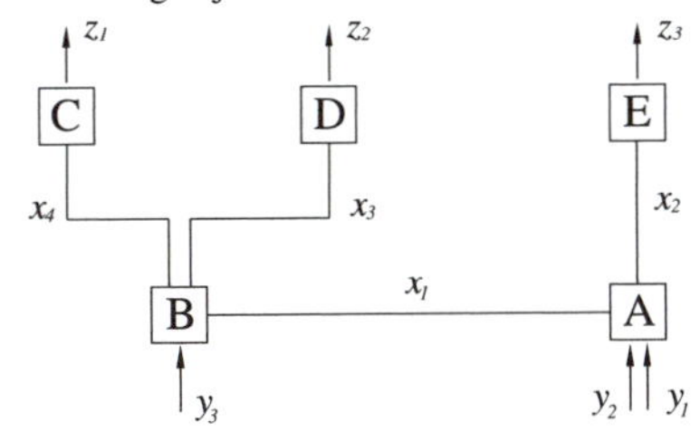

(c) input-output schematic for joints

$$\begin{bmatrix} y_1 \\ y_2 \\ y_3 \end{bmatrix} = \begin{bmatrix} 1/2 & \sqrt{3}/2 & 0 & 0 \\ \sqrt{3}/2 & -1/2 & 0 & 0 \\ -1/2 & 0 & \sqrt{3}/2 & \sqrt{3}/2 \end{bmatrix} \cdot \begin{bmatrix} x_1 \\ x_2 \\ x_3 \\ x_4 \end{bmatrix}$$

$$\begin{bmatrix} z_1 \\ z_2 \\ z_3 \end{bmatrix} = \begin{bmatrix} 0 & 0 & 0 & 1 \\ 0 & 0 & 1 & 0 \\ 0 & 1 & 0 & 0 \end{bmatrix} \cdot \begin{bmatrix} x_1 \\ x_2 \\ x_3 \\ x_4 \end{bmatrix}$$

(d) input-state, output-state relations

Figure 2.8 Simple tension-bar system

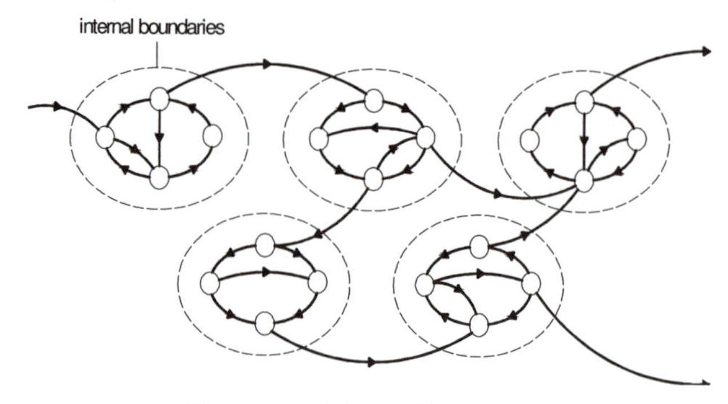

(a) system with natural internal boundaries

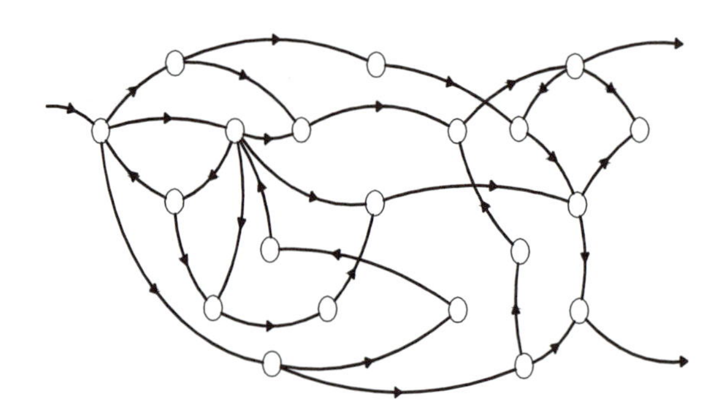

(b) system without natural internal boundaries

Figure 2.9 Component boundaries

The "natural" boundaries thus occur where there are fewest input-output connections among the sub-components. Even when it becomes necessary, for analysis or design, to break a complex system into arbitrary components, it is advantageous to choose the boundaries in such a way that the resulting inputs and outputs among the components are as few and as simple as possible.

Black-box components

In order to achieve an overview of the functioning of a system component it is not always necessary to study simultaneously the operation of each sub-component. In fact, by avoiding detail and by concentrating on the overall input-output relations among the various components, it is often possible to achieve a good overview of the behaviour of the system. When the detailed internal functioning within a component is ignored, and the component is regarded simply as a device, which transforms inputs into outputs, it can be thought of as a **black box**.

An example of a black-box representation of a component is the use of equations to calculate the run-off from a catchment in terms of the rainfall history. A very simple model treats the time history of the rainfall as input, and applies to this a delay and attenuation in order to represent the run-off as a second time history, as indicated in Figure 2.10.

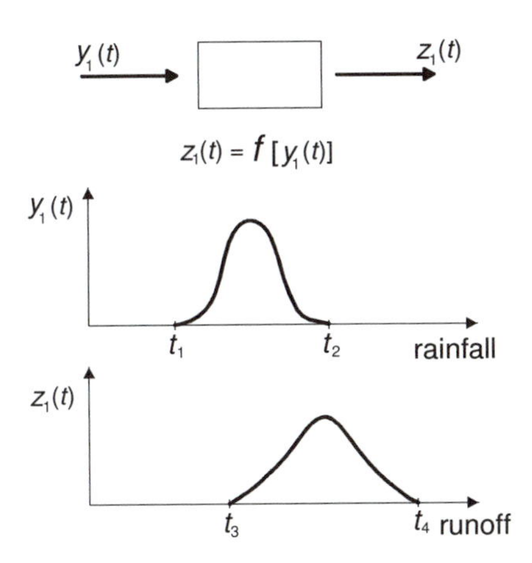

Figure 2.10 Black-box analysis: rainfall-runoff relations for a catchment area

A highly simplified and idealised treatment of a complete system can often be obtained by lumping closely linked components together and replacing them by black boxes with simple inputs and outputs. Such simplified treatments are often very valuable, for example at the preliminary stage of an analysis or of a design, when an accurate consideration of the functioning of the system is not needed. The idea of black-box components is used in electronic engineering for an arrangement of transistors, resistors and capacitors, assembled to fulfil a specific function. Such a device, designed to detect fish, is shown in Figure 2.11.

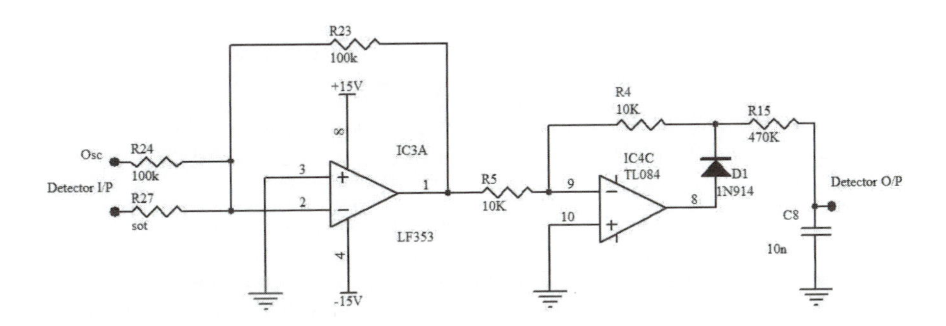

Figure 2.11 Electronic circuit for a fish detector. (Source: EngTest, University of Adelaide)

Control systems

Engineering systems usually operate with the purpose of achieving a specified goal. If the goal can be quantified, it may be possible to introduce a **control device** as a component to control the performance of the system so that it becomes goal seeking. For example, the control device may be linked to a sensor that measures the condition of the system and sends the relevant information to the control device. The actual condition (or state) can then be compared with a desired or target condition, and a control signal can then be sent to another component in the system to initiate changes that move the system progressively towards the target condition. This arrangement, shown in Figure 2.12, is a **feedback control system**, or **closed-loop control system**.

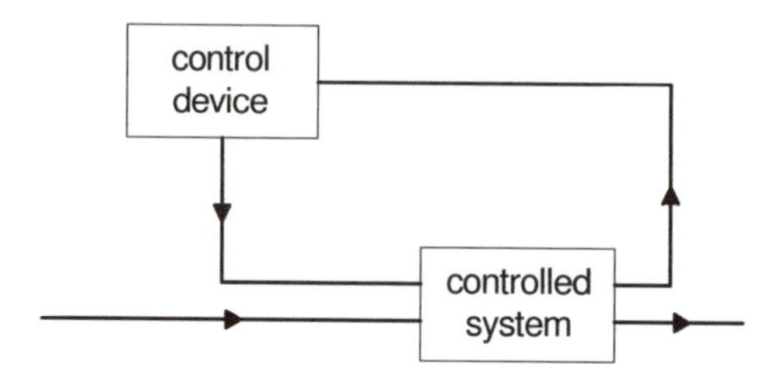

Figure 2.12 Closed loop control system

A simple example of a control system is an air-conditioned room (Figure 2.13). The desired air temperature is pre-set. The thermometer senses the temperature in the room, and the control component switches the conditioning unit on when the difference between the measured and the target temperatures exceeds a specified margin.

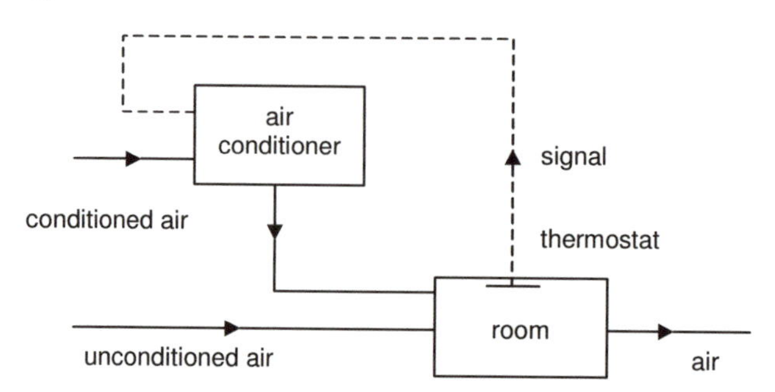

Figure 2.13 Air-conditioned room

Another example is the traffic control lights at a road intersection, where sensors are used to detect and measure the volume of traffic flowing into and out of the intersection in all directions. A simple control device calculates the optimal green times for flow of traffic in each direction, and a control signal is sent periodically to change the lights. While such localised traffic control systems are still used occasionally, it is far more common to use more sophisticated control systems to deal with the traffic in large sectors of a city, each of which includes a number of intersections. The traffic density is measured continuously in key streets and intersections and the information is sent back to a central computer system which has been programmed to calculate a sequence of signals to control local flows and minimise overall flow time through the road system.

A simpler, but less effective arrangement for a control system is shown in Figure 2.14. The feedback control loop has been omitted and so the arrangement is called, somewhat illogically, an **open loop control system**. As an example we may consider a set of traffic lights at an intersection that operates without the traffic flow counters. The preset timing of the lights is chosen to allow for, say, normal peak hour traffic flow. Obviously such a system would not be able to deal with unexpected or unusual events, such as traffic accidents or heavy rain.

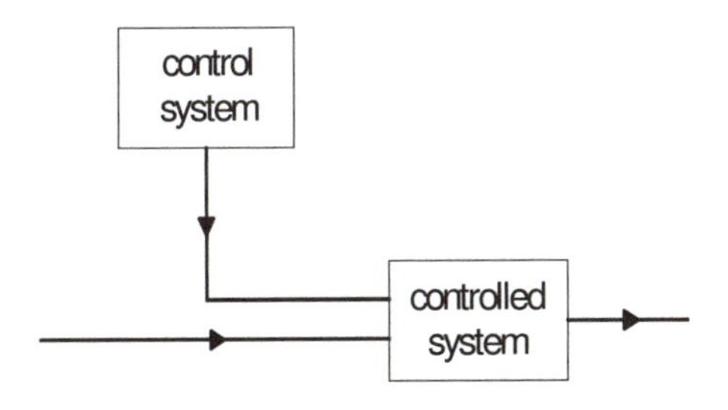

Figure 2.14 Open loop control system

The state of a system

The concept of system **state** is used in analytic studies of the behaviour of dynamic systems, that is to say, systems that change progressively in time. The state of such a system can be represented by a set of key variables, called the **state variables**. The values of the state variables at any given instant in time completely define the condition of the system and so allow its state at succeeding times to be evaluated. The state at a particular time instant thus takes account of all previous states and all inputs up to this instant.

When a medical practitioner investigates the state of health of a patient by measuring blood pressure, heart rate, condition of the eyes, chemistry of the blood and urine (counts of red and white blood cells, levels of cholesterol, triglycerol, uric acid, and sugar), the values represent a primitive set of state variables which indicate, very approximately, the state of health of the patient. The patient is of course a highly complex biological system, much more complex than the simplified state model being used in the health check.

The idea of system state contrasts sharply with the black-box representation previously described, because, in evaluating the state of the system, we look at the internal functioning of the system and not just at the inputs and outputs. The treatment of system behaviour as a sequence of changing states will be considered further when we look at methods of analysis.

From a formal mathematical viewpoint, the state of a system can be represented using a multi-dimensional space in which each state variable has an axis allocated to it. If there are n state variables then the state space has n dimensions. The state of the system is represented by a point in state space. The behaviour of the system over a time period thus traces a line in state space. Although state space concepts are used mainly for dynamic systems, the idea of state can be seen in Figure 2.8, where the state of the tension-bar system can be represented by the four forces in the bars, x_1, x_2, x_3 and x_4. The state variables in this case depend purely on the system inputs, which are the forces y_1, y_2 and y_3. An example of state space analysis for a simple linear dynamic system is presented in the appendix to this chapter.

Equilibrium and stability

Equilibrium and stability are vitally important concepts in engineering, and indeed in many other fields of study. These important ideas are best defined using state space concepts. A system is in an **equilibrium state** if there is no tendency for the system to move away from that state unless it is acted on by external inputs, that is to say, by external disturbances. An equilibrium state is thus represented by a point in state space. Depending on the complexity of the system, there may be more than one equilibrium state, or no equilibrium state at all. Furthermore, a system may be held in a state of pseudo-equilibrium by control inputs that maintain the state of the system at or close to a desired "equilibrium" point in state space.

This definition of system equilibrium is general, and is applicable in fields other than engineering, including economics and biology. An example of an engineering system in equilibrium is a building structure supporting vertical loads on its floors and resisting horizontal wind loads. The state variables of the system include the internal forces that are induced in the component members in response to the loads, and also the deformations and deflections that result from the loads.

We speak about the **stability** of a system in relation to its equilibrium states. A system is in a stable state of equilibrium if it returns to that equilibrium state after it has been disturbed. The equilibrium state is unstable if, following a disturbance, the system moves progressively away from that state. A system is in a state of neutral equilibrium if, following a slight disturbance, it remains in the new state.

A building structure, consisting of slabs supported on beams and columns and carrying vertical floor loads, can be tested for stability by applying a set of horizontal loads at the various floor levels and then removing them. The test may be carried out mathematically on a theoretical model, in the laboratory on a physical model, or may even be carried out on the real system. If the system returns to its original state with the same original internal forces upon the removal of the disturbing horizontal loads, it is in a stable equilibrium state. On the other hand, if the building is potentially unstable the small horizontal forces will induce large lateral displacements which may bring about a state of collapse that begins with buckling failure of the columns.

In reality of course, the response of a system to a disturbance depends on the magnitude of the disturbance. If the disturbance is sufficiently large, a stable system may be damaged or even destroyed and will not then be able to return to its equilibrium state. Thus, a bomb blast or earthquake can bring a stable structural frame to the point of destruction, if it is sufficiently intense. It is therefore common to describe a stable system as one for which the response to a disturbance is not out of proportion to the magnitude of the disturbance.

System coefficients

The main system variables consist of the input and output variables and the state variables. The latter represent the progressive behaviour of the system in time, or, more precisely, the behaviour of the system model. However, various other quantities are needed to construct the model. These are called the **system coefficients**. They represent the properties of the system and its components. System coefficients, in contrast with the system variables, are usually fixed in value and represent the physical characteristics of the system. For example, in the simple structure in Figure 2.8 the lengths of the component members, the co-ordinates of the various joints, and the stiffness properties of the components are all system coefficients that are needed to model the structural system. System coefficients typically include the properties of the materials comprising the system, and its geometry. Unlike the state variables and the output variables, they do not usually change in value in response to changes in the input variables.

Nevertheless, some system coefficients may be found to change slowly in time, not in response to the applied inputs but because the system itself is gradually changing. For example, some engineering materials change their properties in time. The modulus of elasticity of concrete increases slowly with time, because of complex chemical changes that occur in the material. This is an important coefficient in the modelling of concrete structures. Progressive changes in system coefficients can occur if there is deterioration, as for example when there is corrosion of steel, abrasive removal of layers of material in a pipe, or scouring and erosion of the bed of a stream. If one or more of the system coefficients vary significantly in time, then it may be advisable to restructure the model of the system, treating the varying quantities not as coefficients but as system variables. This is the case when stream erosion is so severe that a stream bed moves progressively over time, thus changing the geometry of the system. When this occurs it becomes necessary to move on from a fixed bed model to a movable bed model to deal with erosion.

Optimization

An engineering system has to have a clearly stated set of goals and objectives if it is to be designed, created and operated. Identification of the goals and objectives is thus an important step in the processes of planning and design. In all situations the aim of the engineer is to achieve the best possible solution, or possibly the most economical solution, for any constraints that have to be satisfied. The term **optimization** refers to the process of finding the best, or **optimal**, solution.

Optimization techniques can sometimes be employed to find the best solution to a mathematically formulated problem, irrespective of its physical nature. In optimizing a system or a process it is necessary not only to identify the goals and objectives but also to devise a measure of effectiveness to quantify these goals. The process of optimization is otherwise useless. Specific optimization techniques are discussed in some detail in Chapter 13 of this book.

The error of sub-optimization

It is important to understand that optimization of the overall system is *not* usually achieved by independently optimizing each of its individual components. Indeed, such an approach can lead to serious malfunction of a system if it is reasonably complex. The optimization of even one of the components in isolation may be detrimental to the overall system. A special term, **error of sub-optimization**, is used to refer to this important idea. The term describes the process of optimising one or more system components to the detriment of the performance of the overall system.

Examples of sub-optimization should make the matter clear. In the design of an aircraft, the aim is to achieve optimal overall performance. If the engines are designed to achieve maximum performance without due regard to their fitting into the airframe, to their weight, and to the way they work in conjunction with the airframe, then the result will be less than optimal so far as the aircraft is concerned. Indeed if the engines were designed disregarding the other components it is unlikely that the aircraft would fly.

A type of sub-optimization error can occur in the structural design of a building if a minimum weight solution is sought, on the assumption that this will lead to a minimum use of materials and hence to minimum cost. Minimum weight does not necessarily mean minimum cost. Structural systems designed for minimum use of materials are usually extremely difficult to construct, and therefore very costly. The additional costs of construction can far outweigh the savings in materials in a typical office building. On the other hand, minimum weight may be far more important in the design of a space vehicle because of the advantages in regard to payload and fuel use.

The design of a system is always a compromise whereby each component has to be tuned, not to perform optimally in itself, but to contribute to the optimal performance of the overall system. Sub-optimisation errors are prone to occur in large engineering projects when insufficient contact and communication occur among the different groups of engineers working on the different components, which eventually have to be brought together. An important task in project planning and management is to avoid sub-optimization errors.

2.5 ENGINEERING MODELLING

We have seen that a physical system may be represented by both physical and abstract (mathematical) models. A physical model may be a prototype, a scaled-down, or even a scaled-up, version of the real system, and is usually constructed and tested in a laboratory. For example a chemical engineer may create a scaled-down laboratory model of a proposed chemical engineering process. A structural engineer may choose to construct a small-scale physical model of a building in order to observe in a wind tunnel the effects of complex gust loadings. A mechanical engineer may decide to construct full-scale (prototype) models of an axle shaft in order to investigate fatigue life by testing. Physical models that have a similar appearance to the original system are sometimes referred to as **iconic** models. It is also possible to create physical models that do not resemble the original system at all. For example, electrical circuits can be used to represent groundwater flow. Such dissimilar models can be used if the same set of underlying mathematical relations apply to two different physical situations.

The modelling process

In order to create a model (either physical or mathematical) of an engineering system the problem of extreme complexity has to be dealt with. We have already seen how the systems concept of hierarchy can be useful in this regard. However, there is another problem that derives from the inherent complexity of the real world. Simply put, there is usually too much information available about any real system, and most of the information is either irrelevant or of marginal value. It is therefore necessary to separate out the aspects of the system that have an important influence on the behaviour that is of interest, and those aspects which are of marginal importance and may be ignored. An important step in the modelling process thus involves discarding information that is irrelevant to the particular analysis we wish to undertake. The difficulty, of course, is to decide which information to keep and which to discard.

For example, if we wish to model a component of a bridge, say a slab which carries wheel loads and transfers them to a supporting girder, the factors of prime structural importance are going to be the size and shape of the slab and the stiffness and strength properties of the materials. The colour of the components and the surface details of the road on top of the slab will not be of relevance and will be discarded. However, this information would be extremely important if an analysis of temperature effects were to be undertaken. Separating out the important from the unimportant for a specific analysis is necessary both in mathematical modelling and in constructing a physical model.

It is important to understand that there cannot ever be a single "correct" model of a real engineering system. On the contrary, there will be an unlimited number of possible models of varying complexity and accuracy, depending on the degree of idealisation employed in the modelling process. Furthermore, there will rarely be a single "best" model. Usually the more accurate models will be relatively complex, and will take longer to create and it will be more costly to use them in an analysis.

Although simple models are not likely to be accurate, they have the advantage that they can give quick (but approximate) answers and can provide a good overview of the functioning of the system. In fact it is often a real benefit to designers and analysts to have several alternative models available, including a simple, rough model and an accurate, detailed model. Designers can use simple models in the early conceptual stages of the design process and the accurate, more complex, models when they are working out the final design details.

To be acceptable and useful, a model has to satisfy a number of important requirements. It must, first and foremost, capture the essential behaviour of the real system; otherwise it could produce erroneous and dangerous conclusions. A process of **verification** is therefore needed to check the reliability of the model and to determine its likely accuracy, and the order of magnitude of prediction errors.

Modelling and Formula 1 cars

Physical modelling involves much more than simply building something to a convenient scale and then blowing air past it, running waves towards it, or imposing loads on it. The way the system behaves must be understood and tested accordingly. For example, a key design aim for Formula 1 cars is to reduce the drag that they experience when they are being driven at speed. It is relatively easy to build a 1/10th scale model of a car and blow air past it, but the question is: what air speed to use? If it's a 1/10th scale model then perhaps the air speed should also be 1/10 of that expected, say 30 km/h to model the full scale 300 km/h? This is not in fact the case. Studies show that the drag is a function of scale such that if the physical dimensions are reduced by a factor of 10 then the air speed must be increased by a factor of 10! So the correct answer would be 3000 km/h. Unfortunately at this speed there are other factors involved (such as breaking the sound barrier) that spoil the modelling. Thus, there is a desire to test at as large a scale as possible and in some cases this would mean a huge wind tunnel to test a full size car.

Problems with large-scale physical models have led to the development of complex computer models that can deal accurately with many of the processes involved in drag. These provide a viable alternative to physical models. One such model output is shown below(courtesy of Sauber Petronas and Fluent Inc). The flow lines were predicted by a computer model called FLUENT.
http://www.fluent.com/about/news/pr/pr81.htm(Downloaded 23rd January, 2006)

An analytical model can be used to predict the behaviour of a system if the physical principles that govern behaviour are well understood. The model may then be a set of equations that comprise a **mathematical model**. With the advent and widespread use of powerful computers it is common to use computer-based mathematical models to study the behaviour of even quite complex engineering systems. Today, mathematical models are normally cheaper to prepare and use than physical models.

There are various ways of verifying a model but generally this requires comparisons between predictions of the model and the observed behaviour of a real system. If the modelling is being undertaken for the design of a new, not yet existing system, then verification of the mathematical model may nevertheless be undertaken by comparing model predictions with the behaviour of similar real systems that have been previously constructed and tested. Physical models, tested in the laboratory, can also provide data for evaluating the adequacy and accuracy of a conceptual model. In some situations it is possible to calibrate a new model against an existing model that has itself been validated.

Occam's Razor

William of Ockham (or Occam) was an English philosopher who lived in the fourteenth century. He was an empiricist who wrote on science and logic, and argued that logic should be separated from theology and metaphysics.

His statement in Latin: *Essentia non sunt multiplicanda praeter necessitatem*, has become known as Occam's Razor; it has been translated as: *Do not multiply hypotheses beyond necessity.*

In the context of engineering modelling and analysis, Occam's Razor is often put forward as an argument for simplicity. According to this principle, of two equally accurate theories or explanations the simpler is always to be preferred. It follows that we should keep our models as simple as possible, and introduce no more assumptions than are absolutely necessary.

The potential accuracy of any analysis depends on the underlying model. However, an inappropriate use of a model can lead to inaccuracy with misleading and dangerous results. The careful interpretation of the results of every analysis is therefore needed to identify possible differences between the real system and the model. This is an essential step in any analysis. Verification becomes crucially important when we model complex systems, particularly non-linear systems, because some effects, such as positive feedback, are non-intuitive and are difficult to recognise. Unfortunately, the more complex the system, the more difficult it is to verify the model.

Yet another requirement of a conceptual mathematical model is **numerical stability**. That is to say, computational problems should not arise when the model is used to analyse a range of different systems, or a single system with varying coefficients and varying inputs and operating conditions. Numerical instability is one of the most common bugbears of mathematical modelling, and is particularly likely to occur when non-linear system models are used. It leads typically to non-convergence of certain operations, so that the calculations cannot proceed to their end. Even when the computations proceed to a conclusion, numerical instability is likely to lead to serious errors in the results.

2.6 ENGINEERING ANALYSIS

We have already seen that many different types of analysis can be carried out for a given physical system or problem, depending not only on the specific aspects of behaviour being investigated, but also on the type and accuracy of the model being used. We shall look briefly at some of the forms of engineering analysis that are commonly undertaken.

Modelling complex physical systems and processes

The modelling work that has been undertaken in an effort to manage and maintain the River Murray mouth in South Australia provides a good example of the use of a range of models of varying complexity. The photograph below shows the mouth as it was in April 2000, choked by a large influx of sand and with a small number of channels allowing water to flow into the lagoon adjacent to the land.

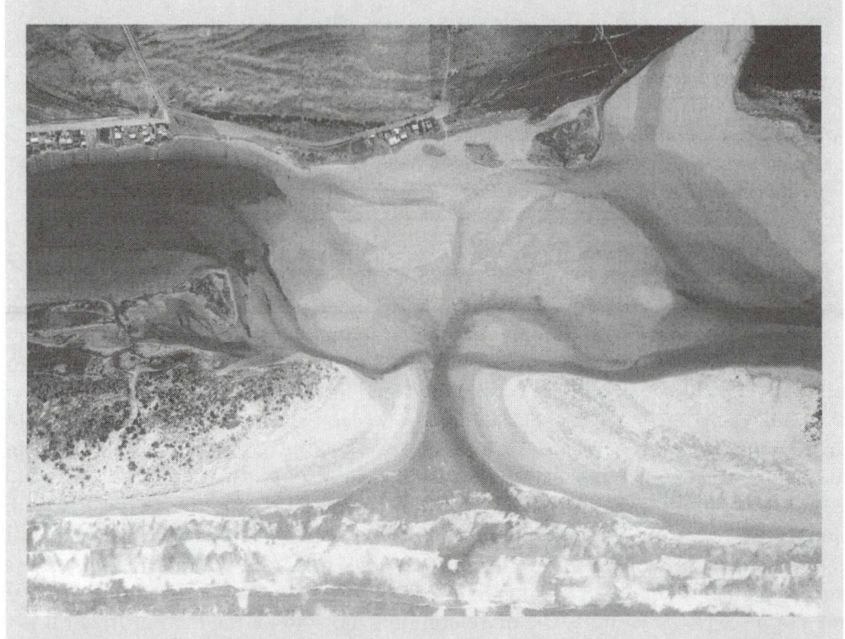

Reduced river flows due to increased irrigation and upstream usage had led to increased deposits of sand inside the mouth area, brought by the tides that flood in twice daily. In an optimisation study of river discharges two numerical models were developed: a simple one-dimensional model that represented the state of the mouth by a single number, the area, and a complex three-dimensional model that allowed a detailed analysis of the mouth, its channels and sandy islands over an area of several square kilometres. While the detailed three dimensional model could be used to evaluate discharge options and to give predictions of which channels would be choked with sediment and which would run freely, it took approximately one day of computer time to simulate one day of operation. Simulations of anything over a month were thus infeasible due to time constraints. On the other hand, the one-dimensional model could not predict what was happening to individual channels but it did allow a 100-year simulation of the mouth size to be undertaken in a matter of minutes and gave a useful prediction of the long-term fate of the mouth of the river.

If the real system displays significant variability in its response to repeated, apparently identical inputs, then it may be necessary to undertake a full

probabilistic analysis. Alternatively, a **statistical analysis** may be undertaken. For example, the coefficients may be treated as statistical quantities. Another possibility is to use a deterministic model but to carry out a **sensitivity analysis** by making a large number of computational runs with systematically varying values of the system inputs. Sensitivity analyses can also be used when the coefficients or inputs are not known precisely, but only statistically. Yet again, a large number of runs might be undertaken with randomly generated values of the coefficients and inputs, in order to look at the variability to be expected.

In contrast, variability might be ignored in favour of a highly simplified deterministic **input-output analysis** for the overall system. This may be appropriate when only very approximate estimates of behaviour are needed. We have already seen that input-output analysis can be carried out by setting up expressions for the output as functions of the input. These may be thought of as expressing a causal relationship with input as cause and output as effect.

Alternatively it may be appropriate to consider the system to be executing a process that transforms the input into output. The detailed modelling of the system usually requires the application of scientific and engineering principles drawn from the areas of knowledge relevant to the particular physical system. The analysis may proceed by considering the behaviour of the various components and the way they interact with each other.

It is a characteristic of real systems that they undergo change with time. If the changes occur rapidly, dynamic effects are likely to be important and so a **dynamic system analysis** will be undertaken and a dynamic system model will be employed. On the other hand, if the relevant changes do not take place rapidly, it may be possible to ignore the time dependent effects and use a **static analysis** or **steady-state analysis**. For example, in modelling a bridge to study the effects of highway traffic, consideration will be given to braking, impact and collision and this will require dynamic analysis; on the other hand, dynamic effects might well be ignored in modelling a small building constructed in a seismically inactive area. **State space analysis** is used primarily to study dynamic system behaviour. An advantage of state space analysis is that the equations of state are in an almost ideal form for computer computation.

The n equations of state for an n-dimensional system show the rate of change of each state variable with respect to time, as a function of the current values of all the state variables. This is usually very easy to program. An example of state-space analysis, with a set of state equations, is given in the appendix to this chapter.

The important concepts of equilibrium and stability have been explained in Section 2.4. It is frequently necessary to investigate the equilibrium condition (or conditions) of a system and its stability in relation to its equilibrium states. Methods of **equilibrium analysis and stability analysis** are available, both for static and dynamic systems.

Computer methods and computer simulation

A computer is normally used to carry out the detailed calculations required in an analysis of a complex engineering system. The use of computers increases enormously the quantity of information that can be generated by the analysis. In fact the form of the analysis, and the form of the underlying model, are sometimes adapted to suit the available computer software. This is because a large number of standardised computer software packages are commercially available in most engineering fields that can be applied to routine problems of analysis and design. The software packages are written to deal with a general class of problem. Thus it is often the case in engineering work that an appropriate software package is first chosen, and then the modelling and analysis decisions are undertaken in a way that is compatible with the software. **Computer simulation** of the behaviour of a system is made by calculating successive values of the state variables (or input-output variables) over an extended period of time for a prescribed set of operating conditions and system coefficients. The aim of computer simulation is to give a detailed picture of the way the real system will behave should it be subjected to the operating conditions chosen for the study.

System simulations can be repeated a large number of times with systematically varied values of the coefficients and inputs in order to observe the sensitivity of the system behaviour to small variations in the system properties. A sequence of such calculations has already been referred to as a sensitivity analysis.

The term **Monte-Carlo analysis** is used to describe repeated analyses undertaken with randomly chosen variations in system characteristics and inputs. Such analyses can be used to determine the effects of random variations on system performance. This in turn provides a means of studying stochastic system effects. The values of the coefficients and inputs are determined by random sampling from frequency distributions which represent variable real-world quantities. The results of the large numbers of analyses can then be used to create frequency distributions to represent system behaviour, thus providing a pseudo-probabilistic treatment of the problem.

In the case of complex non-linear systems, behaviour can be almost impossible to anticipate intuitively, so that modelling and analysis then become critically important. The only feasible approach is to use models that have been developed specifically for computer analysis. State space concepts and state space analysis are very useful in such circumstances. Complex, counter-intuitive behaviour in systems was demonstrated by Forrester (1971), one of the early pioneers in the use of numerical modelling and computer simulation of complex system behaviour. Forrester used the idea of positive and negative feedback loops to show how apparently simple changes can lead to unexpected and drastic outcomes. An example of a positive feedback loop is the effect of birth rate on population. Over time, even a small positive birth rate leads to an exponentially increasing population because of the compounding effect. In contrast, the death rate has a negative feedback effect on population. When several feedback effects occur simultaneously, especially when there are some opposing and some reinforcing loops, the results can be very difficult to anticipate. An example of modelling a system with such feedback effects was given by Dörner (1996), who investigated the problem of controlling excessively bad odours in a small pond. Some of the feedback effects he considered were related to temperature, oxygen, organic and inorganic substances present in the pond, and the anaerobic micro-organism population. Other examples of computer simulation of complex system behaviour, dealing with river mouth silting and global warming, are described briefly in interest boxes in this chapter.

Global Warming: modelling and analysis

The earth together with its oceans and atmosphere form what is now known to be a highly complex system that is acted on by the sun and moon and other celestial bodies. This understanding is relatively new and it is interesting to see how it has developed (and continues to develop) with time.

In the early 19th century Joseph Fourier determined that without the earth's atmosphere normal surface temperatures would be below freezing. In 1859 John Tyndall tested a number of the atmosphere's components and found that while oxygen and nitrogen were ineffective at trapping heat, methane and carbon dioxide were effective. The first two greenhouse gases had been identified. In 1896 Svante Arrhenius suggested that if for some reason carbon dioxide levels were to rise, for example as a result of increased volcanic activity, this could lead to a general warming which in turn would increase atmospheric moisture, and this would lead to further warming. These processes, including the feedback between temperature and atmospheric moisture, might be shown as follows:

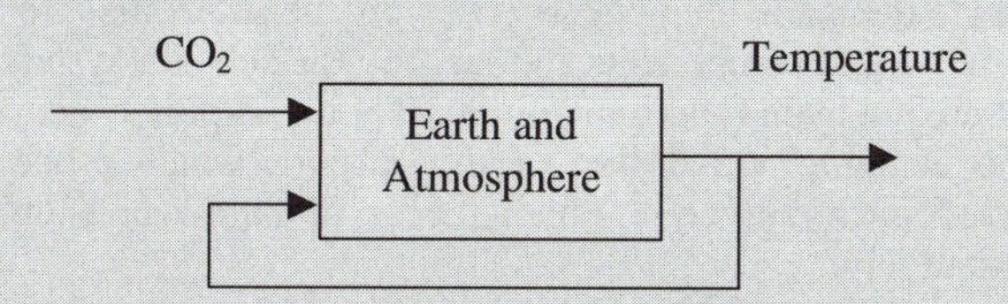

This figure suggests a system description of the earth's atmosphere with inputs of temperature and carbon dioxide affecting the output temperature. This is of course a highly idealised model of the atmosphere, and one that is actually wrong and misleading because it ignores many of the other important processes involved. Other factors that are now believed to be important include cloud cover, ice and snow cover, the oceans and their ability to store heat and absorb carbon dioxide, the ocean currents, the earth's ecosystems that act as both sources and sinks for carbon dioxide, man-made features such as aeroplanes and their exhaust contrails. The list goes on. Many of these have significant feedback effects and it is for this reason that there are so many arguments about global climate change. In all this uncertainty one thing is certain: it would not be possible to model the climate without a systems approach to describe and quantify the actions, reactions and interactions between all the processes at work.

2.7 SUMMARY

Various systems concepts have been introduced in this chapter, including system and process, hierarchy, input and output, state, optimisation, equilibrium and stability. These are used in modelling and analysing engineering systems. The concepts also provide a means for dealing with the complexity that is inherent in engineering systems and indeed in engineering problem solving.

The problem of complexity is handled by means of the process of decomposition whereby a system is broken down into components and sub-components that are simple enough to allow us to study how they behave individually and how they interact with each other.

The ideas of engineering modelling and analysis have also been introduced. A model is a simplified and idealised representation of a real system or process. It may be a physical model or a conceptual or mathematical model. Conceptual models provide the starting point for any mathematical analysis to determine how a system behaves under a given set of operating conditions. The process of engineering modelling has been briefly described and various methods of engineering analysis have been described. As we shall see in Chapter 3, systems concepts and modelling and analysis are essential tools in engineering planning and design.

PROBLEMS

2.1 Consider the airport for a city that caters for both local and international flights:

(a) What are the main components (i.e. subsystems) of the airport system?
(b) Choose several components and identify the sub-sub-systems.
(c) What are the main inputs and outputs for the airport system?
(d) In the planning and design of this airport, which processes should be analysed quantitatively? Which pieces of hardware should be analysed quantitatively?
(e) Which variables could be used to represent the state of the airport system at any time instant?

2.2 Answer the questions listed in Problem 2.1 in relation to a ten-storey city office building.

2.3 Identify and describe the functioning of some of the control systems to be found in the house or apartment where you live, and in the car you drive.

2.4 What would you choose as the state variables when analysing the dynamic behaviour of a diving board above a swimming pool? Which are the system coefficients?

2.5 Figure 2.7 in this chapter represents the layout of a city water supply system, but no outputs are shown. What are the outputs? What happens to these outputs? How does this system link up with other systems that are components of the city?

2.6 Identify possible sub-optimization errors in the design of a road network you are familiar with, and for the building you live in.

REFERENCES

Ashby, W. R. 1965. *An introduction to cybernetics*. London: Methuen.
Ashby, W. R. 1965. *Design for a brain*. London: Methuen.

Bazant, Z. P. & L. Cedolin 1991. *Stability of structures*. Oxford: Oxford University Press.
Chapra, S. C. & R. P. Canale 1998. *Numerical methods of engineering*, 3rd Ed. New York: WCB McGraw-Hill.
Coates, J. F. 2000. Innovation in the future of engineering design. *Technological Forecasting and Social Change*, 64, 121–132.
Dörner, D. 1996. *The Logic of Failure*. New York: Basic Books.
Elgerd, O. 1967. *Control systems theory*. Tokyo: McGraw-Hill Kogakusha.
Forrester, J. W., 1971. *World Dynamics*, Cambridge (Mass): Wright-Allen Press.
McLoughlin, J. B. 1969. *Urban and regional planning, A systems approach*. London: Faber & Faber.

APPENDIX 2A: EXAMPLE OF STATE-SPACE ANALYSIS

To illustrate the development of a set of equations of state, we consider the dynamic analysis of the two-storey frame shown in Figure 2.16. The building is subjected to horizontal loads due to wind or earthquake.

For the purposes of a preliminary analysis the floors are considered to be stiff in comparison with the columns and hence to move laterally under the influence of the horizontal forces, but not to bend. This is shown in Figure 2.16a. The wind forces $y_1(t)$ and $y_2(t)$ act at the first and second floor levels and vary with time. They are the input variables. The floors move horizontally and the supporting columns deflect laterally as in Figure 2.16a.

In order to construct a state model of this system we must first choose appropriate state variables to represent the condition of the system. The horizontal displacements at the floor levels, $x_1(t)$ and $x_2(t)$ are clearly appropriate choices. However these two variables alone are not sufficient to specify the state of the system: initial values of these two variables at some starting time t_0, i.e. , $x_1(t_0)$ and $x_2(t_0)$, do not contain enough information to allow all future states to be calculated when an input history, $y_1(t)$ and $y_2(t)$, is stipulated. In fact we need to include two further state variables. It is convenient to choose the velocities of the floors as the additional variables. Denoting the velocities as x_3 and x_4, we have:

$$x_3 = \frac{dx_1}{dt} = \dot{x}_1 \tag{2.1}$$

$$x_4 = \frac{dx_2}{dt} = \dot{x}_2 \tag{2.2}$$

If initial values of these four state variables are known at the time instant t_0, it is possible to undertake a dynamic analysis to determine all subsequent states of the system for a known history of the applied forces. The equations of state for the simplified system in Figure 2.16 are derived from simple dynamic principles.

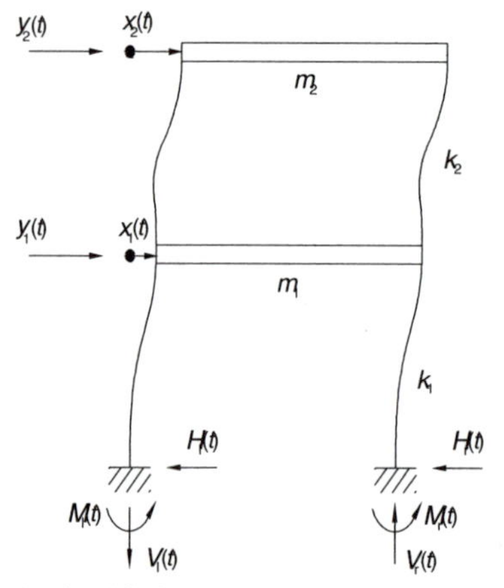

(a) simplified dynamic structural system

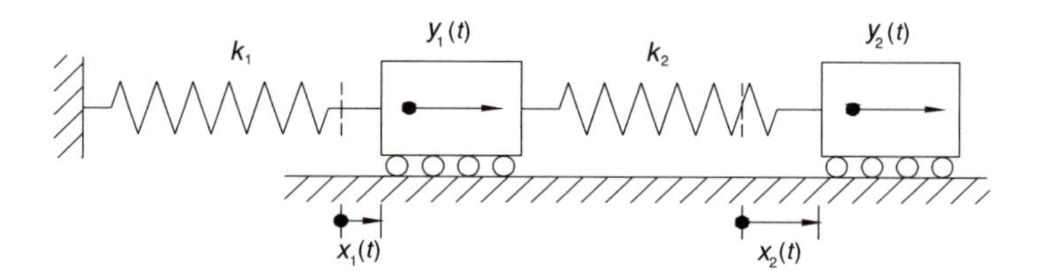

(b) equivalent dynamic system

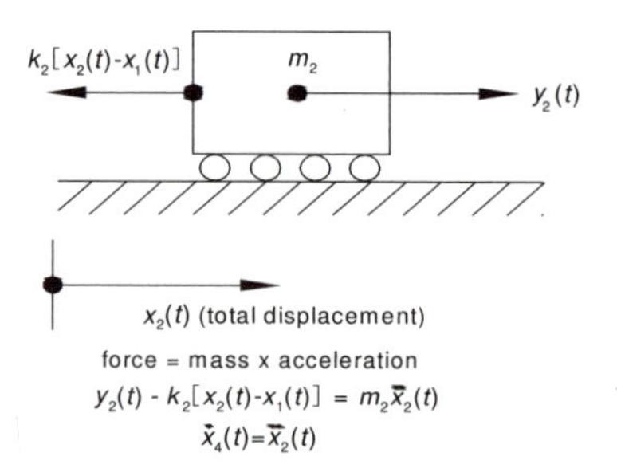

(c) forces acting on top floor

Figure 2.16 State space analysis of a two-storey frame

The top floor is represented as a point mass which is accelerating under the resultant of the two forces acting on it. One of these forces is the applied load $y_2(t)$; the other is the restraining force exerted on the floor by the tops of the second storey columns. The stiffness of the two columns together is k_2 Newtons per mm. This means that a force of k_2 Newtons, applied to the columns at the top floor level, would produce a movement of 1 mm, relative to the bottoms of the columns at the first floor. By equating the resultant force acting on the top floor mass to the mass times the acceleration, we obtain:

$$y_2(t) - k_2[x_2(t) - x_1(t)] = m_2\ddot{x}_2(t) \tag{2.3}$$

From Equation (2.3) and the definition of the fourth state variable, we obtain the following state equation:

$$\dot{x}_4(t) = \frac{1}{m_2}[y_2(t) - k_2(x_2(t) - x_1(t))] \tag{2.4}$$

Similar considerations apply to the mass at the first floor. In this case, however, there are three forces acting on the mass: the horizontal load $y_2(t)$; the force from the second floor columns acting on the mass; and the force from the first floor columns acting on the mass.

Table 2.1 System variables and coefficients for dynamic analysis

Input variables	
$y_1(t)$ (Newtons):	horizontal force applied at first floor level
$y_2(t)$ (Newtons):	horizontal force applied at second floor level
State variables	
$x_1(t)$ (mm):	displacement at first floor level
$x_2(t)$ (mm):	displacement at second floor level
$x_3(t)$ (mm):	velocity at first floor level
$x_4(t)$ (mm):	velocity at second floor level
Output variables	
$z_1(t)$ (mm):	displacement at first floor level ($=x_1(t)$)
$z_2(t)$ (mm):	displacement at second floor level ($= x_2(t)$)
System coefficients	
m_1 (kg):	mass of first floor
m_2 (kg):	mass of second floor
k_1 (N/mm):	stiffness of lower columns
k_2 (N/mm):	stiffness of upper columns

We thus obtain another state equation that represents the dynamics of the first floor mass. The four state equations are thus:

$$\dot{x}_1(t) = x_3(t) \tag{2.5}$$

$$\dot{x}_2(t) = x_4(t) \tag{2.6}$$

$$\dot{x}_3(t) = \frac{1}{m_1}\left[y_1(t) - k_1 x_1(t) + k_2\left(x_2(t) - x_1(t)\right)\right] \tag{2.7}$$

$$\dot{x}_4(t) = \frac{1}{m_2}\left[y_2(t) - k_2\left(x_2(t) - x_1(t)\right)\right] \tag{2.4}$$

These equations of state are in what is called the normal form, whereby the time rate of change (the first derivative with respect to time) of each state variable is expressed as a function of the current values of the state variables and the inputs. The equations of state thus allow us to calculate the incremental change in each state variable over the next small time interval from the current values of the state variables.

The normal form of Equations (2.5)–(2.8) is very convenient for undertaking numerical calculations. Various numerical and analytical techniques are available to solve the equations and hence obtain the values of the state variables at a succession of time instants, from the starting state and the inputs. We are not concerned with the mathematical methods here, and we shall not pursue the example further.

CHAPTER THREE

Engineering Planning and Design

Planning and design are the core activities that drive an engineering project from its inception through to a successful conclusion. They are essentially problem-solving processes. In this chapter we introduce a simple methodology for solving open-ended problems which provides the basis for undertaking engineering planning and design. The main steps involved in planning and design are described in some detail, and the chapter concludes with a brief review of some important related considerations, namely safety and risk, the role of codes and regulations, and legal and social issues. These are discussed further in later chapters.

3.1 TERMINOLOGY

Planning and design, together with management, are used to bring an engineering project from the initial problem recognition phase through to successful implementation. Broadly speaking, **planning** is a deliberative mental process that is undertaken to ensure that a proposed action will be successful. It involves deciding on the goal you want to achieve and then identifying the steps needed to achieve the goal. These steps, when clearly formulated, constitute a **plan**. Planning plays a crucial role in any engineering project. **Design**, as distinct from planning, is undertaken to produce the information needed to create a new system or process. Design work is also undertaken when an existing system or process is to be modified and improved. **Management** refers to the effective use of available resources to achieve a designated outcome. Management is thus the means by which the results of engineering planning and design are brought to a successful conclusion.

These descriptions of planning, design and management can be sharpened if we apply the systems terminology of Chapter 2. An engineering project is thus undertaken in order to change the current state of some part of the physical infrastructure to a preferred new state, either by introducing new systems or processes or by modifying existing ones. The planning for an engineering project therefore begins with a study of the current state of the infrastructure and proceeds to the identification of a preferred future state. It also involves working out the steps needed to change from the present state to the new state. Design is the process of determining the details of any required new systems and processes (or the details of modifications to be made to existing ones) that are needed to achieve the desired changes to the infrastructure. The purpose of management is to bring about the changes effectively, in a designated time frame, and with an efficient use of available resources.

By way of example consider a project that aims to improve the quality and quantity of water which will be needed by a township over the next 25 years. In the initial planning work it is necessary to determine details of the present water supply system, and in particular the quantity and quality of the water currently used as well as details of the delivery systems. The expected demographic changes in the city, including the increasing physical size and changing distribution of population and industry, need to be determined over the planning period so that future requirements and future demand can be estimated in order to identify the desired new system details. Possible new sources of water may have to be investigated, such as the use of groundwater and domestic rainwater tanks, and possible sites for reservoirs and dams. Combinations of these alternatives may also have to be considered. Additional options in the case of inadequate water sources may include the re-cycling of water and the use of pricing policies to curb the demand for water. Modifying and adapting the existing water supply system progressively over the planning period is also an important planning consideration.

Design work for the project would be needed to determine the details of any new components, or modifications to existing components. If, for example, it is decided that a new dam is to be constructed, design work becomes necessary to determine where it is to be sited, the size, shape and other details of the dam wall, and how it is to be keyed into the valley walls at the site. Before such design work can begin, it is necessary to identify and survey the possible sites and to obtain data on rainfall and runoff patterns, as well as relevant geological information. Careful planning is also needed to determine the sequence of steps for constructing the dam wall and other related components. The resulting construction plan for the dam will include activities such as providing access roads to the site, bringing in heavy equipment, excavating soil and rock, shaping and preparing the foundations, grouting unsound regions of bed rock, and progressive construction of the main wall. Design work is also needed for other related components of the delivery system such as pipes, pumping stations, holding tanks, and pressure tanks.

Comparable planning and design work would of course be needed for any other options that may be chosen. The planning and design activities themselves need careful management, as do all the steps in the implementation of the project, with progress monitored regularly to ensure that the work proceeds on time and within budget.

The term **project planning** is applied to the planning for an overall project, while the planning undertaken at sub-project level is referred to as **activity planning**. The planning required to carry out a particular step in an engineering project is thus activity planning. The planning of an activity results in a sub-plan, or component, of the project plan. Using this hierarchical view of planning, Griffis and Farr (2000) go further and define **program planning** as the planning involved at the broadest level where, for example, various related engineering projects are identified and appropriate resources are determined and allocated on a preliminary basis.

Program planning sits at the top of the hierarchy of engineering planning and is used to create a program aimed at improving some aspect of the physical infrastructure. The program would normally include a range of projects to be carried out together or in sequence. For example a flood mitigation program in a river valley might include the construction of dams in the upper reaches, silt dredging near the mouth and the building of levy banks in intermediate regions along the river. Such work would require several different engineering projects.

Strategic planning is undertaken by businesses and other organisations to produce long-term plans needed for their ongoing successful operation and survival. Another form of long-term planning, of interest particularly in situations where the future is uncertain, is **scenario planning**. In this book the focus of attention is restricted to project planning and activity planning, as used in engineering work.

The design activities in an engineering project can also be broken down into tasks and sub-tasks. This breakdown is needed to deal with the complexity inherent in a large infrastructure item such as a dam, a road, a bridge, a building or an airport. Also in the design of a unit that is to be manufactured in large numbers, such as a washing machine, a mobile telephone, or an automobile, the design is undertaken using the hierarchical approach. It is also used in the design of processes, such as one to manufacture and deliver ready mixed concrete to building sites in a city.

It will be appreciated that the design and planning work required for a large engineering project are closely linked to each other and also to management. It is not necessary, or even desirable, to classify all of the interrelated activities in an engineering project into distinct and separate categories of planning, design and management. Indeed, there are very close similarities between these processes because they are, basically, problem solving activities.

3.2 PLANNING AND DESIGN AS PROBLEM-SOLVING PROCESSES

Besides being complex, engineering problems are also ill-defined and open-ended. An ill-defined problem is one that is vaguely or ambiguously formulated. An open-ended problem, on the other hand, is one that does not have a single "correct" solution even when it has been clearly formulated. It is a characteristic of engineering problems that many alternative, potentially acceptable solutions can usually be found, so that the problem is not so much to find "the" solution but rather to find the "best" solution. While systems concepts allow us to deal with the complexity inherent in engineering systems and in engineering problems, they do not provide a means to solve open-ended problems.

Gilhooly (1982) has suggested that a problem has three components: a starting state, a goal state, and a set of procedures that are used to reach the goal state from the starting state. A well-defined problem is thus one for which all three components are clearly specified. Some problems in the fields of mathematics and logic are well defined. Board games (such as chess and backgammon) provide good examples of well-defined problems. For example, the game of chess presents to a player a well-defined problem with a clear starting state, a target end state (checkmate of your opponent) and precisely defined intermediate steps, or moves.

It should not of course be assumed that a well-defined problem is necessarily simple or easy to undertake. The three components of chess are clearly defined; nevertheless it can be a very difficult, challenging and exceedingly complex game. If your opponent is a computer, the chess-playing software probably handles complexity by exhaustively elaborating sequences of possible moves and counter moves. The capability of your opponent in this situation depends on the length of the sequence that the computer can handle.

Planning and design for the Great Pyramid of Giza

Studies of the great pyramid of Khufu (or Cheops) have shown how complex and ingenious the planning and design for this mammoth engineering project must have been. The pyramid, located on the outskirts of present day Cairo in Egypt, was built more than four thousand years ago. It has a 230 m square base and covers an area of 5.3 hectares or 13 acres. Over 140 m high, it is the greatest monumental construction of antiquity. Apart from its special chambers, the pyramid originally consisted of three main components: a massive inner core of stone blocks, a thin surface cladding of white limestone, and an intermediate fill layer between the cladding and core. The surface cladding must have given the construction a spectacular appearance, but it has been vandalised over the millennia. The main inner core contains over 2 million large stone blocks of limestone and granite, mostly weighing between 2 and 6 tonnes. These were sourced from various quarries, then sized and shaped before being transported by Nile barges to the site, where they were moved on rollers and raised to the working level (up to 140 m high) before being placed precisely in position. One theory suggests that a temporary inclined helical ramp was built around the partially completed pyramid and used to move the blocks up to the working level. It is estimated that the project was completed in a little more than twenty years by around 20,000 men working in small teams.

Pyramid **design** was complex and ingenious. In earlier pyramids the stone blocks were given a slight inward inclination to improve stability and reduce settlement. The great pyramid was constructed on bedrock which had been cleared, levelled and shaped to give stability and prevent settlement. The blocks were placed in horizontal layers, generally with larger blocks at lower levels and with smaller blocks in the higher layers. Corbelled ceilings with granite beams were used for special rooms, and stress-relieving chambers were introduced above the burial chamber.

The **planning** for the construction must have been of the highest order, comparable with that needed for a modern monumental engineering project. On average it was necessary to transport, deliver, raise and place about 300 blocks per day. With say a twenty-man team to look after each block, an average on-site workforce of above 6000 would have been needed. However, it seems that the main building work took place during the yearly flooding of the Nile, when agricultural work could not be undertaken. The building site must have then been a hive of activity. Even looking after the needs of the workforce would have been a highly complex exercise in planning and logistics.

The planning, design and management skills evident in the construction of the Khufu pyramid had been acquired by experience and trial and error over generations of pyramid building. The name of Imhotep, a vizier, medical doctor and scribe, is associated with pyramid design and construction, in particular the step pyramid of Saqqara. He is perhaps the first engineer to be known by name. The vizier responsible for the great pyramid is thought to have been Hemiunu.
Sources: dtv Atlas (1974) ; Craig B. Smith (2004)

In the case of engineering problems, the starting states and the target states are both likely to be ill defined, initially. Furthermore, the intermediate steps, needed to proceed from the initial state to the target state, do not necessarily follow simple, well-defined rules.

There are good reasons why engineering problems are initially poorly defined. The need for an engineering project can become evident gradually over time as inadequacies in the infrastructure are experienced and community dissatisfaction grows, eventually with demands for improvements. In such situations the needs of the community are expressed forcefully, but not necessarily precisely and unambiguously. Even when an engineering problem has been clearly formulated and stated, there is going to be a range of alternative approaches available to solve the problem. In other words, the problem is open-ended. For example, road traffic congestion in a part of a city might be dealt with by increasing the capacity of the existing road system. An alternative approach might be to develop an efficient and cheap public transport system, or even to use some social engineering and modify the starting and finishing times for employees of industries and businesses in different regions of the city.

In some situations it may be very difficult to identify the required target state. It may not even be possible to know whether the correct engineering decisions have been made until the project has been completed. The planning and design for a new and improved car model involves a great deal of engineering work, but the introduction of the new model is, in the end, a risky commercial venture which is finally tested in the market place. The open-ended nature of engineering work raises a number of important questions. In particular:

- how do we solve an open-ended problem if there is not going to be a unique, correct solution?
- how can we know if we have found a good solution to an open-ended problem?
- how should we go about finding a good solution to an open-ended problem?

To answer such questions we need to look at problem solving strategies.

3.3 METHODOLOGY FOR SOLVING OPEN-ENDED PROBLEMS

People use a range of problem-solving strategies in their everyday life. One popular method is to try to solve a current problem by applying a method that proved to be successful in the past. This approach can work well in relatively simple situations, provided the problems, past and present, are very similar in nature. However, difficulties arise if there are substantial differences between the present and past situations. If both problems are complex then there are inevitably going to be differences between them.

In engineering work, past practice is often drawn on for ideas for solving current problems. However, it is very dangerous to rely exclusively on this approach. Adaptation is always needed, and great care must be exercised to ensure that the differences between the past and present situations are taken into account. Furthermore, past practice is not relevant in the case of a new type of engineering problem. In many situations there is simply no relevant past practice to rely on and engineers must then create innovative, new approaches.

A simple strategy for solving open-ended problems

When faced with a problem that is both ill-defined and open-ended, common sense suggests that the first step should be to clarify the problem and re-state it in clear and unambiguous terms, in so far as this is possible. If a range of alternative, approaches present themselves when the problem has been clearly formulated, then the initial instinct may well be to choose the most "obvious" one and develop this into a solution. The obvious or instinctively most acceptable solution is likely to be one that has been used previously and that we are therefore familiar with.

On reflection, it becomes clear that instinct alone will *not* lead necessarily to the best, or even to a good, solution. It follows that we should *not* concentrate, initially, on any particular solution, but rather do just the opposite and look for as many different, promising solutions as possible. To allow this to happen, it is important that we formulate the problem in general (not over-specific) terms so that unusual but promising solutions are not excluded.

It is also important to understand that our search for unusual but promising ideas cannot be successful if we rely on an analytic approach. Instead of deductive, convergent reasoning we have to adopt a divergent thinking approach. We must think laterally. Creativity is extremely important because we want to find unusual, non-routine solutions. In Chapter 4 we shall look at the question of creativity and how creative approaches to problem solving can be developed.

If we have been successful in creating a range of different and promising solutions to our problem, the next task is to evaluate them, rank them and hence identify the best approach. These simple thoughts provide the basis for a straightforward problem-solving strategy which can be applied to poorly formulated, open-ended problems. The steps are listed in Table 3.1.

Table 3.1 Strategy for solving complex, open-ended problems

Step	Action
1	Formulate the problem clearly but in general terms
2	Develop a wide range of promising approaches for solving the problem
3	Evaluate and compare these approaches and hence identify the best one
4	Work out the details of the solution, based on the best approach
5	Implement the solution

Methodology for complex engineering problems

The strategy listed in Table 3.1 is appropriate for relatively simple open-ended problems where it is easy to evaluate and compare all of the alternative options. This will be the case in simple activity planning tasks. If, however, we have to deal with a problem where there are many, relatively complex solution options (and this is typically the case in engineering work) then an enormous amount of unnecessary effort would have to be expended in the third step. In such circumstances the strategy becomes very time consuming and inefficient.

For example, if we consider a harbour city and the problem of choosing the most suitable water crossing for the road traffic, there may be a dozen main options, including a bridge at various alternative sites and an under-harbour tunnel with alternative routes. Each bridge site probably requires a different bridge design, just as the various tunnel routes have to be dealt with on an individual basis. If the problem solving strategy of Table 3.1 were applied unmodified, it would be necessary to undertake full, detailed designs for all the options. The design and planning costs in engineering work are always substantial and in this case they would be many times the cost of a single design. This means that the bill for the planning and design work alone could take up a very substantial portion of the budget.

A modification to the problem-solving strategy in Table 3.1 is thus needed for use in engineering planning and design. The costs in time and effort can be reduced enormously if, instead of fully evaluating all of the options before identifying the best one, we use a step-by-step approach. It is far better initially to make only a simple and rough evaluation of each option and then, on the basis of this limited information, immediately eliminate the least competitive ones. These initial evaluations are not costly because they are based on approximate, order-of-magnitude calculations which are nevertheless sufficient to identify the uncompetitive options. A second, more detailed evaluation of the remaining options can then be used in a further cull.

Proceeding step-by-step and progressively eliminating uncompetitive options on the basis of additional information, we can obtain a short list of the most promising options. The best option can then be chosen in a final round of detailed comparisons. In this way, the total amount of design work and analysis can be reduced substantially. The steps in this methodology are listed in Table 3.2.

Table 3.2 Methodology for solving complex, open-ended problems

Step	Action
1	Formulate the problem clearly and in general terms
2	Develop a wide range of promising approaches
3	Choose criteria for ranking the alternative approaches
4	Cull the least promising approaches using simple evaluations
5	Cull progressively, using more detailed evaluations, until a short list remains
6	Choose the best approach from the short list, using very detailed evaluations
7	Develop the best approach into a detailed solution
8	Implement the solution

A big advantage of this methodology is that the most investigative effort is expended on the most promising options. The detailed information thus obtained is not wasted because it can be used in Step 7 to develop the detailed solution.

The need for iteration

It might appear from the foregoing discussion that the methodology can proceed in the simple linear sequence listed in Table 3.2. This most definitely is not so. It will often become clear in a later step that a modification, improvement or correction is needed in one or more of the steps already completed. For example, in Step 4 when a new and unusual, but promising, approach is being evaluated, it might become clear that the original problem statement is unnecessarily restrictive and should be revised. Rather than staying rigorously with the original problem statement it is far better to go back and restate the problem in a more encompassing way, since the overall aim is to obtain the best possible solution. Likewise, when alternatives are being compared and ranked in Steps 5 and 6, it might well be found that an improved problem statement will allow improvements to be made in some of the approaches.

While comparing and ranking the short-listed alternatives it might also become apparent that some approaches have been unnecessarily penalised or even eliminated because of arbitrary, unnecessary restrictions inherent in the criteria. A reformulation of the criteria is justified if this can lead to a better solution to the problem. It may also become clear in Step 6 that some approaches can be modified to achieve a much more favourable evaluation and a better solution to the problem. Iteration is thus an inherent ingredient in the methodology.

It may sometimes be advantageous to undertake several steps simultaneously. For example, it may be difficult to choose good evaluation criteria before any approaches have been proposed. It would then be preferable to undertake simultaneously the search for alternative approaches and the search for evaluation criteria. However, a possible complication to be avoided is unintentional bias in the choice of criteria that may favour an intuitively appealing and favoured approach.

Rather than elaborate now on the steps of Table 3.2, we shall first see how this methodology can be applied to the planning process, and then to design.

3.4 ENGINEERING PLANNING

A methodology for undertaking engineering planning is shown in Figure 3.1. It follows closely the problem-solving methodology listed in Table 3.2 but typical feedback loops are included to emphasise the need for iteration.

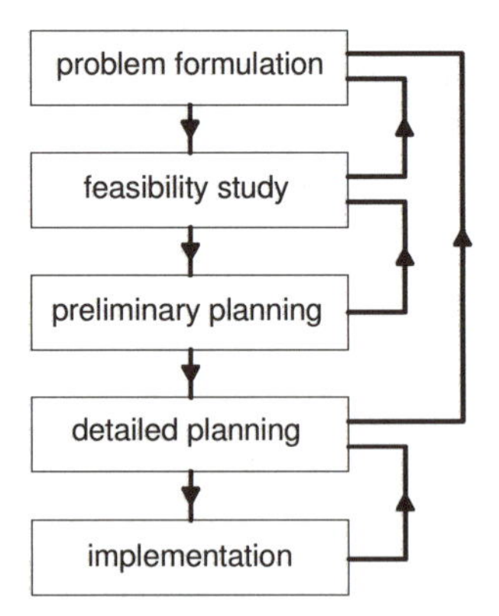

Figure 3.1 The process of engineering planning

There has been a regrouping of the steps in Table 3.2 into five main phases with some new terminology. In particular, the step of choosing the evaluation criteria (Step 3 in Table 3.2) has been combined with Step 1 into the initial

problem formulation phase. The **feasibility study** incorporates Steps 2 and 4, while **preliminary planning** consists of Steps 5 and 6. Step 7 has been called **detailed planning**. These changes have been made to conform to normal engineering usage.

The overall aim of the initial problem formulation phase is to clarify and, if possible, quantify the problem. This is not as simple as it may at first appear because engineering problems are usually many faceted. For example, **constraints** always apply to an engineering project, and limit in various ways the approaches that can be taken. For example there are limits on overall cost, on available time, and on the quantity of materials and human resources that are available for a project. Account needs to be taken of these constraints in the initial problem formulation phase.

The **feasibility study** is the second phase of the process in Figure 3.1. This name implies that the feasibility of the project is to be demonstrated by actually finding a feasible (or workable) approach. Of course, the aim is not just to find any solution, but rather a very good one. It is therefore important to consider a wide range of solutions, including unusual ones. As already mentioned, we try, initially, to avoid focusing on any particular approach, but rather to generate as wide a range of different and contrasting feasible alternatives as possible.

The preliminary comparison of alternatives commences during the feasibility study when the options have been identified. This allows us to test out our evaluation criteria and modify them as necessary. The feasibility study continues until a short list of good, feasible options has been found. If no feasible approach has been found then the study cannot continue. The problem has to be postponed, modified or left unsolved. Reasons for not finding a feasible solution may be related to difficulties in satisfying the constraints that are imposed on the solution. For example, costs may be unacceptably high. Sometimes the technology may not be sufficiently advanced to allow an economical and safe solution to be formulated. If no feasible solution has been found, it may be appropriate to repeat the feasibility study, but with relaxed constraints or less optimistic goals.

In the third phase of the process, **preliminary planning,** further rounds of comparisons and evaluations are carried out, the aim being to identify the best option on the short list. The comparisons now require more detailed analysis and evaluation than previously. If one of the alternatives stands out clearly from the others, then the preliminary planning phase is straightforward and can be quickly and easily completed. On the other hand, if the alternatives are closely matched, this phase may become protracted because of the need for very detailed comparisons of the competing options.

The purpose of the **detailed planning** phase is to take the best option, previously identified, and use it work out the full details of the solution. If a considerable amount of analysis and evaluation went into identifying the best option, then the detailed planning work will be correspondingly reduced. The end of this phase of the planning process is a plan for implementation. However, this is not a final plan, as it is likely to undergo further changes. We have already emphasised that iteration is inherent in the planning process. Indeed, the plan will not be finalised until the project is finally completed. There is an often quoted statement: it is not the plan that is important, it is the planning.

The form of the last phase, **implementation**, depends on the nature of the engineering problem. Implementation may require the construction of new infrastructure systems or the implementation of a new plan. At first sight implementation may seem to be a separate phase of the project, to be undertaken when the planning and design work has been completed. This is not so. Whatever form the implementation takes, it will be necessary to return to and revise some of the earlier planning work. In order to achieve ease and efficiency in implementation, it must be treated as an interlinked phase of the planning process.

In later sections of this chapter we shall discuss in some detail how to undertake the various phases of the planning process. Before doing this, however, we look at the corresponding phases of the design process.

3.5 THE DESIGN PROCESS

The phases of the design process shown below in Figure 3.2 parallel those in Figure 3.1 for planning. However, there are some differences in terminology. In particular, the term **concept design** is used to describe the second phase of the process. This term describes the initial search by designers for a range of alternative approaches. Emphasis is thus placed on finding alternative design concepts or options, rather than on investigating feasibility. The aims of this phase, whether in design or planning, are nevertheless identical: to produce a short list of the best feasible options for solving the problem.

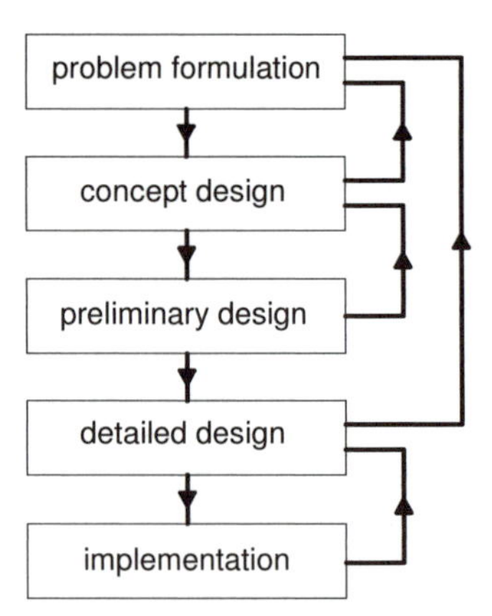

Figure 3.2 The design process

In the structural design of a bridge the concept design phase is undertaken with the purpose of finding suitable alternative structural forms that provide adequate "load paths" for the design loads. These load paths carry the applied loads, such as self-weight, vehicular loads and wind loads, through the different structural components and into the supporting foundations. For a large-span bridge, the alternative structural forms might include an arch, a suspension bridge, a cable-stayed bridge and a multiple span system of beams with intermediate piers.

Initially, however, the aim is to produce a wide range of alternative, promising concepts and then to reduce these down to a short list, using the evaluation criteria developed in the problem formulation phase, with calculations that become increasingly detailed.

It is during the concept design phase, when original and unusual approaches are considered, that the most general, far-reaching and financially most important design decisions are made. As the design proceeds through to the detailed phase, the design decisions become more and more restricted. This means that the possibilities of cost savings are correspondingly reduced.

In Figure 3.2, the terms **preliminary design** and **detailed design** correspond to their counterparts in Figure 3.1. The purposes of these phases are similar in both planning and design and, as we shall see later, they are carried out in the same way.

The purpose of the preliminary design phase is to study the concepts short-listed in the previous phase in sufficient detail to allow the best concept to be identified. If it is difficult to choose from several very good concepts, then successive rounds of calculations and comparisons are needed. These studies may become, in effect, detailed designs. It is not unusual for the preliminary design phase to blend into the detailed design phase, with detailed designs being carried out for several of the very best options. For example, in the design of a dam for a water supply system that was planned for the Hastings district in New South Wales, Australia, 24 sites were identified in an initial study. During preliminary design this was narrowed to two, before the better was chosen for the final design (Thompson, 2005).

In the detailed design phase, the aim is to develop the solution from the previously identified best approach, with enough information to allow full implementation. If one concept has been clearly identified at an early stage as the best, then a fair amount of additional work will be required in the detailed design. On the other hand, the effort needed in the detailed design is correspondingly reduced if the chosen concept has already been investigated in detail.

As previously noted, implementation always needs careful consideration. Unfortunately, poor engineering design almost always results if inadequate consideration is given to the details of implementation.

While Figure 3.2 gives an accurate view of the way engineering design work is undertaken, some designers prefer to describe the design process differently. It is not unusual to treat problem formulation as a part of concept design. Also, the term preliminary design is sometimes used in lieu of concept design. Thus, the design process may sometimes be described as a three-phase sequence consisting of concept design, detailed design and implementation, or preliminary design, detailed design and implementation.

Irrespective of the way the design process is described, the activities of problem formulation, concept design and preliminary design (in the sense explained above) are essential steps in the design process. In this book we have chosen to describe design in terms of the five phases of Figure 3.2. There are two reasons for this: firstly, this highlights the importance of problem formulation in design; secondly the close similarities between the processes of planning and design are emphasised.

What is not shown in Figure 3.2 is how the need for design work is often first recognised during the project planning process. This is when the need for a new system or process is often first identified. Initially the need is not formulated very clearly or precisely, so that problem formulation remains an important step to be undertaken at the commencement of the design process. The following statement by Richard Seymour and quoted by Liston (2003) emphasises the crucial role of problem formulation in engineering design:

> *Something which is often forgotten or misunderstood is that the vast majority of the work involved in design is finding out what the problem really is– and it's rarely what you think it is. More often than not the client is asking the wrong question. Once you've found out what the problem really is, then things virtually design themselves because you've so comprehensively understood the problem that the solution is self-evident.*

3.6 PROBLEM FORMULATION PHASE

Table 3.3 contains a checklist of actions that can assist in the problem-formulation phase of engineering planning and design. The terminology used here comes from Chapters 1 and 2. Some of the actions are information-gathering activities while others are aimed at formulating and, to the extent possible at this stage, quantifying the problem. Although intended specifically for planning and design, these activities can be useful in engineering problem solving generally. We now discuss them in turn.

Table 3.3 Checklist of actions for the problem formulation phase

Action
Identify the problem as part of a larger or wider problem
Identify the relevant engineering system as part of a wider or larger system
Identify the components of any relevant system
Find interest boundaries for the problem
Determine the real underlying needs that are to be addressed
Gather relevant background information
Search for possible side-effects
Identify constraints
Specify objectives and identify possible conflicts among objectives
Specify any performance requirements and operating conditions
Devise measures of effectiveness

Identify the problem as part of a wider problem

It has been seen already how an engineering project can begin as a response to perceived needs in the community, and that the objectives may be ill defined initially. The problem as stated may be vague; alternatively, it may be stated in over-precise terms which imply an "obvious" solution. In the latter case, the implied solution will rarely be the only possible one, and may not be the best one. It is therefore important to begin the problem formulation phase by trying to identify the problem as part of a wider or larger problem.

By way of example, let us return to the problem of providing traffic access between two parts of a city separated by a waterway. Suppose that the cross traffic is already catered for by several bridges but these cannot handle the increasing traffic volumes. The "obvious" problem might be seen as constructing an additional bridge to cater for present and future traffic. Is this really a statement of the problem, or is it statement of a solution to the problem? A broader problem statement would be: how to improve the cross-harbour traffic flow. An advantage of this statement is that it widens the range of possible solutions. It is now possible to consider a tunnel, a range of small peripheral bridges, various forms of water transport, or even air traffic as alternatives. This example was used by Svensson in his book on engineering design in 1974.

In fact, the city of Sydney has been faced with a traffic problem of this nature for many years. In the post-war years there was a single bridge to carry traffic across the harbour between the northern and southern suburbs. Although a second bridge was constructed on the periphery of the harbour at Gladesville, traffic volumes continued to grow and traffic congestion became worse. There were community calls for another large harbour bridge. The next step was in fact to construct a tunnel to address the cross-harbour traffic problems in the medium term. The tunnel, integrated into the arterial roads of the city, was initially considered successful, although the problems of traffic congestion in the city continue to grow as the traffic volumes increase.

It is useful to take our hypothetical example further. Is the real problem simply one of regularly increasing the traffic capacity on all the main routes in response to growing volumes? A broader problem statement might be to reduce or eliminate the current and future traffic congestion. This statement allows more varied approaches to be considered. It allows, for example, a reduction in peak traffic by staggering starting and finishing times for various industries, and the use of flexitime in offices. It also encompasses initiatives such as imposing surcharges on traffic using critical facilities at peak times, and encouraging the increased use of alternative forms of transport such as public transport. A form of congestion tax has been introduced in cities such as Singapore and London to reduce the number of vehicles using the inner city streets.

It will be clear that broadening the problem statement does not disallow the initial "obvious" solution. On the contrary, it opens up the problem to a wide range of alternative, competing solutions and in this way actually tests the adequacy of the "obvious" solution.

Identify each engineering system as a component of a wider system

To assist in identifying the problem as part of a wider problem, it may be useful to identify the relevant engineering system as a component of a larger or wider system. In the above example the broader problem statements have focused on the transport system, of which a bridge is just one possible component.

It can be useful to take the problem-broadening technique further. In the above example of the bridge, is the problem purely one of achieving an efficient city transport system? The costs involved in providing roads and bridges to improve the traffic network of a large city run into thousands of millions of dollars. When such expensive traffic options are considered, other radically different possibilities deserve consideration. It might be possible to achieve a more workable city by reducing the need for cross-harbour traffic. Here we are extending our discussion beyond traffic engineering and into the realm of town planning. We can go further: the use of expensive resources to improve road transport and hence the efficiency of an already congested city can be queried. Even if the road transport system were to be improved, the result is likely to be more city growth, followed by further congestion and yet more calls for improvements to the again-inadequate transport system. An alternative use of the resources could be to introduce decentralisation schemes to encourage the relocation of industry and population away from the city and into regional growth areas. We have now moved on from town planning questions to a consideration of regional planning and various associated political issues.

Identify interest boundaries for the problem

How far should the process of problem broadening be taken? Clearly we could go even further with the bridge argument, beyond the regional planning alternative and into questions of national or even international planning. At some stage the problem broadening argument breaks down. How do we recognise this stage? The questions cease to be relevant when they are so broad that they cease to be relevant to the original problem statement. This occurs when the expanded system that we are considering is so large that the original problem (in this case, the need for a bridge, or some alternative) is not relevant. When this happens, we have clearly gone too far.

The limit, where the problem has been widened to the extent that it is of marginal relevance, is the **interest boundary** for the problem. The relevant solution options are always to be found within this boundary. In Figure 3.3 the interest boundary is shown for the bridge example already discussed. In this case the very large costs suggest that the interest boundary should certainly include the city system and perhaps extend to the surrounding region or even state.

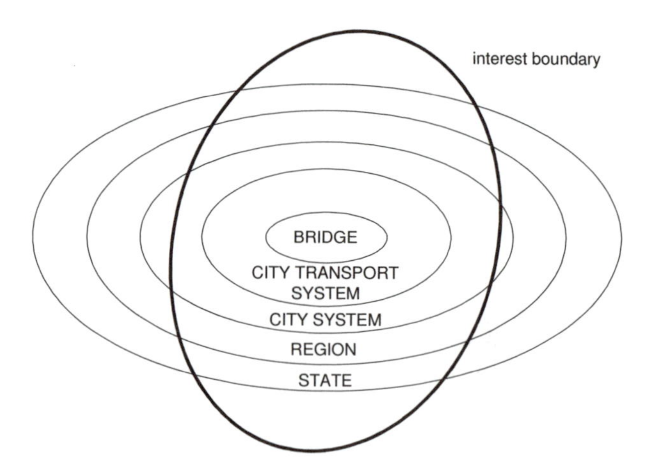

Figure 3.3 Problem-widening: interest boundary for a bridge

The construction of a large bridge in a small country can become an international issue. An example is provided by the K-B Bridge in Palau, a small island nation in the Western Pacific, somewhat less than a thousand kilometres west of the Philippines. The bridge was constructed in the 1970s with aid money when Palau was a protectorate of the United States. At the time, the bridge was the largest prestressed concrete box-girder cantilever construction in the world, linking the two main islands of Korror and Babeldaob. It served a crucial purpose because essential services including the airport and power generation equipment were located on one island while most of the population lived on the other. The bridge collapsed suddenly and apparently without warning in July 1996, shortly after it had undergone extensive refurbishment. Life in Palau was severely disrupted, even though a ferry connection was established between the islands. The construction of a new bridge was clearly beyond the resources of Palau and had to wait until it could be undertaken with international aid. The interest boundary in this case extended well beyond national boundaries.

As another example of an interest boundary, this time a relatively small one but again in the field of traffic engineering, we consider the redesign of a hazardous traffic intersection in a suburban area. The traffic intersection may initially be seen as a part of the local traffic system, consisting of the intersection and the immediately adjoining streets. This can in turn be regarded as part of the traffic system in one region of the city. The interest boundary is indicated in Figure 3.4.

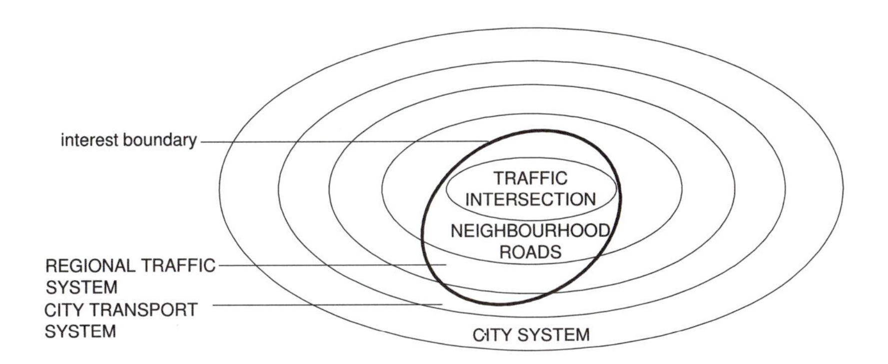

Figure 3.4 Problem-widening: interest boundary for a traffic intersection

In this case the costs of the proposed project are a small proportion of the annual road transport budget. Also, conditions at the traffic intersection are of only marginal relevance in relation to the efficient flow of traffic through other regions of the city.

In summary, the boundary-widening process should be halted while the expanded problem still has an impact on the original system. The interest boundary can be identified in this way.

Determine the real underlying needs as distinct from the stated needs

The real needs that initiate an engineering project are not always those that are initially perceived. The process of problem widening and the identification of the engineering system as a component in a wider system usually lead to a better understanding of the real needs. In the bridge example, the need is not for a bridge *per se*, but for a reduction in traffic congestion, or, possibly, a more efficiently functioning city. When the needs are properly identified the problem can be formulated in a more general and much more useful way.

In situations where the size and scope and influence of a project are limited, it may be much easier to identify the relevant underlying needs. As an example, consider the construction of a swimming pool as part of a house design for a private client. Before the design work is undertaken the real needs of the client have to be clarified. The pool may be needed for exercise, for diving, or for social entertaining and relaxation. It might be required purely for prestige reasons, or for a combination of reasons. In each case a different design may be appropriate. The problem statement needs to be obtained through discussions with the client.

The attempt to identify underlying needs sometimes moves attention away from purely technical problems towards social issues and political problems, especially when the proposed work is large in magnitude and costly. This is almost inevitable if large sums of public money are to be spent. As the problem-widening and system-widening processes are undertaken, the questions inevitably acquire an increasingly political and social flavour.

Gather relevant background information

When work commences on a new project it will often be found that there is a lack of relevant background information. Such information has to be gathered as early as possible, and before the problem formulation phase is completed. The required information may be scientific and technological, but it may also be non-technical and sociological, legal or political in nature. The data-gathering exercise may require the use of libraries, the Internet, textbooks, and databases. Sometimes it will be necessary to gather field data and conduct laboratory tests, as for example when the properties of a foundation material have to be determined.

The type of information potentially needed in engineering work is almost unlimited in scope. When international one-day cricket matches were initially planned, a lack of suitable turf wickets was seen as a potential problem and the design of replaceable, transportable turf wickets was investigated. In this case the background information involved the way the game is played and the laws of cricket. Detailed horticultural information was required on the appropriate types of grass for preparing turf wickets, how to grow the grass, and the depth of soil needed, not only to allow the grass to grow but also to provide the right "bounce". Other information was needed on how to achieve the best type of drainage and compaction, and the required wearing properties of the turf. The structural questions, relating to creating a rigid, transportable structural tray to support the pitch, seem relatively uncomplicated. In unusual projects, the engineer rarely has sufficient knowledge or training in the relevant specialist fields. It is necessary to gather information rapidly and also to work in association with consultants and experts with specialised knowledge.

As a further example, consider the relocation of access roads in a suburban area. There would be an obvious need for technical information concerning the soil properties along the proposed routes as well as the present and expected future traffic volumes. To obtain this information a program of soil testing and traffic counting might be undertaken. Other relevant background information of a quite different nature could relate to the social structure of the community through which the roads are to pass and the location of the roads in relation to natural community boundaries, such as the catchment areas for local schools. The political implications of such a project, measured in terms of votes gained and lost, may also turn out to be important background information, influencing the decision on whether the project proceeds or not.

Search for side effects

Each engineering project is designed to bring about some change in the world in which we live. Although the intended aim is to achieve an overall improvement in the infrastructure and hence satisfy community and individual needs, it is inevitable that there will be side effects. These may be obvious or subtle, desirable or undesirable. They may only become evident after a considerable period of time, when the project has been completed and is in operation. In the case of a large-scale project there will usually be a wide range of side effects, some beneficial and some detrimental.

A side effect of improved car security

In a report in the Australian Newspaper of 5th June, 2006, a rise in the number of "car jackings" was attributed by police to improved "immobilisation technology". The new technology has made the theft of unattended late model vehicles almost impossible. According to the newspaper report, thieves have therefore taken to "holding up drivers across Sydney's wealthy suburbs".

The identification of possible side effects should begin as early as possible during problem formulation but it should also continue into the evaluation stages because some side effects will depend on the chosen solution. It is clearly important that the relevant side effects be identified and allowed for when the costs and effectiveness of each alternative approach is evaluated.

Some disturbance of the environment is inevitable when a large infrastructure project is undertaken, and environmental side effects can become a focus of community attention. When a road or freeway is constructed, either in an urban or rural development, some modification of the surroundings is involved, such as the demolition of existing older buildings. It may involve cutting into a hillside or filling in part of a valley floor in the countryside. Even when a large span bridge is constructed across a valley to lessen the physical effects on the environment of a roadway, there is a visual impact. Of course the impact may be pleasing to many.

Technical side effects may be advantageous or disadvantageous for a project. For example, the provision of lift wells and walls in a multi-storey building design can also improve the resistance of the building to lateral seismic loads. Non-technical side effects can be very significant, such as increased or decreased job opportunities. These may be temporary or long lasting. The influence on the tourist industry of the construction of a monumental, iconic building, or the development of water sports and recreation facilities following the construction of a dam, are examples of non-technical side effects which may have an effect on the outcome of a project. They need to be identified and spelt out in the problem formulation phase.

Identify constraints

It is necessary in the problem formulation phase to identify the constraints that apply to the problem. Constraints arise in various ways. They may be technical, legal, economic, social, environmental and even political in nature. Monetary cost is a constraint that applies in one way or another to every engineering project. Constraints arise when certain side effects are unacceptable or undesirable, such as excessive atmospheric pollution. Technological limitations create other constraints.

Legal constraints apply, for example, to building construction. They limit the maximum footprint size of a building on a building site in a city and also the maximum building height. Other examples of legal constraints are the emission control requirements for automobile engines and the minimum requirements for fire protection and toilet facilities in buildings that house people.

Legally enforceable industry standards are often the means by which constraints arise in engineering design. Limits for noise control and thermal insulation thus apply in the design and construction of apartment buildings. Design constraints may be introduced for the components of a system so that they fit together and work in harmony with each other. Constraints can sometimes be quantified as physical limits or as minimum performance requirements which have to be achieved.

The unavailability of certain resources can result in other constraints. For example, telephone poles were commonly made of wooden tree stems in most parts of Australia in the mid-20th century. However, the lack of trees in South Australia severely constrained the use of timber and led to the early and widespread use of a steel-concrete composite pole, called a Stobie pole after its designer. Stobie poles have been a distinguishing feature of the South Australian landscape for many decades.

Another important constraint is time. Completion deadlines are crucial constraints in the construction of sporting complexes and facilities to be used every four years for the Olympic Games. The timelines for construction are determined by the set dates of the games. Such deadlines are fixed. If they are not met then heavy financial penalties can be enforced.

Physical constraints frequently apply to engineering problems. In the early stages of the design of the Gateway Bridge in Brisbane, a minimum clear height above water level at mid span was required in order to allow shipping to pass under the bridge. On the other hand, a maximum overall height limit was imposed to give air clearance for flight paths to the nearby airport. The initial constraints on minimum and maximum heights clashed for this bridge so that no solution was possible until the constraints were examined more closely and loosened.

In the case of large engineering projects with overt government support, it is not unusual to find that political considerations lead to explicit or implicit constraints on the engineering work. Local sourcing of materials and the creation of employment are frequently negotiated before any engineering work is undertaken and become constraints on the engineering work. Such constraints have to be identified early in the problem clarification phase of the project.

Identifying Constraints: Power supply for Waterfall Gully

As already noted in the main text, much of South Australia's power is carried on the distinctive steel and concrete Stobie poles. However, there are situations where these cannot be used. In the late 1970s the increasing demands for electricity in Adelaide's eastern suburbs led the state electricity body, the Electricity Trust of South Australia (ETSA), to investigate the provision of additional capacity to suburbs adjoining the Adelaide Hills. This required 66kV power lines to traverse difficult terrain over a number of hills.

Various constraints were identified in relation to the location and nature of the proposed work, such as:

Cost and visual appeal: although underground power lines would have been preferable from a visual point of view, the cost of this solution had recently increased significantly and essentially ruled out underground lines as an option.

Legal: planning approval was required from the Department of Environment and Planning and this led to further constraints on what could be installed.

Environmental: the Department of Environment and Planning imposed constraints in terms of visual amenity and the physical environment, including limiting the scope of roads that might be used to access the area.

Social: although the residents of nearby councils were the beneficiaries of the new power supply, they exerted significant social and political pressure on the planners in an effort to maintain their views and the natural characteristics of the area.

Technical: without the ability to drive roads and tracks through the area a solution had to be developed using available technology. The solution was built around the use of a helicopter. This had technical constraints associated with it, because there was a limit to the load that could be lifted, in this case a spare carrying capacity of less than 140kg.

The constraints that were imposed left few viable options. The one that was chosen was to build and partially assemble three steel structures. The components were to be brought to the site by helicopter and bolted together while secured by guy ropes.

Difficulties encountered in assembling the pylons, coupled with other issues, led to the deaths of four workmen who were on one of the structures when it suddenly fell to the ground during an operation designed at securing it in alignment. Three died instantly and the fourth was pronounced dead on arrival at the Royal Adelaide Hospital.

Source: Grabosky (1989)

Define objectives and identify conflicts among objectives

While it is important to clarify the initial problem statement, it can be advantageous to postpone trying to do this until at least some of the background information has been collected. In particular the process of identifying relevant systems as components of larger systems and identifying the problem as part of a larger problem can assist in identifying the objectives.

Furthermore, engineering projects usually have not just one but a number of different objectives, and in such situations some of the objectives are very likely to conflict with each other. In particular the twin requirements of maximum performance and minimum cost are always going to be conflicting objectives. Dorner (1998) has suggested that "contradictory goals are the rule, not the exception, in complex situations." It is always necessary to identify the potential conflicts among the objectives.

For example, a dam can usually be used both to store water and to assist in reducing down-stream flooding. Some dams are also designed for the additional purpose of electricity generation. For effective storage the dam should be kept nearly full, whereas for flood prevention it needs to be nearly empty and ready to accept runoff after heavy rain. The operating policy for the dam has to recognise and allow for the implied conflicts.

It is important for any potential conflicts among objectives to be explicitly recognised. The question of how to deal with conflicting objectives will be discussed in some detail in Chapter 9.

Specify any performance and operating conditions

When a project involves the design and production of a physical system or device it is necessary to determine both the conditions under which it will operate and the level of performance that will be required. For example, if a pipeline is to be used to move natural gas from a production field to a consumption site, the performance requirements for the system will include the maximum and average quantities of gas that will be transported in a time unit, as well as the minimum life of the operation. Such information is also needed for the planning and design of other components of the overall system, such as storage tanks and pumps.

The operating conditions of interest in this example might include the ambient temperatures throughout the year, the magnitudes of any possible seismic action, and the chemical properties of the gas in relation to its effect on the pipe material.

Safety and reliability are also performance requirements which always require very careful consideration. These are discussed briefly in Section 3.12 and will be examined further in Chapter 11.

Devise measures of effectiveness

Measures of effectiveness are needed to evaluate and rank alternative approaches and concepts. They are used in the feasibility study phase of project planning and also in the preliminary and the detailed design phases. When a project has more than one objective, a measure of effectiveness is needed for each specific objective, together with an overall measure that takes account of the separate objectives.

In many situations total cost can be used as an overall measure of effectiveness. This is the case if the various objectives can be stated in terms of equivalent cost. Overall cost is also appropriate if the various objectives can be reformulated as minimum performance requirements, or as constraints that have to be satisfied. Some situations are more complex and more difficult to deal with, for example when measures of effectiveness are needed to take account of matters such as aesthetics, environmental effects, risk of injury and loss of life.

Even when cost is used as the measure of effectiveness, conditions can become complex when the matter is analysed carefully. For example, in the replacement of the superstructure of a bridge, total initial cost might at first appear to be entirely appropriate, so that it would be a simple matter to choose, say, from construction based on steel trusses, steel plate girders, in-situ reinforced concrete girders and precast, prestressed concrete girders. Purely in terms of cost of construction the plate web girder solution may be a clear winner. However when maintenance costs are taken into account for the expected lifetime of the bridge, which may be between 50 and 100 years, the reinforced concrete girder solution may take precedence. But other factors might also influence the decision, such as the need for minimum disruption to traffic during construction. The problem is now to establish a time-cost trade-off. This might lead to the choice of a prefabricated truss system. Yet again, if appearance is important, as well as cost and time needed for construction, the extremely difficult matter of the monetary value of appearance arises. The precast, prestressed concrete solution may now be competitive.

The manner in which the measures of effectiveness are formulated can be enormously influential in determining the direction that a project will take. In some circumstances an engineering project might be carried out without the explicit use of a measure of effectiveness. For example, a design approach might be chosen subjectively by the design engineer without a study of alternatives and without any measure of effectiveness. Such an approach might be taken if the project is small, on the argument that an exhaustive problem formulation study could add more to the cost of the project than non-optimal decision making. However, there are inherent dangers in such an approach which should be evident from the discussions to date. One possibility is that the wrong problem is solved. Errors of judgement easily occur unless an attempt is made to identify measures of effectiveness. In fact the attempt to define a measure of effectiveness is a valuable starting point in design and planning, even for small projects, because it gives fresh insight into the project.

Iterations

The various activities listed in Table 3.3 all occur within the problem formulation phase; nevertheless, they may need to be undertaken iteratively. The type of background information that is required becomes progressively clearer after some attempt has been made to identify side effects and constraints and possible approaches. Likewise, some of the constraints and side effects are more easily recognised after some background information has been gathered. It will also be necessary to return to the problem formulation phase as we proceed through the later phases of planning and design.

3.7 FEASIBILITY STUDY AND CONCEPT DESIGN

The purpose of the feasibility study in planning is to show that the project can be carried out successfully, by developing a short list of promising alternative approaches, any one of which can form the basis of a solution. Likewise, the aim of the concept design phase in design work is to establish a short list of promising concepts, each of which may result in a successful engineering design.

To ensure that all the options in the short list are competitive, it is necessary to begin with as wide a range of alternatives as possible. We have seen how each option can be investigated superficially in order to cull the non-feasible and non-competitive options. A further study, with a somewhat more detailed analysis of each option, leads to a further culling. The culling continues, using progressively more information, until a short list of feasible, promising options remains. A list of steps that may be useful in the feasibility study and in concept design is shown in Table 3.4. The steps are discussed below.

Table 3.4 Checklist of actions for feasibility study

Action
Check available resources
Investigate and quantify the constraints
Develop as many promising concepts as possible
Compare alternative concepts using the measures of effectiveness
Progressively eliminate the non-competitive approaches
Modify the problem formulation as necessary
Modify the measures of effectiveness as necessary
Scrap or defer the project if no feasible approaches can be found
Identify the most promising concept

Check available resources

Resources are used up in the course of an engineering project and a check is needed to ensure that sufficient resources are in fact available. Resource availability (or non-availability) may affect the viability of particular options, so that this check should be undertaken with the various alternative concepts and approaches in mind.

It is necessary to consider human, financial and technical resources, as well as special materials and machinery that may be needed during implementation. Engineering expertise and scientific knowledge are resources, as are relevant trade and craft skills. Specialised design and analysis skills may also be required. Time is a resource because engineering work always has to be completed within a limited time frame.

Investigate and quantify the constraints

The technical and other constraints that have been identified in the problem formulation phase now have to be investigated and, if possible, quantified. Some of the constraints that apply to engineering work are legal-technical, and are quantified in codes and standards and in legislation. Thus, in the design of a large city building, constraints on overall height, minimum services for water, sewerage, fire protection, thermal and acoustic insulation and even vertical transport, are imposed through the relevant ordinances and building acts.

Develop as many promising concepts as possible

This is the key step in the entire processes of engineering planning, design and problem solving. The success of any design or planning work depends on the range, quality and appropriateness of the approaches and options that are generated in this step. As already emphasised, the diversity of the alternatives is important. Innovative and creative new approaches, as well as traditional and proven approaches, should be included in the initial list of alternatives. It is by no means clear whether an innovative new approach or a well-tried standard approach will eventually prove to be best; both types need to be considered.

The technique of problem widening, described in the previous section, should prevent an overly specific problem statement that would otherwise stifle new and unusual approaches. Creativity is central to this activity and is discussed further in Chapter 4, together with techniques that may prove helpful in the search for new and different options.

Compare alternative concepts and progressively eliminate non-competitive approaches

The preliminary comparison of options is made using only simple calculations and preliminary information. This information is usually sufficient to allow the poorer options to be culled.

In the design of a highway bridge to cross a river, order-of-magnitude design calculations for the alternative options are sufficient to provide very approximate sizes for the main components of the structural system. Such calculations can be used to obtain preliminary costs and also to check that the constraints and operating conditions are met. Information on costing can be found in Chapter 8.

As further rounds of comparisons are made, the detail and depth of the analysis and design increases so that more refined comparisons and evaluations can be made.

Modify the problem formulation as necessary

The need for an iterative approach throughout the problem solving process has been emphasised. Modification of the original problem statement needs consideration during the feasibility study. As work proceeds and new and unusual approaches are considered and investigated, the understanding of the problem is inevitably improved. With improved understanding, it is often possible to achieve a better problem formulation. New and unexpected approaches tend to challenge the validity of the original problem statement, particularly in regard to the measures of effectiveness that are used to compare and evaluate the alternatives.

Scrap or defer the project if no feasible approaches can be found

If none of the options are feasible, for example because they do not satisfy constraints relating to time, cost or resources, then the project cannot proceed. In these circumstances it might be decided to cancel the project, or to defer it until technical knowledge has improved to the level needed, or until additional resources become available. Another possibility is to look for new, innovative approaches that will prove to be feasible. Yet another possibility is to reformulate the problem with a different, perhaps more modest set of goals and less severe constraints.

3.8 PRELIMINARY PLANNING AND DESIGN

The purpose now is to bring the search for the best option to a positive conclusion and identify the approach that will lead to the best solution to the problem. Each of the short-listed options is investigated in turn and in sufficient detail to allow comparisons and rankings to be made, using the measures of effectiveness.

Even at this stage, very accurate comparisons are avoided if at all possible because of the cost implications. On the other hand, if alternatives are eliminated on insufficient grounds, the most appropriate alternative might also be incorrectly eliminated. The progressive approach therefore continues until all but one of the alternatives are eliminated.

Although attention is focused on the original options which came out of the feasibility study, the search for new and better alternatives should not be discontinued in the later stages of planning or design. Modifications to existing approaches should be made if this will improve them. As work proceeds, there is a build up in expertise. The increased expertise and additional background information can lead, even at this stage, to further improvements and changes to the original problem statement and to the measures of effectiveness, as well as to new or modified design or planning concepts.

3.9 DETAILED PLANNING AND DESIGN

The aim in this phase is to work out the details of the solution to the extent necessary to allow implementation. For example, at the end of the preliminary design of a reinforced concrete bridge the form of construction and the approximate overall dimensions of the component members will have been chosen. It is now necessary to determine details of the structure, including final member sizes, the amount, type and location of the reinforcement in each member, non-structural fitments, concrete strength, concrete finishes, special road surfaces, handrails, bearing pads and storm water pipes, so that full construction plans and specifications can be prepared.

Accurate calculations are needed in detailed planning and design. Sophisticated methods of analysis may be required, the details of which will depend on the relevant field of engineering. The methods are nevertheless developed from fundamental areas of knowledge including solid mechanics, fluid mechanics and physics and chemistry.

Even in the detailed phase of planning and design, iteration is necessary. It may be necessary to develop alternative trial details for some components, and then use calculation and analysis to evaluate and rank the alternatives. If adjustments are made to the design details with the aim of improving performance or decreasing cost, a new analysis will be needed to check whether the improvements have in fact been achieved. Such iterations continue until an effective and economical design or plan has been achieved, that meets all the design requirements.

It is in the detailed phase of planning and design that optimisation techniques may be employed. If the behaviour and performance of a component lends itself to theoretical modelling, it should be possible to improve the design by mathematically optimising the parameters that define or characterise the component. The process of optimisation is discussed in some detail in Chapter 13, together with various mathematical optimisation techniques.

At all stages of the detailed design it is important to check that the design constraints are not violated. In some situations overly severe constraints may add disproportionately to the cost or detract from the effectiveness of a solution. Even during the detailed planning and design phase it may be advisable to modify decisions made previously in the problem formulation phase so that a return to the problem formulation phase may be necessary.

In the detailed design of the components of a system, consideration must be given to the overall operation and cost of the parent system. If the design of a component has a disproportionate effect on overall cost and effectiveness, then some modifications may be possible for this component and for other neighbouring components in order to achieve an improved overall design. If components are to be manufactured in quantity it may be desirable, depending on the nature and expected cost of the component and the number to be produced, to construct prototypes and test and modify them as an adjunct, or alternative, to the theoretical analyses.

The important final step in the detailed design and planning phase is full documentation, with a permanent record of relevant calculations and analyses and any other investigations that have been used to produce the final plan or design.

3.10 IMPLEMENTATION

Implementation of a plan, a design or a solution to an engineering problem can take many forms, depending on the context of the work and it is not possible to discuss the details of implementation here.

Unfortunately, poor engineering design almost always occurs if inadequate consideration is given to implementation. For example, constructability is a very important criterion that is too often forgotten in the structural design of buildings. An undue focus on optimum design can thus lead to an elegant design with a minimum use of materials, but exorbitant construction costs.

We have specifically mentioned (if only briefly) the implementation phase here in order to emphasise its importance in the overall scheme of planning and design.

3.11 THE SOLUTION-FIRST STRATEGY

The sequence of steps shown in Figures 3.1 and 3.2 follows from the methodology for solving open-ended problems. However, the necessity of iteration, and the advantages of undertaking several steps simultaneously, has been mentioned. In some situations a rearrangement of the sequence shown in Figure 3.1 or 3.2 may be advantageous. It can be argued, for example, that choosing the evaluation criteria *before* the range of alternative approaches has been identified is an overly abstract exercise. An alternative is to choose the evaluation criteria after some or all of the options have been identified. This may be a better alternative in some situations, although it may lead to an unintentional bias in the criteria that favours some approaches over others. Another possibility is to undertake these activities simultaneously. Various rearrangements of the sequence are possible and desirable in certain circumstances.

To emphasise the fact that alternative sequences may be appropriate, we now mention briefly a solution-first strategy, which stands in sharp contrast to the sequence in Figures 3.1 and 3.2. Numerous examples can be found in the history of engineering where an important project has *not* commenced with the identification of a problem but, on the contrary, has started with a potential solution. The task is then to search for an appropriate use for the solution. A good technical idea may arise from some technical or scientific development, or by bringing new, potentially useful knowledge from another field of engineering.

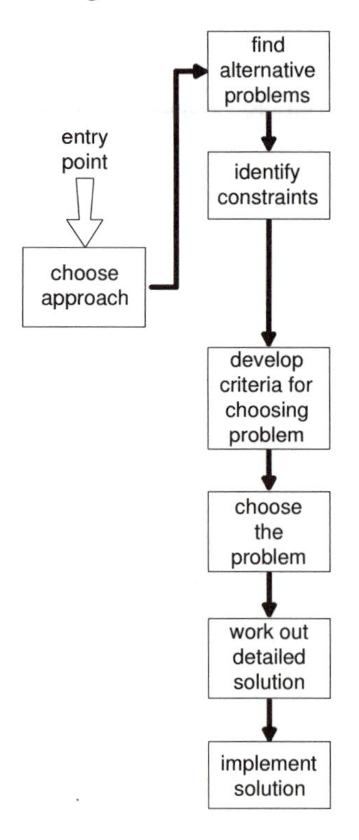

Figure 3.5 Solution-first approach to problem solving

Ideas may arise concerning applications of new materials with unusual and useful properties that have been developed in non-engineering fields. Such scenarios lead to the search for possible applications of new ideas. The solution-first approach shown in Figure 3.5 applies to such situations. A specific solution is the starting point to the procedure, which is simply a rearrangement of Figure 3.2, the steps being similar to those already discussed. This again emphasises the important point that engineering problem solving is an iterative activity. The entry point to the process is not of prime importance.

In a similar vein, Rogers (1983) has argued that answers often precede questions. He has suggested that while organisations face many problems, they typically possess only limited knowledge of a few innovations that can offer solutions. The chance of identifying an innovation to cope with a specific problem is therefore small. However, by commencing with an innovative solution there is a good chance that it can be matched to some problem facing the organisation. According to Rogers, a strategy for organisations is thus to scan for innovations and to try to line up a promising innovation with relevant problems.

Post-it Notes® (the adhesive that wouldn't stick)

The development of the Post-it note pads is a good example of the application of the solution-first strategy. In 1968 a 3M scientist, Dr Spence Silver, discovered a new type of adhesive, one that was quite different from anything currently available but whose properties defied conventional use. Dr Silver tried for five years to generate some interest in the new product, but because it was apparently inferior to existing adhesives he was unsuccessful.

Eventually a company researcher, Art Fry, took notice of the adhesive and its properties and started using it as a bookmark. The advantage here was that it was sticky enough not to fall out, but not so sticky that it left a residue on the page. By using the adhesive in this way it was soon noticed by others around the 3M Company and the idea of a lightly adhesive note pad soon developed.

The Post-it® note was introduced commercially in 1980 and named outstanding new product by 3M in 1981.

Source: 3m (2006)

3.12 OTHER ASPECTS OF PLANNING AND DESIGN

Risk, safety and failure

Once the aims and objectives of an engineering project have been established, the expectations are that they will in fact be achieved. Unfortunately, engineering work is always undertaken in the face of uncertainties which may arise, for example, from extreme natural events such as flood and earthquake, or from human error. Engineering work thus involves an element of risk, in the sense that there is always the possibility of some form of **failure**. It may occur while the project is in the planning and design phases, during implementation, or even after implementation when the project is in the operational phase. An engineering failure may be relatively minor, such as the breakdown of a pump in a water treatment plant. Unfortunately, it may also be catastrophic, with loss of property, injury or even loss of life. Minor failures are not uncommon, but the catastrophic ones, such as the collapse of a building, a bridge or a dam, are fortunately very rare.

It is not uncommon for the actual use of resources during the implementation phase of a project to be greater than was predicted during the planning phase. A project might not be completed within the time schedule allowed. Strictly speaking these are failures and are not uncommon. They lead to additional costs and budget over-runs which may be small or large. There are many other ways in which failure may occur in an engineering project. If there are undesirable and unintended consequences of a project, then there is a sense in which there has been a failure.

The cause of an engineering failure may originate in human error, for example by miscalculation in the design phase of the project, or by errors made during the construction phase. Failure may alternatively be the result of a catastrophic natural event, such as a severe earthquake or a flood, which could not have been reasonably forecast or allowed for during the design and planning phases of the project.

The cause of a catastrophic failure may simply be a very unlikely event, even one that had been foreseen and considered carefully in the planning and design phases. In the original design of the Tasman Bridge in Hobart, Australia, careful consideration was given to the possibility of a ship colliding with the main piers. Although this possibility was considered to be remote, steps were taken in the design to limit the effect of such a collision (New et al, 1967). Nevertheless, the unlikely event occurred in January 1975, when a ship did collide with the bridge. The collapse of the main bridge deck had catastrophic consequences, not only with severe disruption to life in Hobart but also with tragic injury and the loss of twelve lives (Laurie, 2007).

In open-cut mining projects a great deal of planning effort goes into minimising and managing risk. Nevertheless, accidents can and do occur and rescue and recovery operations are therefore included in the planning, design and ongoing management of such projects. Similarly, in petroleum and crude-oil handling facilities all procedures are carefully planned and monitored with the express purpose of preventing fire and explosion. Accident, fire and explosion cannot be prevented completely, and have to be allowed for in the planning and design of such projects. Management procedures are carefully worked out to handle accidents when they occur and to minimise their consequences.

Even with the introduction of ingenious fail-safe concepts, the probability of failure in real engineering systems cannot be reduced to zero. Unfortunately, the incremental cost of reducing risk increases sharply as the risk level decreases. In other words, it is extremely expensive to achieve a marginal increase in safety when the safety level is already high. Eventually a stage is reached in any design situation where the cost of improved safety is impossibly high. Engineering design must therefore deal with calculated risk and, in rare cases, the prospect of malfunction and failure.

Given that an element of risk is inevitable in engineering work, one of the difficult decisions is the choice of the risk and safety levels to be applied in planning and design. If the safety requirements are set too low, the rates of occurrence of failure will be unacceptable to the community. If the levels are set too high, the cost of achieving the safety levels becomes unrealistically high. The consequence of extremely high project costs is that very few important projects can be undertaken. Risk and safety are critical issues in almost all engineering projects. Methods for risk management and for dealing with failure have to be worked out in detail. Risk and safety in engineering are discussed further in Chapter 11.

While engineers understand that absolute safety is not an achievable or a feasible objective, this is not well appreciated by the community at large, or in legal circles. Nevertheless, the risk levels adopted in engineering work have an indirect effect on the entire community, not just in regard to the safety levels achieved, but also in the cost of the engineering infrastructure. It is for this reason that national codes and standards are used to give guidance and set minimum safety and performance requirements in many fields of engineering.

Codes, standards and regulations

Much engineering design and planning work is constrained by legal requirements and regulations. These are typically formulated in national and international codes and standards for the engineering industries. Regulations, codes and standards have various functions.

A prime function is to ensure that minimum safety levels are achieved in engineering work. For example, in the structural design of an office building, a factory or a hospital, design load values have to be chosen to represent the effects of natural events such as wind and earthquake. There are legal standards that prescribe minimum load levels to be used in design calculations. Other codes and standards prescribe how materials are to be manufactured in order to ensure that minimum strength and other properties are achieved. The methods of analysis to be used by designers may also be prescribed. Prescriptions on the way design work is carried out, although sometimes onerous, are aimed at ensuring that minimum levels of safety against failure are maintained. The safety levels are chosen to represent, approximately, the levels of risk that are acceptable to the community. In the choice of safety levels, an attempt is made to achieve a balance between over conservative (and therefore over costly) design on the one hand, and unconservative (and therefore unsafe) design on the other.

Another important function of standards is to achieve uniformity and standardisation in engineering industries and hence to improve efficiency. The common use of agreed dimensions and tolerances for simple items such as nuts and bolts leads to enormous savings and improved efficiencies in industries both nationally and internationally.

Standards are formulated in various ways. They may be prescriptive or performance based. Prescriptive standards stipulate the way a product is to be designed and manufactured, whereas performance standards set out minimum acceptable levels of performance for the product that must be met. Standards are sometimes brought in by an industry on a voluntary basis, rather than through government agencies.

The extent of standardisation varies greatly from one engineering industry to another. For example, the design work of structural engineers is regulated by various codes and standards, while the design work undertaken in water engineering projects, such as the design of irrigation channels and systems, is subject to less regulation. Nevertheless, basic data, such as information on rainfall and runoff intensities, are indirectly regulated by design guidelines.

Unfortunately, government regulations can sometimes become complex and cause difficulties to planners and designers. For example, in a project to improve water quality running off the Salt Creek catchment in California, Fowler and Rasmus (2005) report that engineers had to obtain permits from the following: the California Coastal Commission, The US Army Corps of Engineers, the California Department of Fish and Game, the San Diego Regional Water Quality Control Board, the Orange County Flood Control District and the South Coast Water District.

Legal considerations

Engineering projects are conducted within the legal framework provided by national laws. Civil law concepts usually apply to engineering work, although criminal law may be relevant in special cases of negligence. Legal liability is always an important consideration when an engineering project is undertaken through contracting and sub-contracting. Legal aspects of planning and design are considered further in Chapter 10 of this book.

Aesthetic, social and environmental considerations

Environmental effects are of importance in almost every type of engineering project. Chapter 9 deals with some of the most important environmental considerations.

We have already seen in Chapter 1 how engineering work impacts on society. We have also seen how engineering projects may have adverse side effects, as well as the targeted benefits. The social side effects of engineering projects, both large and small, are not always obvious. For example, the widening of a street to improve traffic flow can disrupt a local community by cutting it into two parts, one on either side of the widened road. Such social effects should be identified and allowed for during the project-planning phase.

Aesthetic considerations may not be foremost in the minds of engineers when they commence work on a new project. Nevertheless many engineering accomplishments provide a continuing visual impact and become part of the daily life of the community. Monumental structures such as bridges, towers and buildings are prime examples, although the aesthetics of manufactured goods, such as cars, can provide ongoing pleasure, or irritation, to the community. The importance of aesthetics in engineering is unfortunately not always appreciated.

3.13 EXAMPLE: PLANNING FOR A CITY WATER SUPPLY SYSTEM

To provide an example of how the steps listed above in Tables 3.3 and 3.4 can be used in the early phases of a project, we now discuss briefly the planning of the future water supply for a small coastal city in a very dry climate. The city of Adelaide in South Australia is chosen as the focus for the discussion. The aim here is not to provide valid, long-term conclusions in regard to the needs of this city. The emphasis is on the process, not the results.

The existing water supply system currently supplies about 200,000 ML per year for a population of around one million people. We look at the problem of planning the supply of water for the next 25 years.

Problem formulation

The steps in problem formulation, listed in Table 3.3, may be applied to this project.

Identifying the problem as part of a wider problem

Given that there is already a serious lack of water throughout the region, the problem of planning for the future water supply in Adelaide has to be considered as part of the wider problem of providing water for the entire State of South Australia. An attempt to look at the Adelaide water supply problem without reference to the water supply problems of neighbouring areas, including industrial cities such as Port Augusta and Port Pirie, the South Australian wine industry and other rural industries, would be pointless. The increasing demand for water in remote regions of the state, generated by extractive industries and mining operations, is also relevant and needs to be taken into account.

The geography of southern Australia and its lack of river systems makes it necessary to identify the problem as part of an even wider problem: the management and use of water throughout the Murray-Darling river system, which extends through three states in the eastern region of the Australian continent and supplies the majority of the water used in South Australia. The apparent change in climate that has occurred in recent years in the region (but also in many other parts of the world) suggests that there are global aspects to this "local" problem.

The water supply system for the city also has to be looked at as one among many system components that make up the physical infrastructure of the city. Other components of particular relevance are the sewerage and storm water systems, which are users of water, but are also possible suppliers of additional water. The parks and gardens and other public areas place significant demand on water. The transport system and the energy generation and supply systems are other components of the city system, as are industrial regions and the suburbs where people live.

Relevant background information

Quantitative data on the current state of the water supply system and its desired future state in twenty-five years time clearly have to be gathered. We find that, in addition to the 200,000 ML of water used each year in the city, a little under 100,000 ML is used additionally in neighbouring rural areas including the Adelaide Hills. On the demand side, about 45 per cent of city water is used by suburban households, about 28 per cent goes to primary production, but only 10 per cent goes to the commercial and industry sectors. Community use (including parks and gardens) makes up the remainder, about 17 per cent.

The above figures, and those following, have been obtained from a planning document recently prepared by the South Australian Government entitled "Water

Proofing Adelaide", which is available from the Department of Water, Land and Biodiversity Conservation of the State of South Australia.

Looking specifically at households, we find that about 40 per cent of domestic water goes into gardens and other outdoor uses, such as swimming pools. Bathrooms (baths and showers) take up 20 per cent, with toilets at about 11 per cent, while laundries and kitchens use around 16 and 11 per cent, respectively.

On the supply side we find that although the water for the city comes from a variety of sources, the two main ones are, firstly, the reservoirs and catchment areas in the Adelaide Hills, and, secondly, water pumped from the River Murray. In a "normal" year these sources provide roughly 60 and 35 per cent respectively of the water for the city and surrounding rural areas, but in a "dry" year the component from the Adelaide Hills reduces to between 10 and 20 per cent. The additional water is obtained by extra pumping from the Murray. Minor other sources of water for Adelaide include groundwater, rainwater tanks and storm water. These figures give a broad picture of the current supply and demand situation in and around the city. Similar figures for the rest of the state, and indeed for south eastern Australia provide additional important background information. For example, of the water available from the River Murray, about ten per cent goes to towns and cities while the rest is used in irrigation.

Background information is also needed on the desired "target" state of the system in twenty-five years time, and at intermediate times. Such information regarding future demand necessarily depends on expected demographic changes to the city and state, and cannot be as reliable as the data on the current state. It could therefore be advantageous to look at several alternative scenarios, including optimistic, pessimistic and most likely, in order to obtain a range of figures for the target state.

The document "Waterproofing Adelaide" suggests a demand of about 230,000 ML in the city in 2025. This unexpectedly low value might include some allowance for a curb on demand as the result of pricing policies. Even so, it is worrying when compared with the predicted supply figures for 2025 from present sources, which are about 190,000 ML in a dry year and 250,000 ML in a normal year. One reason for the relatively low supply figures is the reasonable assumption that climate change will continue and will, for example, reduce the available water from the Adelaide Hills by ten per cent.

It is clearly important to look for possible new sources of water in the feasibility study. Given the unavailability of new, plentiful and cheap sources of water in the surrounding hills, wider possibilities have to be considered. In anticipation of the steps in the feasibility study, we can see that background information will be useful on processes such as water desalination and recycling of used water, as well as methods for improving the efficiency of use of the existing sources of supply. Background information will be useful on more radical options such as the transport of icebergs (or iceberg water) from the adjacent Antarctic Ocean and the use of measures to modify the location and quantity of rainfall. Another option requiring careful background information is the possibility of adjusting water usage in the south east of the country, perhaps through the marketing and trading of water and water rights.

Other relevant background information relates to the quality requirements of water when used in different ways. Clearly the quality of water needed for drinking and cooking is not the same as for irrigating public parks and gardens or for flushing toilets.

Underlying needs

While a useful picture of future domestic, industry and rural water needs can be built up for the planning period, the evident lack of water raises important questions that highlight the differences between needs and requirements. Specifically, considering the relatively high domestic use of water, with 40 per cent going to gardens and other outside applications, we need to look at questions such as whether a distinction needs to be made between the quality and cost of water when used for different purposes. Such questions arise when we seek to identify the underlying needs, as distinct from demands, for water. This leads to further strategic questions, for example regarding the possible supply and use of water of varying quality with pricing policies related to cost of supply. At present, almost all water is supplied at the one quality level. The issue of water trading clearly focuses attention on the difference between needs and demands. Such questions are relevant when we undertake the feasibility study. Our purpose here is not to try to answer such questions, but to illustrate the process.

An important step in the planning process could well be to place important questions before the community for discussion and debate. In the year 2006 in various country communities in Australia questions were being looked at, such as the acceptability of recycling water as a way to boost the capacity of town water supplies.

Side effects

Increasing pollution of the waterways in the Adelaide Hills and pollution and salinity problems in the River Murray are already prevalent. Worsening pollution is an obvious side effect to be expected in the present project. The side effects of depleting underground water through aquifer pumping are not well appreciated in the community, but have been investigated. They include increased salt in the aquifer water and overall deterioration of quality. This is a side effect to be considered, should there be an increased use of underground water supplies. Overuse of river water generally means reduced flow, with adverse effects such as silting, algae bloom and other adverse environmental effects on the biota. Cessation of river flow and closure of the river mouth is a real possibility. It has already been observed on the Murray and the local community is well aware of such possibilities.

Other side effects include increased salinity in the river water of the Murray due to increased pumping, and the need for correspondingly more expensive treatment procedures to achieve adequate water quality. Another consideration is the inhibiting effect of poor quality water on industrial expansion, city development and the rural and wine industries.

Constraints

Constraints derive from state and federal regulations regarding water quality, environmental impact, land resumption, health regulations, city planning and zoning, and water use. World Health Organisation standards for water quality and health also act as constraints. Current intrastate agreements on the allocation of water from the Murray-Darling basin impose severe constraints on the available options for the city of Adelaide. In regard to some of these constraints, consideration has to be given to of modifying existing laws and regulations and the re-scheduling of state and commonwealth agreements.

Objectives

In any statement of objectives, account has to be taken of the demands and needs for water, and the likely limits on supply. Water scarcity and interstate and intrastate competition for this limited resource also have to be considered in the statement of objectives. A rather general statement is as follows:

> *To match, as closely as is economically and technologically feasible, the quantity and quality of water provided against the realistic needs of the various uses (domestic, industry, etc) over the planning period.*

This formulation treats both demand and supply as variables. On the other hand it skirts the problem of differentiating between need and demand. The prediction of realistic figures for demand versus need thus becomes a central task to be undertaken in this project. The above statement does not stipulate that water will always be available without restriction, for example in times of severe drought, although some minimum level of supply is clearly implied.

The linked concepts of technological and economic feasibility allow for changes and improvements in technology over the planning horizon. For example the economic feasibility of desalination plants is likely to improve substantially in the next decade, assuming that there is continued development and improvement in efficiencies in the relevant technologies.

Performance requirements and operating conditions

The water supply system has to be able to deliver adequate quantities of water of suitable quality progressively through the planning period, and not just in 2025. A wide variation in conditions in this period must be expected, including effects such as progressive climate change and alternating drought and flood throughout the state and in the upper reaches of the Murray-Darling system. Drought and flood are even possible at the same time, in different parts of the system. In taking account of these possibilities, it will be necessary to specify minimum limits on the quantities of water to be supplied on a daily, weekly and yearly basis. Quality limits also have to be specified. Separate statements of the requirements for domestic use, industrial use, etc may be needed. Above all, the performance requirements have to be realistic and achievable.

Measures of effectiveness

Total cost in dollar terms is one of the prime measures of effectiveness, but certainly not the only one. Another measure is needed in regard to the reliability of supply, which can be measured, for example, by the number of days per year when restrictions occur. Another measure may be needed for the quality of the water supplied, for example taking account of the average and peak salinity levels. Adaptability of the supply system to unexpected changes in demand over the medium term (in other words, system robustness) is another possible factor for inclusion in the measure of effectiveness.

A combined measure of effectiveness might be made in terms of the dollar cost per unit of water supplied, using penalty costs to take account of reliability and adaptability. Such a measure of effectiveness can be broadened to allow for water being supplied at several different quality levels according to different uses. This leads to a separate measure of effectiveness for the different water quality levels considered. A single overall measure of effectiveness, again in dollar terms, can nevertheless be obtained by differential costing of the different quality levels. The present comments are intended to show how a measure of effectiveness might be developed. They are not meant to indicate preferred options.

Feasibility study

We now look very briefly at the steps listed in Table 3.4, as they apply to the present example.

Available resources

In addition to the main resource, water, which we have already discussed, it is necessary to consider the financial resources that will be available, and the technical resources and expertise that will be needed to undertake the project. When each specific option is considered, such as the construction of a new dam or the construction of a desalination plant, the specific resources and expertise will have to be investigated.

Investigation and evaluation of constraints

Some of the constraints to the problem have already been mentioned, such as the regulations and laws concerning water quality. It is necessary in the feasibility study to investigate and study them to observe how they will affect the possible options and solutions.

Develop as many promising concepts as possible

Some of the possible options for dealing with the problem have already been foreshadowed in the previous discussion. Consideration has to be given to increasing the use of the available but presently unused water resources, such as the storm water which flows into the sea. An improved efficiency in the use of other existing resources is another important option, for example by reducing the losses that occur due to evaporation from open water surfaces and leakage from reticulation pipes. Another option is legislation to require the installation of domestic rainwater tanks. Further options that are more costly also have to be considered, such as desalination, recycling and reuse of brown water. More radical approaches, which may appear at first sight not to be feasible, include the harvesting of icebergs from the Antarctic Ocean. The possibility of actively changing the climate to increase rainfall falls into this category.

A more obvious and potentially very effective approach is to curb the increase in demand for water, and possibly even reduce demand, through pricing policies coupled with education programs to alert the community to the real costs of water and to methods for effectively reducing its use.

Technical evaluation is needed to determine the relative costs, the feasibility and the quantities of water that can be delivered by each of these approaches. It is extremely unlikely that any one approach will be adequate in itself. It is far more likely that a range of the cheaper and more effective measures will have to be used so that, together, they produce the outcome sought

While some of the approaches listed might not be feasible today, we must remember that the planning horizon is twenty years, and the need for water will not stop then. With improving technology and new scientific discoveries, we can confidently expect that some of the options that today seem unlikely will become quite realistic over time and may well be introduced at later stages as the century progresses.

Comparison of alternatives and elimination of the non-competitive options

In this project we are dealing with a scarce resource, water, and, as already mentioned, a mixed strategy is likely to be appropriate. The employment of some options and not others will be according to the relative costs and efficiencies, which will undoubtedly vary over the planning horizon.

The example being discussed is not typical of many engineering projects, in that the solution here will *not* be found by progressively eliminating all but one from a list of options. The nature of the problem means that a mix of options will be required, with only the clearly the uncompetitive ones eliminated.

An evaluation of viable options can lead to their ranking in terms of initial set-up cost, cost per water unit delivered, and total amount of water deliverable. Such a ranking could be used to make decisions on the mix of options to be employed initially and at subsequent stages during the planning period. Regular updating of the list, especially when new developments and discoveries are made, will allow decisions to be made on when specific options should be taken up. At the beginning of the century, unlikely options such as desalination, rain making and harvesting icebergs should not be discarded and forgotten; they may well become attractive in time, as conditions change, and as the options initially employed are exhausted.

PROBLEMS

3.1 What do you see as the main engineering problems to be faced in order to establish a permanent human settlement on the moon?

3.2 How would you undertake a feasibility study for the problem of creating a lunar base to house between 10 and 25 persons in a reasonably comfortable environment for living and for undertaking scientific work? List the steps you would use in undertaking this feasibility study. Describe each step in a short paragraph and mention what you see as the key considerations. For example, if one of the steps is to obtain background information, what kind of information would you want to obtain?

3.3 What are the main engineering problems to be solved in establishing a small permanent Antarctic base for, say, twenty scientists? Describe briefly the steps to be taken in the concept design for such a base.

3.4 Consider the energy supply and distribution system of the city where you live. How would you undertake a planning study to provide adequate energy over the next thirty years? What background information would you want to gather in undertaking this study? List several alternative sources of energy that might be used to increase supply to meet demand. What other energy options would you consider in your study?

3.5 In constructing a building it is usual to begin from the foundations and work progressively upwards, floor by floor. Is this the only way to undertake building construction? Can you think of any situations where construction follows another sequence? Under what circumstances might it be advantageous to develop an alternative construction sequence?

REFERENCES

American Society of Civil Engineers 1986. *Urban Planning Guide*. New York: ASCE.

Chadwick, G. 1978. *A Systems View of Planning*, 2nd Ed, Oxford: Pergamon Press.

Department of Water, Land and Biodiversity Conservation, South Australia, 2005. *Water Proofing Adelaide, a Thirst for Change*.

dtv 1974. *Atlas zur Baukunst, Band 1*, Munich: Deutscher Taschenbuch Verlag.

Dym, Clive L. & Patrick Little 2000. *Engineering Design, A Project-Based Approach*. New York: John Wiley & Sons.

Fowler, B. & Rasmus, J. 2005. Seaside Solution. Civil Engineering, Magazine of ASCE, 75 (12), 44–49.

Grabosky, P.N. (1989) Electricity Trust of South Australia: Fatal Accident at Waterfall Gully. Wayward Governance: Illegality and its Control in the Public Sector. Australian Studies in Law, Crime and Justice Series, 161–171. Web reference: www.aic.gov.au/publications/lcj/wayward/ch10.html (Downloaded 24th January, 2006)

Griffis, F. H. & J. F. Farr 2000. *Construction Planning for Engineers*, McGraw-Hill International Editions, Engineering Series.

Hall, P. 1980. *Great Planning Disasters*. London: Weidenfeld and Nicolson

Hyman, Barr 1998. *Fundamentals of engineering design*. New Jersey: Prentice-Hall.

Laurie, Victoria, 2007, Time Capsule: Tasman bridge disaster claims 12 lives. The Weekend Australian Magazine, 6–7 January, 2007, p. 6.
Lee, Colin 1973. *Models in Planning*. Oxford: Pergamon Press.
Neilson, A. M. 1991. Sydney Harbour Tunnel – Why? *Proceedings of the Seventh Australian Tunnelling Conference*. Institution of Engineers, Australia. Also in Tunnelling and Underground Space Technology, *Elsevier Science*, 6(2), pp. 211–4.
New, D.H., J.R. Lowe & J. Read, 1967, The superstructure of the Tasman Bridge, Hobart. *The Structural Engineer*, 45, pp. 81–90.
Rogers, E. M. 1983, *Diffusion of Innovations*, 3rd Ed., The Free press, Macmillan Publishing.
Saaty, T. L. 1990. *The analytic hierarchy process: Planning, priority setting, resource allocation*, 2nd Ed., Pittsburgh: RWS Publications.
Spinner, M.P. 1997. *Project Management Principles and Practices*. New Jersey: Prentice Hall.
Smith, Craig B. 2004. *How the Great Pyramid was Built*. New York: Smithsonian Books, (Imprint of HarperCollins).
SSFM Engineers, Inc, 1996. *Preliminary Assessment of Korror-Babelthuap Bridge Failure for US Army Corps of Engineers*. Honolulu, Hawaii, 21 October.
Svensson, N. L. 1974. *Introduction to Engineering Design*. Randwick (NSW): University of New South Wales Press.
Voland, G. 2004. *Engineering by Design*. New Jersey: Pearson Prentice-Hall.
Yee, A. A. 1979. Record span box girder bridge connects Pacific Islands. *Concrete International*, June, 1(6).
3M (2006) www.3m.com/about3m/pioneers/fry.jhtml(23 January, 2006)

CHAPTER FOUR

Creativity and Creative Thinking

Creativity plays a crucial role in engineering and in particular in the planning and design of engineering projects. In this chapter we find that the brain works in some quite surprising ways; that much of its success comes from the use of simple heuristics (or search rules), and from hard-wired features that allow it to make reasonable decisions quickly. However, this form of thinking is convergent in nature and does not provide a sound basis for the divergent thinking we need for the creative aspects of planning and design. It is therefore important to develop techniques to enhance a divergent and creative approach to problem solving. We review some of the available techniques in the context of engineering problem solving.

4.1 INTRODUCTION

We have seen in Chapters 2 and 3 that engineers are often charged with the responsibility of solving complex open-ended problems; ones that have no single obvious solution and ones that will benefit from a novel and creative approach. This may seem easy initially, but, as we will see in this chapter, our brains have evolved to be very efficient at making quick decisions that are often quite appropriate, using processes that have developed to ensure the survival of the species. However, these are not the sort of processes needed for creative thinking.

Before we look at the brain and its workings it is worth defining clearly the types of thinking that are important to engineers. On the one hand there is convergent or deductive thinking, which aims to apply logic and take in just as much information as necessary to develop a solution to a particular problem. We have seen in Chapter 2 that this is the sort of thinking that occurs when people are trying to solve an analytical problem, or make a decision by narrowing down the alternatives. Gilhooly (1982) quotes a finding that scientists working on the analysis of lunar samples as part of the Apollo mission tended towards convergent thinking: when presented with an open-ended question, they would quickly transform it into a more tightly defined one. Divergent or inductive thinking, on the other hand, is aimed at solving problems where there might be a large number of possible solutions or where there are no obvious solutions. Divergent thinking deliberately takes in as much information as possible, without too much concern about relevance, and tries to develop solutions by making links that are not necessarily obvious or (apparently) sensible.

Take as an example of divergent thinking the case of the Swiss engineer, Robert Maillart (1872–1940). In 1901 he constructed a bridge at Zuoz in Switzerland and noted a couple of years later that a section of concrete webbing near the main supports was cracking. The obvious solution was to repair and strengthen it, but Maillart instead investigated the behaviour of the structure and found after some study that the presence of cracks in those locations was not having a detrimental effect on the bridge's performance. On this basis he modified his design to take advantage of this finding, leading to a much more efficient section (Ferguson, 1999). Many must have observed the cracking at Zuoz, but only he was able to think about it in quite a different manner and come to a quite different and highly creative solution.

And so begins a look at creativity (or divergent thinking) in engineering – what it is, how it can be recognised, how it can be promoted, and why it is important in the planning and design of projects. It is worth noting at this stage that much of what we know about creativity comes from work undertaken by highly creative individuals and that when we consider creativity we are looking not just at the solutions to engineering problems but also to the problem formulation. As Findlay and Lumsden (1988) have argued this "may be just as much a part of the creative process as its solution".

4.2 CREATIVITY DEFINED

Creativity is a somewhat elusive concept. Bohm (1996) believes that it is impossible to define in words, but others have tried. One that picks up many of the elements is by Gardner (1993) who defines the creative person as one "who regularly solves problems, fashions products, or defines new questions in a domain in a way that is initially considered novel but that ultimately becomes accepted in a particular cultural setting." He goes on to suggest that "initial rejection is the likely fate of any truly innovative work". Gardner was looking for much more than general creativity, and was attempting to pick out individuals who had shown significant creative output. As a matter of interest, he selected Albert Einstein (physicist), T.S. Eliot (writer), Sigmund Freud (psychologist), Mahatma Gandhi (politician), Martha Graham (dancer), Pablo Picasso (painter) and Igor Stravinsky (composer) for his list from a wide variety of fields and justified each based on his definition.

It is proposed that, for engineering planning and design, creativity must demonstrate novelty, value and eventual acceptance in the posing or solution of engineering problems. The requirement for eventual acceptance means that it is not possible to dismiss ideas at the time they are generated. It is worth remembering the numerous examples of ideas that were ridiculed when first presented, only to be accepted some time later when it was agreed they came "before their time".

The aim of this chapter is not to turn everyone into an Einstein or Goddard, but to demonstrate the need for a creative and open mind in engineering and to show how it can be achieved in practice (by practise). In the next section we look at the brain and its workings with the aim of showing just how constrained human thinking often is, and how much of what we do is governed either by features that are hard-wired into our brains or by the unconscious application of simple rules or heuristics. In the later sections a number of procedures are outlined to demonstrate how it is possible to coerce the brain to ignore these rules, and to work in a way that promotes creativity.

4.3 THE BRAIN AND ITS WORKINGS

As recently as two thousand years ago little was known about the brain, and serious research into it really only began in the first half of the 20th Century. The fact that many parts of the brain are named after their shape rather than their function gives an indication of just how little was known. For example, in the brain there are areas called the pons, amygdala and hippocampus named after bridges, the shape of almonds, and seahorses, respectively (Rose, 1998). It is only quite recently that it has been possible to determine some of the actual functions of some of the various brain regions using techniques such as functional magnetic resonance imaging (fMRI), which is able to measure oxygen uptake rates as a measure of neural activity, and positron emission topography (PET) which can measure, in addition to oxygen uptake rates, the release of neurotransmitters such as dopamine (Knutson and Peterson, 2005). Based on these studies it is now known, for example, that the hippocampus is important for short term memory and the ability to make mental maps, while the amygdala handles fear and other emotions.

The basic building block of the brain is the neuron. Humans start with around one hundred billion neurons, but the number declines with age. Neurons (see Figure 4.1) have a cell body with input (dendrite) and output (axon) channels that connect to other cells. The number of input and output channels per neuron can be in the thousands. The contact between neurons is made at synapses and each neuron has between a thousand and ten thousand synapses. A section of brain the size of a grain of sand contains about one hundred thousand neurons, 2 million axons and one billion synapses (Ramachandran and Blakeslee, 1998).

Figure 4.1 A neuron in a rat hippocampus showing the cell body and a small number of the connections to adjoining cells. Note that the staining technique highlights single neurons and so does not show the density of connections to adjacent cells. © Used with permission of Synapse Web.

The connections in the brain develop from birth (Klawans, 2000) and are strengthened by use. The infant brain starts with a high level of redundancy in the form of parallel paths. In fact, one of the strengths of the human brain is based on its parallelism which is measured as fanout, the number of direct contacts a neuron makes with other neurons (Holland, 1998). Typical computer elements have a fanout of about 10 whereas human central nervous system neurons have a value between 1,000 and 10,000. It is believed that this explains much of the potential for quite complex behaviour. Until more details are available on exactly how neurons store information it is not possible to determine a measure of brain power based on a simple neuron count but a comparison with other animals is instructive. Camazine (2003) points out that honey-bees, with approximately one million neurons, have the capability to navigate by the sun, fly to a food source, make what appear to be decisions, communicate with other honey-bees, and perform other complex activities. Boden (2004) tempers any enthusiasm for reading too much into these sorts of abilities with a description of the hoverfly which, although capable of quite complex behaviour, has much of it hard-wired into its brain and nervous system. With this understanding she believes that the hoverfly's intelligence "has been demystified with a vengeance".

There is still debate about the exact mechanism that the brain uses to store and process information but it would appear that the synaptic connections are heavily involved. When messages are transmitted the synaptic gap (approximately 1/40,000 mm) is reduced, making further communication along that path easier. In this way pathways are established leading to patterns that the brain can use. Pathways that are not used retain a large synaptic gap and are therefore less likely to be used. Researchers in computing are using this model to add some "intelligence" to the internet. According to Brooks (2000), a mechanism for making active world wide web links has been developed so that links that are used most commonly are given prominence and those that are not used for a time are discarded. In this way the internet may be able to optimise data collection by highlighting the most common paths and making them easier to follow (a convergent thinking approach). Whether this will actually assist people gathering information is yet to be seen. There could be an argument for not wanting to follow all the paths that everyone else on the web has followed but allowing a few less-travelled ones to come up in searches (a divergent thinking approach). One way to achieve this might be to allow searchers to specify an equivalent acetylcholine level for the search. Acetycholine in the brain is important in strengthening synapses. It is in particularly high levels during rapid eye movement (REM) sleep and it has been suggested (Phillips, 1999) that this might be a reason that dreams during REM sleep can be so bizarre.

One of the findings that has come from research is the level to which the brain manipulates raw data before presenting it to the consciousness. As an example, consider the sense of sight. It has been discovered that when humans "see" the information is in fact passed to approximately 30 different areas of the brain, with each working on a specific aspect of the vision (Ramachandran and Blakeslee, 1998). For example, there are separate areas for colour, depth, motion, slopes of lines, edges, and outlines. According to Goleman (1995) studies have identified neurons that only fire in response to certain facial expressions such as a threatening opening of the mouth, a fearful grimace, or a docile crouch. There is also a group of neurons that fire in response to "seeing" the shape of the back of a monkey's hand! (Popper and Eccles, 1977) What this means is that many decisions are effectively made using hard-wired sets of neurons rather than conscious thinking. We do not look at someone's face and actually have to think about what it is showing – as soon as our eyes gaze on the face neurons fire in response to the aspects they have evolved to react to and the brain is fed the information that the face is showing, for example, a threatening opening of the mouth. This results in extraordinary speed, but at a cost: it's all automatic, and we don't necessarily realise that.

It is not commonly known, but every eye in every human has a significant defect: a blind spot. The fact that there is such a problem was first discovered centuries ago by a researcher studying the optics of the visual system, and is due to the geometry of the lens and retina. That it should be so well hidden says more about the brain than the eye. The brain actually covers for the eye and "fills in the missing details" automatically. An example of a test picture that can illustrate the effect is shown in Figure 4.2. To demonstrate the effect: hold the diagram close to the face with the left eye looking directly at the circle and the right eye closed, then move the page away from the eye. At a certain distance the cross should disappear from the peripheral vision and be replaced by the general background.

Once the effect has been observed, just stop and think for a second. The brain is actually "making up" some of what it reports it is seeing! If this is true, then people cannot necessarily believe what they are seeing, because everything that they pick up is filtered and modified by the brain, mostly in a harmless way, but modified nonetheless.

The human brain is the result of approximately 200,000 years of evolution since *Homo Sapiens* split from earlier ancestors (Gould, 1991). For most of that time humans were living in a more primitive world where many of the decisions were ones of life and death and where survival led to an emphasis on quick reasonable solutions rather than slower well thought out ones. Cohen and Stewart (1995) make this point in relation to avoiding black and yellow stripy things. Often they might not be tigers, but it is not worth having a complicated and time-consuming procedure to work this out because this will not aid survival!

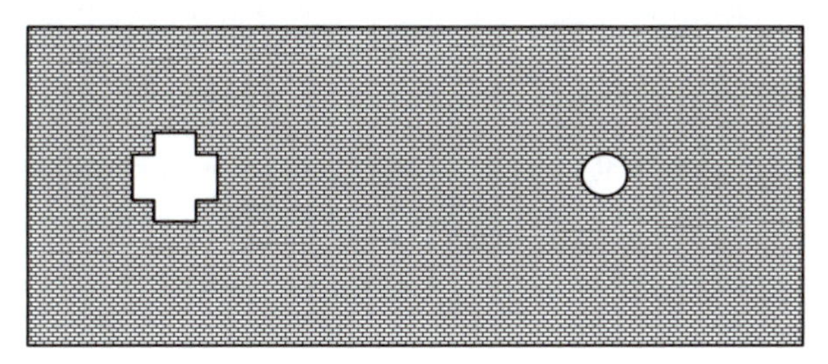

Figure 4.2 Example of a diagram that can be used to illustrate the optical blind spot. Hold it close with the left eye looking directly at the circle and the right eye closed, then move the page away until the cross disappears.

We now turn to conscious thinking and find that there are also issues with this activity, especially where creativity is concerned. Most high school graduates majoring in mathematics and science would consider themselves to be well trained in thinking. However, much of the training has been to develop skills of a very specific kind. There are various terms for the traditional type of thinking: convergent, analytical, vertical, and logical are just a few. Application of this type of thinking means that people start at a particular point and move one step at a time, justifying each step as they go and narrowing down the options until it leads to a clear solution. This type of thinking is particularly true in mathematics, for example, where there is generally a correct way of arriving at an answer. Even if there is more than one way, each follows a logical sequence of steps that lead on from the last. Although this type of thinking is important, and particularly so in engineering, there is another type of thinking that is equally important; it is creative, divergent or lateral thinking. De Bono (1970), classifies the key differences which he has identified between vertical and lateral thinking and these are listed in Table 4.1.

A science-mathematics trained student might look to the right hand side of characteristics and be horrified at some of the entries. For example, "does not have to be correct" and "welcomes intrusions" seem like odd ways of tackling serious engineering problems, but these are in fact ideal when one considers that the brain has not been designed to tackle tasks where one wants lots of solutions and where the time taken to arrive at the solution is not an issue.

4.4 THE BRAIN AND HEURISTICS

According to work by a number of researchers (e.g. Gigerenzer and Goldstein, 1996; Borges et al., 1999; Cross and Jackson, 2005) much of the convergent thinking we do can be described based on the application of very simple rules (heuristics). The heuristics are interesting for two reasons: firstly to see how real decisions are made, and secondly, to highlight the need to force the brain to act otherwise in situations where quick and dirty decisions are not wanted. These heuristics will be referred to again in the context of decision-making in Chapter 12.

Table 4.1 Key characteristics of the types of thinking by de Bono (1970).

Convergent/Vertical	Divergent/Lateral
• moves forward in logical steps	• generation of new ideas
• selective	• generative
• rightness matters	• richness matters
• analytical	• provocative
• sequential	• jumps
• correct at every step	• does not have to be correct
• use of negative to block pathways	• no use of negative
• follows most likely paths	• follows all paths
• finite process (expects answer)	• probabilistic
• excludes that which is irrelevant	• welcomes intrusions

Recognition heuristic

We start the description of the recognition heuristic with a simple example. In one experiment (Reimer and Katsikopoulos, 2004) a group of US university students was asked to say which city was larger, San Diego or San Antonio. The correct answer was given by 66% of the group. When a similar group of German university students was asked the same question, 100% of them selected the correct answer. So how is it that German students seem to know more than US students about American cities and their size? The explanation given was in terms of the recognition heuristic (Goldstein and Gigerenzer, 2002). In this type of question with only two alternatives, if someone recognises only one of two options, then they will assume that that one is likely to be the correct answer. In this case, it was likely that the German students had heard of San Diego but not San Antonio and so selected the former. The recognition heuristic is often referred to the "less is more effect". The less one knows, the better the chance of getting the correct answer! The method will of course only work in situations where there is a systematic bias in the options, but is important nonetheless. Of course, once people know more than the basic minimum that particular heuristic breaks down and people move on to other more complicated ones.

Minimalist heuristic

The next stage in decision-making is to use a very small piece of additional information and the minimalist heuristic. Continuing the search for the larger of two cities: if one is forced to choose between two about which a little is known, it might be good to ask which has an airport, or a national soccer team – then choose that one. Again, this way of thinking gives a quick and easy answer and will be reliable under a wide range of situations.

One of the minimalist heuristics is "Take The Last". In this, a problem is tackled using the scheme that was used to solve a similar looking problem last time. This may seem a sensible approach and will often work but it can have its problems. In one experiment reported by Gilhooly (1982), a group of people were given three water jugs of different volumes and asked to make up a particular volume of water by manipulating the quantities. For example, when given 18, 43, and 10 litre jugs and asked to make 5 litres people found, after some manipulation, that the answer was to fill the 43 litre jug, then use it to fill the 18 litre jug and then the 10 litre jug twice leaving the required 5 litres. Having got good at this and other similar problems they were then given 28, 76, and 3 litre jugs and asked to make 25 litres. The subjects had difficulty because they were intent on using all three jugs because that had been necessary previously.

Take the best heuristic

The next stage in the decision-making process is to gather additional data, and to sift through it in order until one choice comes up as the best. At this stage the search stops. According to Gigerenzer and Goldstein (1999) the "Take the Best" heuristic works as well or better than more complicated decision-making methods such as linear and multiple regression in selected situations, and is faster as well. This conclusion was based on the study attempting to determine the larger of two cities where nine pieces of information were available for each including, whether it was a capital city, or had a soccer team, an intercity train or a university. It has been suggested that this is the process that humans go through when searching for a mate. They set up a list of requirements and search until someone matches enough of them, at which time they move to the next phase.

Availability heuristic

One of the aspects that has fascinated people studying human behaviour is that way people assess danger and its relative likelihood. Shark attacks or aeroplane crashes, for example, are relatively rare yet people will overestimate their perceived danger from such an event. This has been put down to what is called the availability heuristic where "individuals estimate the frequency of an event, or the likelihood of its occurrence, by the ease with which instances or associations come to mind" (Wänke et al., 1995). As Tversky and Kahneman (1973) point out, the ease of recall should be related to relative frequency (and often is) but it should be understood that this short-cut is being used because there are times when it doesn't work. An example of where it works was given in an experiment by Tversky and Kahneman (1973) where subjects were given 7 seconds to estimate how many flowers, or four legged animals or Russian authors they would be able to list in 2 minutes. Their estimates and subsequent performance was highly correlated and the authors suggested that a very quick assessment allowed people to make an accurate prediction of their recall ability.

However, the heuristic is open to abuse and can perform poorly in some circumstances. Examples of where it did not work were also described by Tversky and Kahneman (1973). In one experiment, subjects were asked to estimate the relative frequency of words starting with the letter 'k' compared to words with 'k' as the third letter in typical text. Most could think of more words starting with 'k' so estimated that these were more frequent. In fact, words with 'k' as the third letter outnumber words starting with 'k' 2:1 in typical text. In a second experiment subjects were asked to estimate (very quickly) the product: 1 x 2 x 3 x 4 x 5 x 6 x 7 x 8. Others were asked to estimate (very quickly) the product: 8 x 7 x 6 x 5 x 4 x 3 x 2 x 1. The median guess for the first group was 512, while the median for the second was 2,250. The true answer is 40,320. Tversky and Kahneman suggest that in estimating the product, subjects looked at the first few terms and extrapolated based on that. They were unable to appreciate the exponential nature of the sum, particularly those who were considering where 1 x 2 x 3 would be going.

The use of the availability heuristic has important ramifications for engineers. As an example, Kates (1962) studied the decision-making process in relation to the provision of flood protection works:

> *A major limitation to human ability to use improved flood hazard information is a basic reliance on experience. Men on flood plains appear to be very much prisoners of their experience ... Recently experienced floods appear to set an upper bound to the size of loss with which managers believe they ought to be concerned.*

First instinct fallacy

When students take multiple choice question exams and tests conventional wisdom states that one should go with the answer that was chosen based on first instinct. In fact, some guides aimed at helping students recommend this course of action. However, it is very likely wrong despite people's opinion to the contrary. Why this is so can be explained very well by looking at how the brain reacts to being right and wrong and what regret might follow a wrong decision.

In a series of experiments with university students undertaking multiple choice exams Kruger et al. (2005) tested student experience and came to a number of conclusions:

- the majority of answer changes were from incorrect to correct, and most people who changed their answers improved their test scores;
- changing the correct answer to an incorrect answer was likely to be more frustrating and memorable than failing to change an incorrect answer to the correct answer;
- the majority of changes were from wrong to right but student's intuitions said the opposite;
- there is more regret associated with switching from a correct to an incorrect answer than failing to switch from an incorrect answer to a correct answer;
- there is a memory bias: students overestimate how many times they have switched from correct to incorrect and underestimate how often they failed to switch from a wrong answer;
- people who were observing the selection of answers from multiple possibilities were more critical of those who switched from a right answer than of those who did not switch from a wrong answer; and finally
- the first instinct fallacy strengthens in the face of mounting personal evidence to the contrary.

The important point in all of this is to realise that when humans make decisions they are pre-programmed to make "fast and frugal" ones (Gigerenzer and Goldstein, 1996). While this is good for survival in a more hostile world, or appropriate when pressed for time, it must be realised that if a team is developing a solution to an engineering problem it is important to be careful to avoid taking one of the quick solutions that the brain generates so easily, and to work in a more methodical way. It will not feel natural. It will be fighting against everything the brain has ever wanted to do, but it should be done.

4.5 CREATIVITY AND THE EUREKA MOMENT

When Francis Crick and James Watson realised that one of the implications of DNA's double helix was the potential to explain the passing on of genetic information it seemed to be one of the many breakthroughs based on a brilliant flash of inspiration by another of the geniuses that fill the pages of history. But, it is not quite as simple as that. Crick's sudden insight occurred in 1952, but by that time he had been studying the problem of the structure of DNA for over a year – thinking of little else, discussing little else, working on little else. It may have been a brilliant flash but he had prepared himself well for it.

The more one studies people that are regarded as creative geniuses the more it becomes evident that there is much more to creativity than the sudden flash of inspiration. Archimedes in his bath, Newton and the apple under the tree, Darwin and evolution: in each case the real story is much more complicated than the popular folklore, and in each case the discovery came after years of hard work.

The Original Eureka!

Many know that Archimedes discovered something while overflowing a bath and ran down the street shouting Eureka, but the details are not always as clear. Koestler (1964) gives a good description.

Eureka is Greek and translates as "I have found it!". The use of it to describe some sort of mental breakthrough dates from Archimedes who had been given the task of determining whether a crown given to his king was in fact pure gold, or gold mixed with a lesser metal. Knowing the density of gold the task was therefore to decide the volume of the crown and to check its density. A lower density would indicate that the gold had been mixed with a lighter element. Archimedes pondered the task of deciding the volume. If he could melt it down he could measure its volume but that was not an option. Whether it is apocryphal or not, the story goes that one day as he was getting into his bath, Archimedes noticed that as he got in his body displaced the water and it overflowed. He realised that if he submerged the crown it would displace its own volume and he would be able to measure it accurately. He then set off down the street overjoyed at his discovery.

There appears to be a general agreement that it takes approximately 10 years of consistent work before significant outcomes eventuate, and the fact that there is little apparent progress does not mean there is no progress. According to Bronowski (1978) "the discovery is made with tears and sweat (at any rate, with a good deal of bad language) by people who are constantly getting the wrong answer". Gilhooly (1982) and Miller (2001) catalogue the stages involved in a creative leap:

- preparation, where the problem solver familiarises himself or herself with the problem by engaging in "conscious, effortful, systematic and usually fruitless work on the problem";
- incubation, where the problem is set aside and no conscious work is done on it;
- illumination or inspiration, where a break is made which, even if not the complete solution, leads to an advance; and
- verification, where conscious work is done to test the idea.

Of the stages, there are a couple that require further examination. The incubation stage is an interesting one and highlights a mode of thinking that many believe to be crucial to human thought: that done by the subconscious. Both Albert Einstein and the French scientist Henri Poincaré, for example, were firm believers in the power of this type of thought and deliberately structured their work to make the most of breaks from conscious thought. While sitting not thinking may sound appealing, it should be noted that it presupposes that work has been done beforehand so that the brain has something to work with! The act of creativity is most clearly associated with the third stage, where things come together in a novel and useful way. A key point that is made by a number of researchers is that the new idea does not appear from nowhere, but comes from the making of novel connections between information already stored in the brain. On this point, Poincaré believed that new mathematical ideas must be "beautiful" and that the brain actually used this beauty to identify the potentially valid ideas from the mass of others (Poincaré, 1908). This need for existing knowledge underlies many of the methods designed to enhance creativity, as we shall now see.

4.6 TECHNIQUES FOR STIMULATING IDEAS

At this stage it is worth reflecting on what we have found so far about brains and thinking. Brains like short-cuts and have developed a whole range of them to allow us to make quick and often quite good decisions. Brains also hide much of their inner workings and are far more pre-programmed to work with anticipated situations than we realise.

Engineers and scientists are generally well trained in convergent thinking; making use of a minimum of information to come to a solution quickly. They may even resist divergent thinking and the sorts of operations necessary to carry it out. And yet, in many engineering planning and design problems this is exactly the thinking that is called for. Therefore, it may be necessary to use a series of seemingly odd and somewhat artificial procedures to force the brain to act in a manner that suits these types of problems. It will not feel natural. It will not make sense. It will, however, with practise allow the development of some quite innovative solutions that no level of convergent thinking could ever approach.

For many problems that are presented to engineers there will be an obvious "common sense" solution that most will see quite quickly. The value that can be attributed to the idea that led to the common sense approach is therefore minimal. However, if one engineer comes up with quite a different solution, one that no-one else had thought of, its value is much higher. That's the aim of creative thinking: to come up with those different ideas. Much of the time they may not lead to a better approach, but every now and again they will and those times are worth striving for.

If any further encouragement was required to promote creative thinking it might come from knowing how many ideas it takes to come up with a truly successful one. Stevens and Burley (1997) carried out such a study by investigating the relationship between new ideas in an industrial setting and how many progressed on to success as commercial products. Based on a wide range of views and data they suggest that for every 3,000 raw ideas about 150 are developed to the stage where a patent application is filed. Of these applications about 112 are issued with patents of which only nine have commercial significance. Of those, only one leads to major commercial success. It seems we need ideas, and we need lots of them.

One of the common aims of the various methods is to force the brain to look at things in a different light and to make connections that it would not normally make. According to Findlay and Lumsden (1988) "the creative process is derived from the establishment of new links among already existing elements". In the following sections a number of established and accepted methods for enhancing creativity are outlined. They are not in any particular order and it should be noted that each method will have situations where it is suited and some where it is not. Therefore, they should be read with an open mind. Nothing stifles creativity more effectively than a sneering cynic.

Attribute listing

Attribute listing works best where there is an existing solution or product and a better one is being sought. As Svensson (1974) has pointed out, this may be with the aim of eliminating defects from a product, improving its operation or reducing costs. The steps involved are: list all the components or elements of the existing solution; list all the attributes of the component (e.g. weight, size, colour, shape, material etc.); and then systematically change each attribute in every conceivable way.

As an example, taken from Miller (2000), consider an improved design for a heating system radiator using attribute listing. The elements of the existing product, their present attribute and other possibilities are listed in Table 4.2. There are two complementary ways the information at this stage can be used. Firstly, if improvements are required to the existing designs then some of the possible attributes can be evaluated while leaving the others fixed. For example, it may be beneficial to consider the effect of changing the colour of the product. Superficial changes of different shades of white or cream are possible but what about a paint that changes colour as it heats? This may enhance the design and improve safety by reducing the chance of accidental burns or scalds. There are similar options that could be tried for all the attributes. Alternatively, is there a quite different finish that does not involve paint? Or might the finish make the product essentially invisible?

Table 4.2 Product attributes and other possibilities for a house radiator.

Element	Present attribute	Possible attributes
material	metal	plastic, ceramic, paper, wood
shape	rectangle	circle, oval, sphere, ellipsoid, tube
mobility	fixed	mobile, float, sink, roll
liquid	oil	coloured water, gas, liquid metal
opacity	opaque	transparent, semi-transparent
finish	smooth	rough, carved, regions of texture
colour	white	red, blue, black, fluorescent, variable

Secondly, combining all present and possible attributes leads to 5 x 6 x 5 x 4 x 3 x 4 x 6 = 43,200 different possibilities and these could be evaluated. One of these is the current configuration, and many will not make any sense, but some surprising and challenging alternatives may come from this, and may prove valuable. For example, some of the variations include:

- plastic sphere that rolls around the floor filled with oil;
- transparent plastic tubing filled with coloured water;
- black metal tubing filled with hot oil;
- plastic tube that is mobile and filled with oil; it comes in a variety of colours.

The challenge is to see if and how these can lead to something new and viable. If nothing else, the various combinations can be used to foster discussion, and to explore what is really required for a solution. In any case, the main aim has been fulfilled in that the designer has been forced away from seeing all home radiators as flat metal components fixed to a wall. At the early stages of the design process the rule should be quantity not quality of concepts. Another important consideration is that sometimes ideas that come up using this technique can be better applied in a quite different area. For example, consider the last of the four examples listed, the mobile plastic tube filled with oil. While this may not be suitable for warming a room it may be ideal as a hand warmer or a foot warmer for those very cold nights. It was not what was asked for but it is worth being open to the unexpected. The mobility idea could provide the germ of further ideas: why for example should one heat a whole house if it is possible for a person to carry a heat source around and so be comfortable anywhere?

Morphological synthesis

Morphological synthesis (also called matrix analysis) is similar in some ways to attribute listing in that it involves the generation of unusual combinations of system functions, however, it is more focussed on generating completely new ideas rather than just improving existing ones. The steps involved are: describe the problem by its systems; list the major system parameters; list the alternative ways of satisfying each system parameter; draw up a matrix with each system parameter as one dimension; and then consider all possible combinations of ways of satisfying each parameter.

For example, if it is desired to establish an innovative mode of public transport in a major city then morphological synthesis can be used to generate alternatives. If the three major system parameters are considered to be the carriageway, energy source and method of support the results can be generated as shown in Table 4.3.

Table 4.3 Morphological synthesis used on public transport.

System parameter	Possible solutions
carriageway	rail, road, monorail, water, air
energy source	liquid fuel, electricity, gas, solar, wind, animal
method of support	wheels, air, magnetic, wires, rail

The method has generated 5 x 6 x 5 = 150 combinations, which include:

- an electric powered monorail with magnetic levitation;
- gas powered vehicles travelling over water with air suspension;
- solar powered buses with wheels on a road.

Many of the 150 may be infeasible; some will already exist; others will not make any sense at all. However, if there is one that is worth further consideration, or one that sparks a new idea of some sort, then the time spent using the method has been well spent.

Setting a minimum quota

One of the main problems in developing novel or creative solutions is the tendency to apply one of the brain's shortcut heuristics and go with the first viable solution that comes to mind. In many cases this may appear to work, but it is a convergent thinking technique and it is possible that other equally viable (and possibly superior) solutions may be overlooked. To overcome this it is considered good practice to set a minimum quota of ideas which must be generated. In this way the tendency to follow the best idea as soon as it occurs is overcome. De Bono (1970) suggests a limit of around four or five as a minimum. Better advice is to estimate how many solutions there are, and then seek double or triple that number.

As with many of the methods that are used to develop creativity, the setting of a minimum quota works on a number of levels. Firstly, it forces a number of solutions to be generated. Secondly, and perhaps more importantly, it forces the engineer to think about the problem more deeply and to consider much more carefully what might be considered a solution.

The idea of setting a quota does not actually help generate new ideas, but it forces the search to continue and is certainly worth the effort. McCormick (2004) warns against the practice of setting rituals where a certain number of ideas must be generated for a "correct" solution, so it is important to keep in mind the importance of the exercise rather than the idea that a certain number of ideas will produce the correct answer.

Challenge assumptions and implied solutions

One of the issues with problem-solving is the tendency to impose limits on the solution due to assumptions which are made even without realising it. For example, consider for a moment the task of inventing a new design for a toothbrush. Even with no further instructions it is likely that one would assume that the solution being pursued would have to be portable, cheap, hand-held, able to use standard toothpaste, taste and odour free, able to be purchased easily, and instantly recognisable as a toothbrush. With all these constraints in place it is no wonder that any innovative design that is unlike what currently exists is virtually impossible. But are these really necessary? A truly innovative approach might come from altering the question to "design a new way of protecting teeth" and this in turn could lead to the search for a much wider solution than the current toothbrush affords. Whatever the outcome, it is important to realise that humans are very likely to self-impose creativity-limiting assumptions as part of a convergent approach.

Suspended judgment

The purpose of thinking creatively is not to be right, but to be effective. In general thinking there is a tendency to judge each and every aspect of new ideas and to try and discard them on any particular or perceived flaw. History, however, provides ample examples of people who were quite wrong about aspects of their work or thoughts and yet made great advances. According to Kolb (1999), Kepler, the astronomer who formulated the three famous rules of planetary motion developed them based on a question that came to him in the middle of a lecture to undergraduates: why are there six planets? From that poor starting position the excellent work that demonstrated the properties and consequences of the planets' elliptical orbits followed. In fact, in the field of astronomy good results have come from ideas and calculations containing basic errors. According to Thornton (2000), Neptune was discovered based on calculations that showed there was something (possibly a new planet) affecting the orbit of Uranus. The calculations contained an error but, as luck would have it, one that had little effect at the particular time the search for the cause of the disturbance was undertaken. Had the search been undertaken at a different time, the error could have made it a futile one.

According to de Bono (1970), there are several advantages with suspending judgment on ideas:

- an idea will survive longer and may breed further ideas;
- other people will offer ideas which they themselves may have discarded but which others can make something of;
- ideas can be accepted for their stimulating effect;
- ideas which may be judged wrong in the current frame of reference may survive long enough to show that the frame of reference was wrong.

The way in which suspension of judgment is used (de Bono, 1970) is:

- one does not rush to judge or evaluate an idea, one explores;
- some ideas which are obviously wrong can be used to explore, not why they are wrong but how the idea could be useful;
- even if one knows that an idea will eventually be thrown out one keeps it for as long as possible to get as much out of it as possible; and
- instead of forcing an idea in a given direction, one follows the idea to see where it will lead.

Marconi and Radio Transmission

There have been occasions when people have done things which were impossible or wrong, and it turned out that despite everything being against it the final result was what they were seeking. Guglielmo Marconi succeeded in transmitting radio waves across the Atlantic believing that the waves would follow the curvature of the earth. He had been working on transmitting radio waves and had been able to achieve it over increasing distances. First in metres, kilometres, and across the English Channel. He was convinced he would be able to do a transatlantic transmission so he set up a transmitter in England and set up an aerial held aloft by a large kite in Newfoundland. He waited. People told him radio waves would travel in straight lines so he would not be able to do it. But he did it, and in that respect was correct. The people were also right when they said radio waves travel in straight lines, but did not understand that the earth had an ionosphere from which the waves could bounce. If Marconi had listened to others he would not have made the breakthrough.

Gilhooly (1982) reports on a study where subjects were asked to generate solutions to a particular problem for 5 minutes. One set of subjects was asked to work in a conventional manner, evaluating ideas as they went. In the other they were asked to defer judgment. The ideas were then evaluated by a team of judges. The first group produced 2.5 "good" ideas (on average) while the second produced 4.3 "good" ideas. The higher productivity was explained in terms of suspending judgment.

Brainstorming

Brainstorming is a technique that was developed in the 1940s and 1950s by a businessman, Alex Osborn (Gilhooly, 1982). When it was first introduced it took off quite quickly, and has now been applied in industry and evaluated in the laboratory. According to Gilhooly (1982) "numerous studies support the hypothesis that groups using brainstorming produce more ideas than similar groups that work along conventional lines". There are also reports of not only more ideas, but better ideas too, coming from brain storming sessions.

A brainstorming session usually involves 6–12 people. Two officials are needed; a chairperson and a recorder. The chairperson introduces the topic, enforces the rules and is responsible for maintaining a continuing flow of ideas. The recorder makes a note of all ideas. The topic must be formulated in such a way that it is not too vague. Since the brainstorming session will be straining the limits of creativity starting with a vague problem could lead to an over-wide collection of ideas. At the same time the problem must not be over-formulated so that some exploration around the topic is possible. This comes with experience. The rules of a brainstorming session are (de Bono, 1970; Perkins, 2000):

- all ideas must be recorded;
- no idea may be criticised or evaluated during the session (suspend judgment);
- it is permissible to build on the ideas of others; and
- the emphasis is on quantity of ideas; quality follows from quantity.

It is better to record the ideas on a whiteboard or similar so that all participants can see them. At the start of the session the chairperson introduces the topic and sessions tend to last around 30 to 45 minutes. As a way of enforcing the rules and maintaining the flow of ideas the chairperson should:

- stop people trying to evaluate or criticise other people's ideas;
- stop people all talking at once;
- make sure the recorder has the idea down;
- fill in gaps by offering ideas;
- suggest different ways of tackling the problem;
- define the central problem and keep pulling people back to it;
- end the session either at a designated time or when things are flagging; and
- organise the evaluation session.

Culp and Smith (2001) suggest that it may be beneficial if, at the start of a session, everyone spends a couple of minutes writing down their own ideas with no discussion. Then the session starts with the ideas being read out in turn. This is designed to ensure everyone gets an initial word in and that the more dominant personalities do not take over the session.

Often, in the course of a brainstorming session the flow of ideas can wane and there is a prolonged silence. A "trigger" is a very useful device to restart the ideas. A number of triggers have been suggested. These include:

- a checklist;
- wildest fantasy;
- juxtaposition.

One checklist is provided by the acronym SCAMPER and is listed in Table 4.4. The use of "wildest fantasy" as a trigger encourages all participants to think of the wildest idea. Some of these may lead to other more practical solutions. 'Juxtaposition' involves the following steps:

- select three words at random;
- discover a connection between each word and the problem being studied.

Water Supply in the Developing World

According to New Civil Engineer (2002), the World Summit on Sustainable Development held in Johannesburg in 2002 developed an aim to halve the number of people without access to proper sanitation and to have all people with access to clean water by 2020. In a surprising twist the World Bank and the United Nations called for no new infrastructure but a program of education and water supply improvements. Although large schemes seem the logical way to go, experience has shown that unless there is full subsidy of the schemes they are rarely sustainable and the result is a flawed system of broken pipes and pumps and a lack of local skilled personnel to effect repairs.

The role of the engineer in these countries will need to change, along with their perception of what constitutes success.

The best words to use are familiar ones such as water, meat, cup, or pen. For example, in trying to suggest ideas for the design of a vehicle which does not use fossil fuels a group has run out of ideas. They decide to use juxtaposition as a trigger, and select the words pencil, water and chicken. The word pencil then may lead to the suggestions of using wood as a fuel, making the vehicle long and thin to reduce drag, using graphite as a fuel or lubricant, and a vehicle which walks on stilts. The other words will also generate discussion in a similar way.

Table 4.4 The checklist SCAMPER

Substitute	Who? What? Other material?
Combine	Units? Purposes? Ideas? Other place?
Adapt	What else is like this? What can I copy?
Modify	Change meaning? Shape? Colour?
Put to other uses	What is different about this use? What is similar?
Eliminate	~ parts of
Rearrange	Interchange components? Other layouts?

There must be no evaluation of ideas in the session. That is a task for later. Statements that must be discouraged include: "that would never work because", "it is well known that", and "how would you get to do that". The ideas generated should be evaluated several days later. Evaluation is better carried out without consultation between participants. One possibility is to give several people the list and ask them to identify the best 10%. These could then go on and be further evaluated and developed. In the evaluation session which follows some time later the central tasks are to:

- pick out the ideas which are directly useful;
- extract the ideas that are wrong or ridiculous;
- list functional ideas and new ways of considering the problem;
- pick out those ideas which can be tried out with relative ease even if they seem wrong at first;
- pick out those ideas which suggest that more information could be collected in certain areas; and
- pick out those ideas which have already been tried.

Although there is supposed to be some benefits in having a number of people in a group feeding each other ideas and benefiting from the group dynamics there is also evidence that a single-person brainstorming session can be beneficial. Here the main benefit comes from the deliberate suspension of judgement and the idea that "anything goes". A number of studies have shown that a nominal group made up of a number of people working individually can combine outputs to produce a better set of solutions than a group of the same size working as one (Gilhooly, 1982).

Reading and contemplation

A common theme among many individuals that have made significant breakthroughs is the place reading and study played in their lives. Charles Darwin was a voracious reader and his natural selection breakthrough came to him following his reading of a work by Thomas Malthus on population dynamics. Albert Einstein's theory of relativity came after reading David Hume's *An Enquiry Concerning Human Understanding* and he made no secret of the benefit he had obtained from reading it (Edmonds and Eidinow, 2006). Isaac Newton and numerous other scientists of note were all well known for their reading that covered a wide range of topics from science to philosophy. The power of reading also comes to lesser mortals. James Ellis was a leader in the field of code-breaking and code-making. Singh (1999) suggests that one of the reasons for Ellis' success was his breadth of knowledge:

> *He read any scientific journal he could get his hands on, and never threw anything away. ... if other researchers found themselves with impossible problems, they would knock on his door in the hope that his vast knowledge and originality would provide a solution.*

The same must be true for much of life and life's experiences. A wider knowledge base provides different ways of looking at things and a much greater chance of seeing something from a different perspective (Ferguson, 1999):

> *More important to a designer than a set of techniques (empty of content) to induce creativity are a knowledge of current practice and products and a growing stock of firsthand knowledge and insights gained through critical field observation of engineering projects and industrial plants.*

The benefits of reading come not only from the additional knowledge that may be gained, but also from the stimulating effect that reading has on the brain. As has already been noted, creativity comes from making novel connections between pieces of information stored in the brain, and the active thinking that goes on as one reads assists in this process.

More than a bit player: Howard Hughes Snr.

The movie *The Aviator* (starring Leonardo DiCaprio and Cate Blanchett), tells the story of the early years of Howard Hughes Jnr. who made a name for himself producing Hollywood movies, building and flying aeroplanes, setting speed and distance flying records, and generally making the most of his money, youth and talent. The source of his early money was based entirely on a single invention that his father, Howard Hughes Snr. had patented: a drill bit for the oil industry that was particularly good at drilling through rock.

Hughes Snr. had been in the mining industry, but was lured to Texas with the discovery of oil near Beaumont in 1901. He worked in a number of locations and by 1907 had a partner in a small business although Bartlett and Steele (2003) note that he was "usually too independent-minded to work with anyone". In 1908 Hughes met a millwright, Granville Humason, who had designed a new drill bit and who showed him a wooden prototype. Humason had come up with the idea for the drill one morning as he ground his coffee and had shown it to a lot of miners but no-one had been interested. Hughes offered him $150 on the spot for it, which was accepted. Hughes then rushed to work up a patent application and had the device patented in the US and overseas. The final drawing was carried out on the kitchen table. Again, according to Bartlett and Steele (2003):

He emerged from the family dining room with the Archimedean cry of 'Eureka' and the picture of a bit that had no less than 166 cutting edges.

And the rest, as they say, is history. Hughes went on to be awarded over 70 patents and founded a company that generated tens of billions of dollars. The interest in the story is to consider what it was that made Hughes Snr. successful? According to the biographers he was a stickler for details. They also contend that he would have no doubt been aware of the previous attempts to patent such a drill bit and would have recognised the significant breakthrough it represented. Perhaps the real success was based on a good notion of what the idea was really worth and the need to protect that with good patents. Hughes spent at least $1500 on that first patent and worked hard to maintain and protect it.

Fostering creativity

While one of the aims of this chapter is to promote creativity in individuals, a related aim is to ensure that those who are creative will have their ideas received in a positive environment. People should be able to recognise and appreciate creativity, and remember that it will not always be logical, immediately useful or even sensible under current thinking. According to a co-worker (quoted in Singh, 1999) of James Ellis, the cryptographer mentioned earlier:

> *He was a rather quirky worker, and he didn't really fit into the day-to-day business of GCHQ. But in terms of coming up with new ideas he was quite exceptional. You sort of had to sort through some rubbish sometimes, but he was very innovative and always willing to challenge the orthodoxy. We would be in real trouble if everybody in GCHQ was like him, but we can tolerate a higher proportion of such people than most organisations. We put up with a number of people like him.*

Creative individuals are not necessarily those that are highly intelligent (there is only a moderate correlation between creative ability and IQ) and so many will not stand out in an academic sense. In many cases their ideas will be of little value. Note above that the workers with Ellis "had to sort through some rubbish sometimes", but in the end having creative people around is well worth it because when virtually everyone comes up with the same solution the truly creative person will be the one who has a different idea that no-one else had thought of or considered. That's when they come into their own. And that is why everyone should try to be more creative and to appreciate and nurture those who are. Goel and Singh (1998) have a similar message when they urge businesses to "understand that creative ability of most people is fragile and could be seriously suppressed by destructive criticism" and to design a company structure that "enhances rather than detracts from useful creative activity".

4.7 SUMMARY

The brain is a complex, power-hungry, secretive, manipulative and vital organ that defines human beings. It gives us the *sapiens* in *Homo sapiens* and is the product of 200,000 years of evolution since our species split from earlier ancestors. The brain has developed a particular skill in coming up with reasonable solutions to many problems in the minimum of time, often without any formal decision-making process, often using a large number of hard-wired techniques.

If the brain is to be used for creative thinking a conscious effort must be made to force the brain away from its natural methods of solution and towards ways that ensure that unusual or unexpected results can follow. There are a range of methods that can be employed including attribute listing, morphological synthesis, the use of a quota, brainstorming and extended reading and thought. Each takes practise and time, but it will be time well spent.

PROBLEMS

4.1 Apply attribute listing to the problem of trying to generate at least 50 new ideas for a different type of ball point pen.

4.2 Use morphological synthesis to generate at least 20 new ideas for a new way of drying clothes in a typical suburban house.

4.3 In 2 minutes write down as many uses that can be thought of for a paperclip. Start now!

4.4 Use a one person brainstorming session to devise ways of reducing fuel consumption for the average family car. In 10 minutes try and generate as many ideas as possible.

4.5 Assume the government has put out a tender for development of a mobile kitchen that could be used for the army in exercises in remote and arid areas. The unit must be suitable for preparing hot meals for a group of 20 soldiers. Generate at least five quite different solutions to the problem.

4.6 Suggest five creative solutions to the problem of an existing bridge that vibrates excessively when people run over it.

4.7 The people who prefer not to define creativity often resort to the argument that they can recognise it when they see it. Consider your group of friends and acquaintances and identify creative ability in them. What are the main distinguishing characteristics?

4.8 If the Eureka moment comes after much concentrated study and thought what are the implications for students wishing to master a subject at university?

4.9 If truly creative people really do see life differently and think about problems and solutions in very novel ways, what fraction of the population do you think they should comprise? What are the issues to be considered?

REFERENCES

Bartlett, D.L. and Steele, J.B., 2003, *Howard Hughes, His Life and Madness.* (Penguin), 687pp.

Boden, M.A., 2004, *The Creative Mind. Myths and Mechanisms.* 2nd Edition, (Routledge), 344pp.

Bohm, D., 1996, *On Creativity.* (Routledge), 153pp.

Borges, B., Goldstein, D.G., Ortmann, A. and Gigerenzer, G., 1999, Can Ignorance Beat the Stock Market? In: *Simple Heuristics that Make Us Smart.* (Eds. G. Gigerenzer, P.M. Todd and the ABC Research Group), (Oxford University Press), pp. 75–95.

Bronowski, J., 1978, *The Origins of Knowledge and Imagination.* (Yale University Press), 144pp.

Brooks, M., 2000, Global Brain. *New Scientist*, 166(2244), pp. 22–27.

Camazine, S., 2003, Self-Organizing Systems. Encyclopaedia of Cognitive Science, www.cognitivescience.net, 4pp.

Cohen, J. and Stewart, I., 1995, The Possibilities of Evolution In: *The Collapse of Complexity*, (Penguin), pp. 97–133.

Cross, F.R. and Jackson, R.R., 2005, Spider Heuristics. *Behavioural Processes.* 69(2), 125–127.

Culp, G. and Smith, A., 2001, Understanding Psychological Type to Improve Project Team Performance. *Journal of Management in Engineering*, ASCE, 17(1), 24-33.

de Bono, E., 1970, *Lateral Thinking*. (Penguin), 260pp.

Edmonds, D. and Eidinow, J. (2006) *Rousseau's Dog. Two Great Thinkers at War in the Age of Enlightenment.* (Faber and Faber), 405pp.

Ferguson, E.S., 1999, *Engineering and the Mind's Eye.* (The MIT Press), 5th printing, 241pp.

Findlay, C.S. and Lumsden, C.J., 1988, The Creative Mind: Toward an Evolutionary Theory of Discovery and Innovation. *Journal of Social Biology*, 11, pp. 3–55.

Fraser, A., 2003, Don't Let 'Em Pick Brains for Breakfast. *The Weekend Australian*, 15–16 November, 2003.

Gardner, H., 1993, *Creating Minds.* (Basic Books), 464pp.

Gigerenzer, G. and Goldstein, D.G., 1996, Reasoning the Fast and Frugal Way: Models of Bounded Rationality. *Psychological Review*, 103(4), pp. 650–669.

Gigerenzer, G. and Goldstein, D.G., 1999, Betting on One Good Reason. In: *Simple Heuristics that Make Us Smart.* (Eds. G. Gigerenzer, P.M. Todd and the ABC Research Group), (Oxford University Press), 75–95.

Gilhooly, K.J., 1982, *Thinking. Directed, Undirected and Creative.* (Academic Press), 178pp.

Goel, P.S. and Singh, N., 1998, Creativity and Innovation in Durable Product Development. *Computers & Industrial Engineering*, 35(1–2), pp. 5-8.

Goldstein, D.G. and Gigerenzer, G., 2002, Models of Ecological Rationality: The Recognition Heuristic. *Psychological Review*, 109(1), pp. 75–90.

Goleman, D., 1995, *Emotional Intelligence.* (Bloomsbury), 352pp.

Gould, S.J., 1991, *Bully for Brontosaurus.* (Penguin).

Holland, J.H., 1998, *Emergence from Chaos to Order.* (Oxford University Press), 258pp.

Kates, R.W., 1962, Hazard and Choice Perception in Flood Plain management. *Department of Geography Research Paper No. 78*, University of Chicago.

Klawans, H., 2000, *Strange Behavior. Tales of Evolutionary Neurology.* (W.W. Norton and Company).

Knutson, B. and Peterson, R., 2005, Neurally Reconstructing Expected Utility. *Games and Economic Behaviour*, 52(2), 305–315.

Koestler, A., 1964, *The Act of Creation.* (A Laurel Edition), 751pp.

Kolb, R., 1999, *Blind Watchers of the Sky.* (Oxford University Press), 338pp.

Kruger, J.; Wirtz, D. and Miller, D.T., 2005, Counterfactual Thinking and the First Instinct Fallacy. *Journal of Personality and Social Psychology*, 88(5), 725–735.

McCormick, R., 2004, Issues of Learning and Knowledge in Technology Education. *International Journal of Technology and Design Education*, 14, pp. 21–44.

Miller, L., 2000, Idea Generation. *Engineering Designer*, Sept–Oct 2000, pp. 26–27.

Miller, A. I., 2001, *Einstein, Picasso. Space, Time and the Beauty that Causes Havoc*. (Basic Books), 357pp.

Perkins, D., 2000, *The Eureka Effect. The Art and Logic of Breakthrough Thinking*, (W.W. Norton and Company), 292pp.

Phillips, H., 1999, They Do It With Mirrors. *New Scientist*, 166(2243), pp. 27–29.

Poincaré, J.H. (1908) L'Invention Mathématique, L'Enseignement Mathématique. 10, 357–371. Translated by editor as Mathematical Invention. In: Musings of the Masters. An Anthology of Mathematical Reflections, Ayoub, R.G. (Ed.), 17–30.

Popper, K. and Eccles, J.C., 1977, *The Self and Its Brain*. (Routledge), 597pp.

Ramachandran, V.S. and Blakeslee, S., 1998, *Phantoms in the Brain*. (Fourth Estate), London, 328pp.

Reimer, T. and Katsikopoulos, K.V., 2004, The Use of Recognition in Group Decision-Making. *Cognitive Science*, 28, pp. 1009–1029.

Rose, S., 1998, Brains, Minds and the World. In: *From Brains to Consciousness, Essays on the New Sciences of the Mind*. Ed: S. Rose, (Penguin), pp. 1–17.

Singh, S., 1999, The Code Book. (4th Estate).

Stevens, G.A. and Burley, J., 1997, 3,000 Raw Ideas = 1 Commercial Success! *Research Technology Management*, 4093, pp. 16–27.

Svensson, N. L., 1974, *Introduction to Engineering Design*, (NSW University Press), 129pp.

Thornton, C., 2000, *Truth from Trash*. (MIT Press), 204pp.

Tversky, A. and Kahneman, D., 1973, Availability: A Heuristic for Judging Frequency and Probability. *Cognitive Psychology*, 5, pp. 207–232.

Wänke, M.; Schwarz, N. and Bless, H., 1995, The Availability Heuristic Revisited: Experienced Ease of Retrieval in Mundane Frequency Estimates. *Acta Psychologica*, 89, pp. 83–90.

White, M., 2002, *Rivals. Conflicts as the Fuel of Science*. (Vintage), 417pp.

CHAPTER FIVE

Project Planning Techniques

The execution of large engineering projects involves the coordination of many activities. Careful planning is required to ensure that a project is completed on time and within budget. In this chapter, techniques to assist in the planning of large projects are considered. The techniques considered include the critical path method (CPM) and the Gantt Chart. The use of these techniques to assist in scheduling activities so that the project finishes within the minimum feasible time is demonstrated. The possible rescheduling of activities taking into account limitations on certain critical resources is also considered.

5.1 INTRODUCTION

Project planning deals with the interrelationships between, and the timing of, the various activities that comprise a project. Detailed planning of a project entails identification of the activities that comprise a project, estimation of the duration of each activity, identification of the precedence relations between activities (i.e. which ones need to precede others) and development of an organisational network or schedule that represents this information accurately. Such an organisational network can be used to provide the following information:

1. the minimum time to complete the project if all activities run on time;
2. the activities that are critical to ensure that the project is finished in the minimum time;
3. the earliest start time and the latest finish time for each activity, if the project is to be finished in the minimum time; and
4. the amount of time by which each activity can be delayed without delaying the project as a whole.

An organisational network can also be used to examine the likely timing of resources (human resources, cash and equipment) required over the duration of the project and whether limits on these resources are likely to cause delays in the project completion. Furthermore, an organisational network can be used to identify which activities should be accelerated if the project needs to be completed in a shorter time than the current estimate.

Before we look at the development of organisational networks, and in particular the critical path method, it is worth giving a brief historical overview of the topic so that its importance in engineering planning can be fully appreciated.

5.2 HISTORICAL BACKGROUND

Although we tend to think of project planning as a recent development, the construction of structures such as the great pyramids of Egypt and the Americas, Stonehenge and the statues on Easter Island undoubtedly required considerable organisational and project planning skills. Unfortunately there are no records of how these projects were planned and managed.

One of the first modern techniques developed to assist in project planning is the Gantt Chart. This was created in 1917 by Henry Laurence Gantt (1861–1919), an American mechanical engineer. The first Gantt chart was developed for planning the building of ships during the World War One. A Gantt Chart is a graph in which the horizontal axis represents time. All activities are listed down the page with each activity having a horizontal bar representing the planned timing of its completion (see Figure 5.7).

In the late 1950s several techniques were developed to assist in the planning of complex projects. These include the critical path method (CPM) and the program evaluation and review technique (PERT). CPM was developed in the late 1950s by Morgan Walker of E.I. Du Pont and James E Kelly of Remington Rand Univac Corporation and was first used to schedule maintenance shutdowns in chemical processing plants. PERT was developed by Booz, Allen & Hamilton and the US Navy with the aim of coordinating the many thousands of contractors who were working on the Polaris missile program (Griffis and Farr, 2000; Shtub et al. 2005). Both techniques represent all activities involved in a project as a network of arrows and nodes. Calculations can then be carried out to determine the information listed in Section 5.1 using the network and the durations of all activities. The basic difference between the two techniques is that CPM assumes that the durations of all activities are known whereas PERT represents the durations of activities as random variables with optimistic, pessimistic and most likely estimates of their durations. In the 1980s, Gantt Charts were modified to include link lines between tasks so that they could also include certain attributes of CPM.

This chapter contains a description of CPM and Gantt Charts. As noted above, Gantt Charts were developed before CPM, however, they are introduced after CPM in this text in order to allow a more logical development of the concepts. PERT will not be discussed in detail in this book. The interested reader is referred to Burke (2001) or Shtub et al. (2005) for more details of this technique.

5.3 THE CRITICAL PATH METHOD

As mentioned above, CPM represents all of the activities that make up a project as a network of arrows and nodes. Two different types of notation are commonly used. These are called activity on node (AON) and activity on arrow (AOA) notation. Either notation may be used to represent the precedence relationships between activities. To illustrate the two notations, consider a simple project consisting of five activities (A, B, C, D and E). Suppose that Activity A must be completed before activity B can commence, Activity C must be complete before Activity D can commence and Activities C and D must be complete before Activity E can commence. An organisational network for the project using activity on arrow notation is shown in Figure 5.1 (a) while the network using activity on node notation is shown in Figure 5.1 (b).

As shown in Figure 5.1 (a) when using activity on arrow notation the arrows represent the activities and the nodes are used to represent the precedence of activities. The nodes are numbered (in an arbitrary fashion), so that each activity has a unique designation in terms of its start and end nodes. For example Activity A is designated 1-2, Activity B is designated 2-4 and so on. Starting at node 1, Activities A and C have no preceding activities, so both may commence immediately. Activity B cannot commence until activity A is completed, so its start node is the end node of Activity A. For example, Activity A may represent the construction of footings for a house and Activity B the construction of the walls. Clearly, Activity B cannot commence until Activity A is complete.

Similarly Activity D has its start node (3) as the end node of Activity C. Now, why do both Activities B and D end at the same node? This is purely a notational convenience that also achieves a compact network, as Activity E requires them both to be completed before it can commence. There are other equally valid ways to draw this diagram that will be described later. Finally, as shown, Activity E can only start after both Activities B and D are complete and it finishes at Node 5 (the completion of the project).

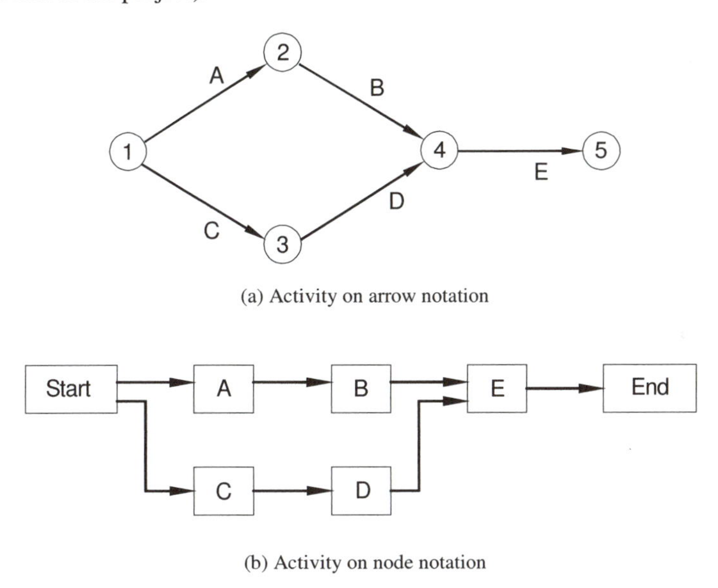

(a) Activity on arrow notation

(b) Activity on node notation

Figure 5.1: Two types of network notation

Figure 5.1 (b) shows an organisational network for the same project using activity on node notation. In this case each activity is represented by a box. There is also a box for the start and finish of the project. The precedence relationship between activities is represented by arrows. From the information provided above, Activities A and C do not need any activities to be completed before they can start. Therefore, they can both commence at the beginning of the project as shown in

Figure 5.1 (b). Activity A must be completed before Activity B can commence, so an arrow goes from A to B. Similarly Activity C must be completed before activity D, so an arrow runs from C to D. As both C and D must be completed before Activity E commence, they both have arrows running into Activity E. Finally the project can end once Activity E is finished.

Activity on arrow (AOA) notation will be used for the remainder of this chapter as it is more commonly used in practice.

There are four basic rules to observe when constructing AOA organisational networks. These are:

1. the network must have one starting node and one finishing node representing the start and finish of the project (respectively);
2. each activity is represented by a single arrow in the network;
3. before an activity can start all activities leading to its starting node must be complete; and
4. there can be, at most, one arrow between any pair of nodes in the network.

Rule 4 is required to ensure that each activity is uniquely defined by its starting and ending nodes. On occasions this may require the use of **dummy activities**. These are artificial activities of zero duration that are used purely to maintain the logic of the network. An example of the use of dummy activities is shown in Figure 5.2.

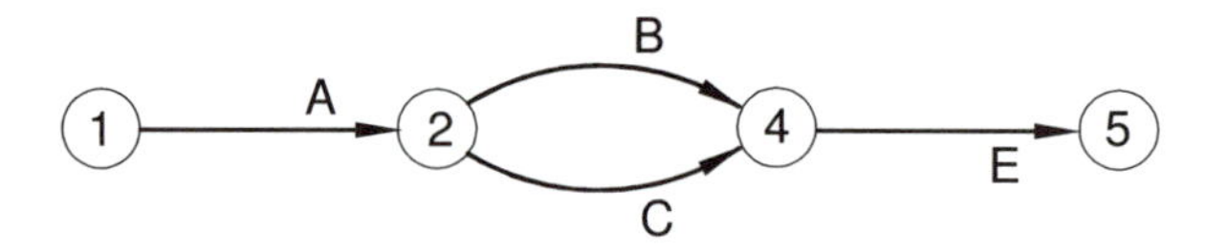

(a) Network that does not satisfy Rule 4

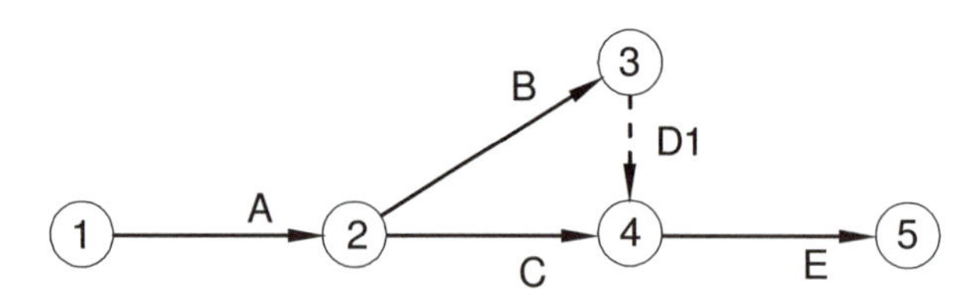

(b) Modified network using a dummy activity

Figure 5.2: Two ways of representing the same organisation network using AOA notation

For the project shown, Activity A must be completed before both Activities B and C can commence. Activities B and C must both be completed before Activity E can commence. Figure 5.2(a) shows both Activities B and C passing from Node 2 to Node 4. This satisfies the precedence rule (Rule 3 above), but does not satisfy Rule 4. Figure 5.2 (b) shows an alternative representation in which a dummy activity, D1, has been introduced. This network satisfies all of the rules given above.

Sometimes dummy activities are needed to represent the precedence logic in a network rather than specifically to satisfy Rule 4. An example is where there are four activities that form part of the project. These are designated A, B, C and E. Activity A must be complete before Activity C can commence, but Activity E requires both Activity A and B to be complete before it can commence. A suitable organisational network for this project is shown in Figure 5.3. Note that a dummy activity, D1, has been introduced to satisfy the precedence logic.

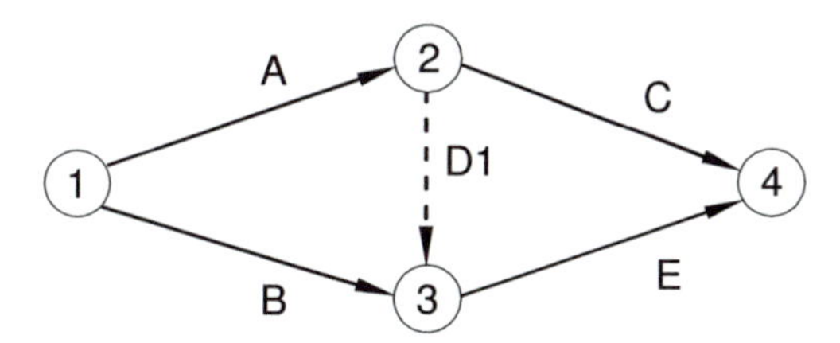

Figure 5.3 Use of a dummy activity to represent precedence logic

Application of the Critical Path Method

The CPM process for project planning involves the following steps:

1. draw up a list of all activities that form part of the project;
2. estimate the time to complete each activity (called the **duration** of the activity);
3. identify the precedence relationships i.e. for each activity, which other activities must be completed before it can commence;.
4. draw the network;
5. analyse the network to identify the earliest start time (EST) and latest finish time (LFT) for each activity;
6. hence identify the critical path(s) for the network;
7. use the EST and LFT to estimate the latest start time (LST) and earliest finish time (EFT) for each activity;
8. use the EST, LST, EFT, LFT and duration for each activity to estimate its total float, free float and interfering float.

The critical path method will be demonstrated by applying it to a case study. This is the construction of a single span highway bridge over a creek. Simplified design drawings of the bridge are shown in Figure 5.4.

Step 1: Draw up a list of all activities that form part of the project.

A list of activities has been drawn up by an engineer with experience in construction. This is given in Table 5.2. The plan involves fabricating the steel girders and handrails off-site and then transporting them for installation on site. Concrete will be produced off-site and delivered to the site ready to pour. All other activities will take place on site.

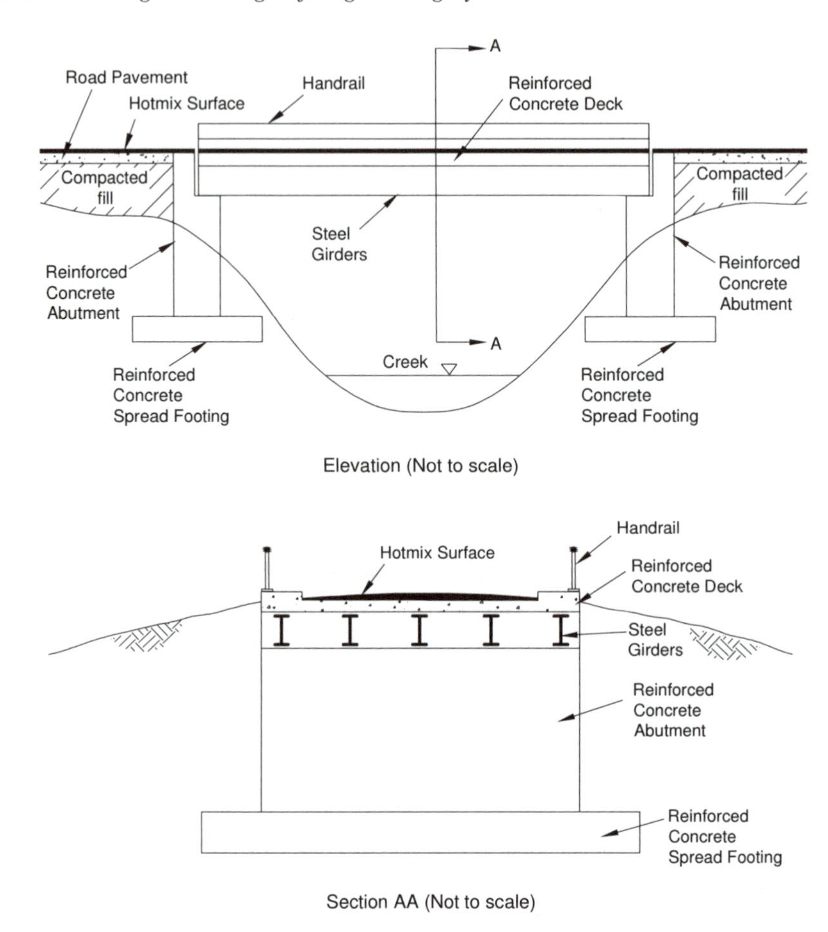

Figure 5.4 Simplified design drawings for a highway bridge

Step 2: Estimate the duration of each activity.

The durations estimated by an experienced engineer are given in Table 5.2. These are based on the assumption that adequate equipment and human resources are available on-site to complete the activities without delays. In this simple example many simplifying assumptions have been made. For example, in reality, the construction of the foundations, abutments and embankments on each side of the creek would be considered as separate activities, as they may be scheduled to occur simultaneously or in sequence, depending on the available resources. In this case, these items are assumed to be undertaken on both sides of the creek simultaneously. A further practical consideration is the planning involved in transporting major items of equipment to the site and scheduling their usage. Such considerations have been ignored in this simple example in order to demonstrate how organisational networks are developed and analysed.

Table 5.2: Activities involved in the construction of a highway bridge

Activity	Description	Duration (days)	Preceding Activities
A	Establish site office and transport equipment to site	3	None
B	Remove topsoil	4	A
C	Excavate foundations	3	B
D	Order reinforcement and have it delivered to the site	5	None
E	Place formwork and reinforcement for footings	2	C,D
F	Pour concrete footings	1	E
G	Cure concrete footings	7	F
H	Place formwork and reinforcement for abutments	2	G
I	Pour concrete abutments	2	H
J	Cure concrete abutments	14	I
K	Fabricate steel girders off-site and deliver to the site	21	None
L	Fabricate steel handrails off-site and deliver to the site	10	None
M	Place steel girders	3	J, K
N	Place formwork and reinforcement for concrete deck	3	M
O	Pour concrete deck	1	N
P	Cure concrete deck	14	O
Q	Install handrails	2	L, P
R	Place and compact fill on approaches	21	A
S	Construct pavement on approaches	10	R
T	Place hotmix on approaches and on the bridge deck	3	P, S
U	Replace topsoil and revegetate the embankments	5	R
V	Paint the handrails	3	Q
W	Clean up the site	7	T, U,V

Step 3: Identify the precedence relationships.

The relationships determined for this project are given in Table 5.2.

Step 4: Draw the network.

The drawing of the network is a relatively straightforward task once steps 1–3 have been completed. In reality, steps 1–4 usually involve several iterations as the process of drawing the network can facilitate thoughts about the activities involved and their inter-relationships and dependencies. The network is shown in Figure 5.5.

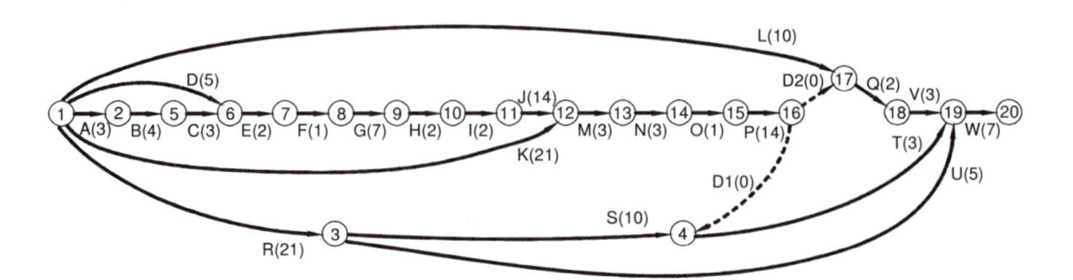

Figure 5.5 Organisational network for construction of the highway bridge (the durations of activities are given in brackets)

Note that the network satisfies the four basic rules given in Section 5.3. Also note that two dummy activities have been added. The dummy placed between Nodes 16 and 4 ensures that Activity T (place hotmix on approaches and on the bridge deck) can't commence until Activity P (cure concrete deck) is complete. The other dummy between Nodes 16 and 17 ensures that Activity Q (install handrails) can't commence until Activities P (cure concrete deck) and L (fabricate steel handrails off-site and deliver to the site) are complete. Two dummies are required in this instance because Activity Q does not depend on the completion of Activity S (construct pavement on approaches) being complete and Activity T does not depend on Activity L being complete. In reality, some work can be undertaken on the abutments before the concrete of the foundations has reached its design strength so the duration for the curing of the concrete in the foundations has been set at 7 days.

Step 5: Analyse the network to identify the earliest start time (EST) and latest finish time (LFT) for each activity

The earliest start time (EST) for an activity is the earliest time that it can commence assuming that all preceding activities and the project as a whole start on time. The latest finish time (LFT) for an activity is the latest time that it can finish without increasing the minimum completion time for the overall project. The EST and LFT are shown on Figure 5.6 with the EST shown in the left hand box and the LFT in the right hand box.

The ESTs are calculated in the following way: Begin at the first node (i.e. the node that has no arrows leading into it). In this case this is Node 1. Set the EST at this node to be 0. For each subsequent node the EST is determined by examining all activities leading into the node. For each activity determine the sum of the EST of its start node plus its duration. Take the largest of these values to give the EST at the new node.

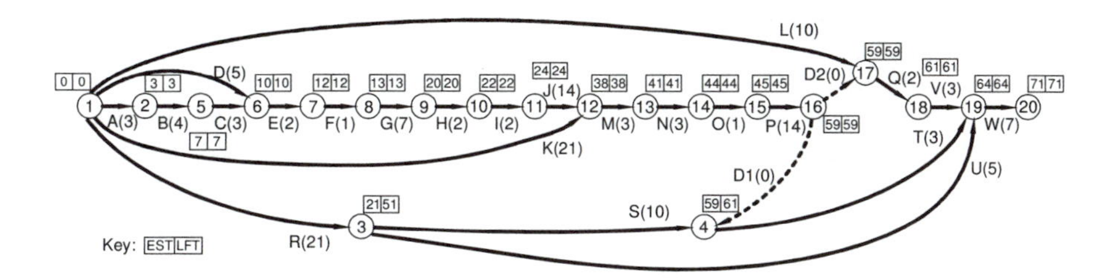

Figure 5.6 EST and LFT for activities involved in the highway bridge construction

For example, Node 2 has only 1 activity (A) leading into it. Clearly the EST at Node 2 is the EST at node 1 plus the duration of Activity A (3 days), thus giving an EST of 3 days at node 2. Similarly, the EST at Node 5 is the EST at Node 2 (3 days) plus the duration of Activity B (4 days) giving 7 days. Node 6 has 2 activities leading into it (C and D). The EST at Node 6 is the larger of the EST plus duration for these two Activities. For Activity C, its EST is 7 days and its duration is 3 days giving a total of 10 days. For Activity D, its EST is 0 days and its duration is 5 days, giving a total of 5 days. The larger of these two is 10 days which is, therefore, the EST at Node 6. This clearly must be the case, because no activities starting at Node 6 may commence until all activities feeding into this node have been completed. That is why it is the larger of the two times.

Mathematically, EST at Node j (EST_j) can be determined using the following equation:

$$EST_j = \max_i\{EST_i + D_{ij}\} \text{ all } i \in I_j \tag{5.1}$$

where, D_{ij} = the duration of Activity *ij*; and I_j is the set of all starting nodes that have activities that finish at node *j*.

In Figure 5.6 the ESTs have been determined on nodes. This can be carried out, because all activities commencing at a node must have the same activities preceding it and, therefore, will have the same EST.

The determination of ESTs requires the nodes to be considered in a certain order. For example, the EST at node 4 requires the EST at nodes 3 and 16 be determined first. This can be accomplished by starting at the node that has no activities leading into it and working one at a time to nodes that only have activities with known ESTs coming into them. This can be carried out by observation when working simple examples (such as this one) by hand and is programmed into software used for analysing more complex organisational networks.

Latest Finish Times (LFT) are determined in a similar fashion by working backwards in time in the following manner: Start at the final node (the one with the largest EST. The LFT at this node is the same as the EST. Then work backwards one node at a time and considering all Activities that commence at the node. The LFT is the **smallest** of the LFT of any Activities that start at the node **minus** their duration. For example, the LFT at Node 20 equals its EST of 71 days. The LFT at

Node 19 is the LFT of Activity W (71 days) minus its duration (7 days) and is therefore 64 days. The LFT at Node 4 is the LFT of Activity T (64 days) minus its duration (3 days). It is therefore 61 days.

There are two activities leading from Node 3. Therefore the LFT at this node is the smaller of the LFT minus the duration for each of these activities. Activity S has a LFT of 61 days and a duration of 10 days and therefore a difference of 51 days. Activity U has a LFT of 64 days and a duration of 5 day and therefore a difference of 59 days. The smaller of the two (51 days) then becomes the LFT at Node 3. Mathematically, the LFT at Node i can be determined using the following equation:

$$LFT_i = \min_j \{LFT_j - D_{ij}\} \qquad \text{all } j \in J_i \qquad (5.2)$$

where, J_i = the set of all ending nodes that have activities that commence at Node i.

As for the EST, the determination of the LFT involves careful selection of the sequence of nodes to be considered. A partial check on the calculations of EST and LFT is provided in that the LFT should be 0 at the first node. Although, if this value is obtained it doesn't necessarily mean that all values of EST and LFT are correct.

Step 6: Hence identify the critical path(s) for the network

The critical path is the set of all activities that cannot be delayed without delaying the entire project. There is always at least one critical path from the first node to the last node. There may be more than one critical path in some cases. The minimum time to complete the project is given by the length of the critical path. This is also given by the EST and LFT at the last node.

In the example illustrated in Figure 5.6 the critical path consists of the following sequence of activities: A, B, C, E, F, G, H, I, J, M, N, O, P, (D2,) Q, V, and W. The minimum time to complete the project is 71 days. It can also be seen that the EST and LFT is the same for all nodes on the critical path.

Step 7: Use the EST and LFT to estimate the latest start time (LST) and earliest finish time (EFT) for each activity.

The latest start time (LST) for an activity is the latest time that it can start without increasing the minimum completion time for the overall project. The earliest finish time (EFT) for an activity is the earliest time it can finish if it and all preceding activities start at their earliest start times.

It should be clear that the LST for each activity is simply its LFT minus its duration. Likewise the EFT for each activity is simply its EST plus its duration. LSTs unlike ESTs are not necessarily the same for all activities starting at the same node in the network. Likewise EFTs are not necessarily the same for all activities that finish at the same node in the network. The EST, LST, EFT and LFT for all activities in the example are given in Table 5.3.

It should be clear from the definitions that the EST and LST are equal for all activities that are on the critical path. Similarly, the EFT and LFT are also equal for these activities. This is verified by the values given in Table 5.3.

Step 8: Use the EST, LST, EFT, LFT and duration for each activity to estimate its total float, free float and interfering float.

The **total float** (TF) for an activity is the amount of time that the activity can be delayed from its EST without affecting the time to complete the overall project. The total float must be zero for all activities on the critical path. For other activities it may be determined by computing the difference between the LST and EST. Alternatively the TF may be determined by computing the difference between the LFT and the EFT for each activity. The TFs for all activities in the example problem are shown in Table 5.3.

Table 5.3 Important times and floats for all activities

Activity	Duration days	Critical (Y/N)	EST days	EFT days	LST days	LFT days	TF days	FF days	IF days
A	3	Y	0	3	0	3	0	0	0
B	4	Y	3	7	3	7	0	0	0
C	3	Y	7	10	7	10	0	0	0
D	5	N	0	5	5	10	5	5	0
E	2	Y	10	12	10	12	0	0	0
F	1	Y	12	13	12	13	0	0	0
G	7	Y	13	20	13	20	0	0	0
H	2	Y	20	22	20	22	0	0	0
I	2	Y	22	24	22	24	0	0	0
J	14	Y	24	38	24	38	0	0	0
K	21	N	0	21	17	38	17	17	0
L	10	N	0	10	49	59	49	49	0
M	3	Y	38	41	38	41	0	0	0
N	3	Y	41	44	41	44	0	0	0
O	1	Y	44	45	44	45	0	0	0
P	14	Y	45	59	45	59	0	0	0
Q	2	Y	59	61	59	61	0	0	0
R	21	N	0	21	30	51	30	0	30
S	10	N	21	31	51	61	30	28	2
T	3	N	59	62	61	64	2	2	0
U	5	N	21	26	59	64	38	38	0
V	3	Y	61	64	61	64	0	0	0
W	7	Y	64	71	64	71	0	0	0
D1	0	N	59	59	61	61	2	0	2
D2	0	Y	59	59	59	59	0	0	0

The **free float** (FF) for an activity is the amount of time that the activity can be delayed from its EST without affecting the starting times of subsequent activities. Once again the FF is zero for all activities on the critical path. For non-critical activities, the FF is determined by subtracting its EST and its duration from the EST of its ending node. This can be calculated for all activities using the information contained in Figure 5.6. The FFs for all activities are given in

Table 5.3. For example the FF for Activity K is based on the EST at Node 12 (38 days) and the EST at Node 1 (0 days). The FF is the difference between these (38 days) less the duration of the activity (21 days) giving 17 days.

The **interfering float** (IF) for an activity is simply the difference between its TF and FF. These values are also given in Table 5.3. The use of interfering float could delay subsequent activities although it will not affect the time to complete the overall project.

The TF for Activity K is all FF, as the use of this time will not affect the timing of subsequent activities. On the other hand, the TF of Activity R (30 days) is all IF as it will reduce the float available to Activities S, T and U if it is utilised by delaying Activity R.

Summary

The construction and analysis of organisational network can provide very valuable information that is needed to manage major engineering projects. This information includes the following: The minimum total time required to complete the overall project, the activities that cannot be delayed without affecting the total time to complete the overall project, a summary of activities that need to be completed before any particular activity can be started, the earliest start, earliest finish, latest start and latest finish times for all activities and the float time for all non-critical activities (i.e. by how much time they can be delayed without affecting the time to complete the overall project or without affecting the timing of subsequent activities.

Because of its great value, it is common to use an organisational network for all major engineering projects. This often involves the use of specialist software products such as Microsoft Project that can facilitate the development of organisational networks.

5.4 GANTT CHARTS

As noted in Section 5.2, a Gantt Chart is an organisational network that represents the timing of activities that make up a project. In a Gantt Chart all activities are listed down the page with each activity having a horizontal bar representing the planned timing of its completion.

Figure 5.7 is a Gantt Chart for the construction of the highway bridge considered in Section 5.3. The shaded bars in Figure 5.7 represent the activities occurring over time (based on their EST) while open bars represent total floats for the corresponding activities. Each activity that has float can be rescheduled within the times represented by the open bar without delaying the overall project (assuming that all other activities run to schedule).

One advantage of a Gantt Chart compared to a critical path network is that the former shows which activities should be running at a particular time (by noting which shaded bars are intersected by a vertical line through the corresponding time). On the other hand it is not usually possible to draw the Gantt Chart without first analysing the relationships between activities using a critical path network in order to determine the EST, LFT and floats for all activities. Furthermore, it is not easy to depict the precedence relationship between activities in a Gantt Chart for complex projects. Although, in theory, vertical lines can be drawn from the end of preceding activities to subsequent activities, this can become very messy and hard to follow for real projects. Nonetheless, Gantt Charts are commonly used to track the progress of projects (often in combination with critical path networks).

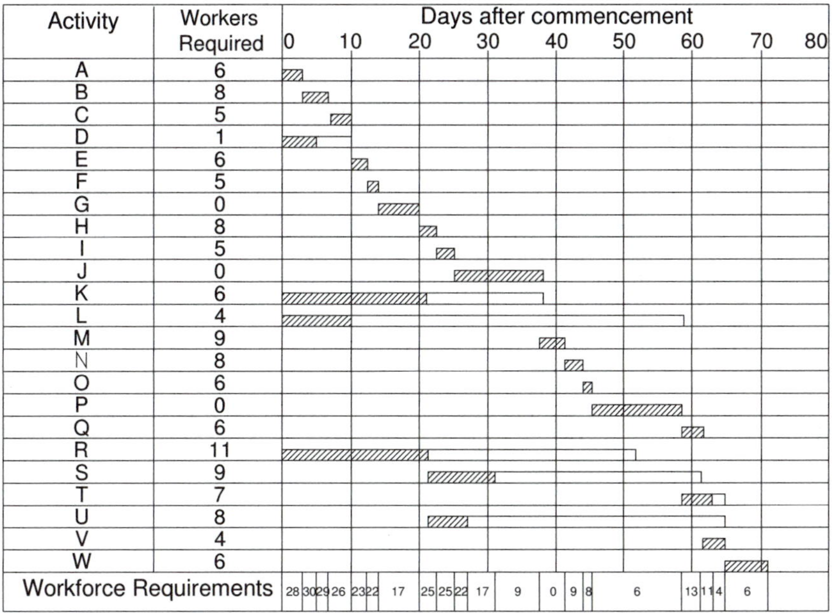

Activity	Workers Required
A	6
B	8
C	5
D	1
E	6
F	5
G	0
H	8
I	5
J	0
K	6
L	4
M	9
N	8
O	6
P	0
Q	6
R	11
S	9
T	7
U	8
V	4
W	6

Figure 5.7: Gantt Chart for the construction of a single span highway bridge.

5.5 RESOURCE SCHEDULING

In drawing the organisational networks to date it has been assumed that the only factor that constrains the start and finish of activities is the precedence relationships i.e. certain activities must be completed before a particular activity can commence. In practice, the availability of critical resources can constrain the timing of activities. These critical resources may be human resources such as the total workforce, skilled labour in particular areas (e.g. steel riggers or electricians) or critical items of equipment (e.g. cranes, bulldozers, graders). Project managers must schedule activities taking these critical resources into account, and, in appropriate cases acquire additional resources by purchase, lease, rent or redeployment from other projects.

A Gantt Chart can be used to indicate the allocation of critical resources on a project (or set of projects). A Gantt Chart together with a critical path network can be used to reschedule activities so that a project can be completed within the resource constraints.

Example

The example of the construction of the single highway bridge (considered earlier) will be used to demonstrate the use of a Gantt Chart to keep track of resources. The critical resource considered is the total labour force. For simplicity it is assumed that all workers on the project are interchangeable and have the same skill set. In reality, there are many different tradesmen and professionals employed on engineering projects. The principles demonstrated in this section can be used for projects with multiple resources.

Table 5.4 shows the estimated workforce requirements for the activities that make up the highway bridge project. Note that Activities G, J and P involve allowing the concrete to cure and do not require significant worker involvement.

The information given in Table 5.4 has been summarised in the bottom row of the Gantt Chart (Figure 5.7) to show the workforce requirements on each day of the project assuming that all activities start at their EST. From the bottom row in Figure 5.7 it can be seen that the workforce for the project (based on EST) starts at 28 workers and builds up to a peak of 30 workers but drops to 9 workers by day 26 and, in fact is 0 workers for some periods of the project. This is unlikely to be optimal. The company carrying out the construction would want to move its workers between projects so that all are fully employed and that no project runs overtime. A common approach is to attempt to "level" or smooth the resources required for any particular project. This can be carried out by a trial-and-error process by moving non-critical activities within their allowable times so as to smooth the workforce requirements as much as possible.

An alternative approach is to reschedule the activities so the workforce used on the project stays within a defined cap. For example, suppose that the maximum workforce available for this project is 20. It is desired to reschedule activities so that the workforce requirements stay within this limit and, if possible, the time to complete the overall project does not increase beyond the minimum 71 days. A heuristic approach to reschedule activities is presented by Meredith et al. (1985). The steps presented by Meredith et al. (1985) are as follows:

1. start with the first day and schedule all activities possible, then do the same for the second day and so on;
2. if several activities compete for the resource, then schedule first the one with the smallest float;
3. then, if possible, reschedule activities not on the critical path in order to free resources for the critical path activities.

This heuristic will be applied to the highway bridge project using the data contained within Table 5.4 and Figures 5.6 and 5.7. For simplicity it will be assumed that once an activity commences it cannot be interrupted in order to redeploy workers to other activities. The heuristic can be extended to the case where interruption of activities can occur, although there are usually inefficiencies in redeploying workers between activities due to the startup time of activities and the need to re-familiarise the workers with the task.

Table 5.4 Workforce requirements for the activities involved in the construction of the highway bridge

Activity	Description	Workers Required
A	Establish site office and transport equipment to site	6
B	Remove topsoil	8
C	Excavate foundations	5
D	Order reinforcement and have it delivered to the site	1
E	Place formwork and reinforcement for footings	6
F	Pour concrete footings	5
G	Cure concrete footings	0
H	Place formwork and reinforcement for abutments	8
I	Pour concrete abutments	5
J	Cure concrete abutments	0
K	Fabricate steel girders off-site and deliver to the site	6
L	Fabricate steel handrails off-site and deliver to the site	4
M	Place steel girders	9
N	Place formwork and reinforcement for concrete deck	8
O	Pour concrete deck	6
P	Cure concrete deck	0
Q	Install handrails	6
R	Place and compact fill on approaches	11
S	Construct pavement on approaches	9
T	Place hotmix on approaches and on the bridge deck	7
U	Replace topsoil and revegetate the embankments	8
V	Paint the handrails	4
W	Clean up the site	6

The process commences at the start of the first day. Five activities could start (Activities A, D, K, L and R). Activity A is on the critical path and so will be scheduled. The total floats for the remaining activities are given in Table 5.5.

Table 5.5 Total floats for non-critical activities that could start on day 1

Activity	Workforce Requirement	Total Float (days)
D	1	5
K	6	17
L	4	49
R	11	30

As Activity A requires 6 workers, there are 14 available for other activities. The order of priority is Activity D, K, R and L. Activities D and K require a total of 7 workers thus leaving 7. This is not sufficient for Activity R so it will not be started at this time, but L can be, so the scheduled Activities on day 1 are A, D, K and L with a total workforce requirement of 17 workers as shown in Figure 5.8. This process has been set up in the first rows of Table 5.6 and will be repeated as we work though the time schedule of the project.

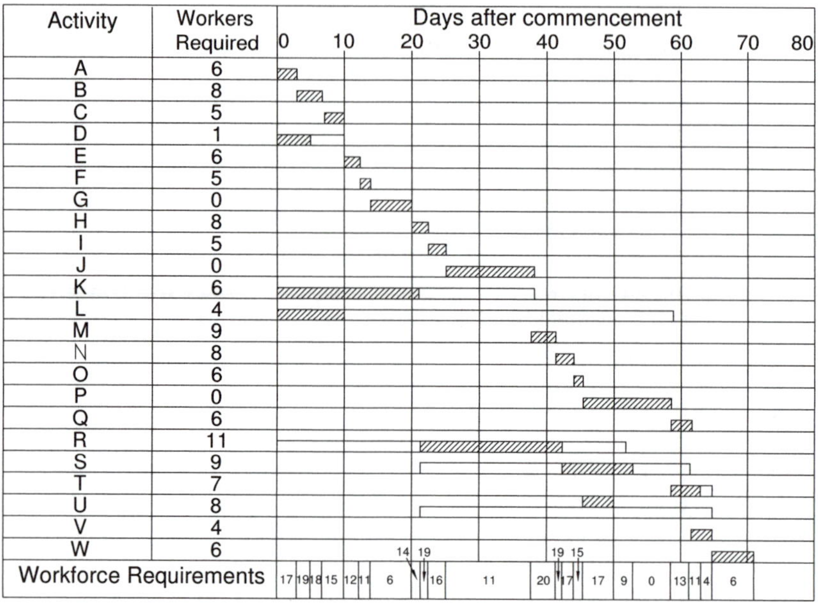

Activity	Workers Required	Days after commencement (0 10 20 30 40 50 60 70 80)
A	6	
B	8	
C	5	
D	1	
E	6	
F	5	
G	0	
H	8	
I	5	
J	0	
K	6	
L	4	
M	9	
N	8	
O	6	
P	0	
Q	6	
R	11	
S	9	
T	7	
U	8	
V	4	
W	6	
Workforce Requirements		17 19 18 15 12 11 6 14 19 16 11 20 19 17 15 17 9 0 13 11 4 6

Figure 5.8: Revised Gantt Chart for construction of the highway bridge

At the start of day 4, Activity A is completed and Activity B (a critical one) is due to start. Activities D, K and L are running and require a total of 11 workers. Activity B (requiring 8 workers) Activity R (11 workers) could be started. As Activity B is on the critical path it is scheduled to start immediately. The total workforce at this time is 19. This process has been continued for the total project. It is summarised in Table 5.6. The revised Gantt Chart is shown in Figure 5.8.

It should be noted that the project can still be completed in 71 days with a workforce of 20 workers compared with a peak of 30 workers using the EST (shown in Figure 5.7). Also note that the organisation network (Figure 5.6) needs to be considered when carrying out the scheduling, for example, Activity S cannot start until Activity R is complete.

Rather than blindly following the steps outlined above, some judgement has been applied in developing the above schedule. For example, Activity R could be scheduled to start at the end of day 13. However, it if did start then Activity H (which is on the critical path) could not start at the end of day 20 as only 3 workers would be available (Activity K and R would be active). This would delay the completion of the project as a whole. Therefore, Activity R has been delayed until the end of day 21.

Table 5.6 Scheduling of Activities to match a resource constraint

Time (end of day)	Activities in Progress	Workforce Committed (Available)	Possible New Activities	Workers Required	Total Float (days)	Activities Scheduled
0	None	0 (20)	A	6	0	A
			D	1	5	D
			K	6	17	K
			L	4	49	L
			R	11	30	
3	D, K, L	11 (9)	B	8	0	B
			R	11	27	
5	B, K, L	18 (2)	R	11	25	None
7	K, L	10 (10)	C	5	0	C
			R	11	23	
10	K	6 (14)	E	6	0	E
			R	11	20	
12	K	6 (14)	F	5	0	F
			R	11	18	
13	K	6 (14)	G	0	0	G
			R	11	17	
20	G	6 (14)	H	8	0	H
	K		R	11	10	
21	H	8 (12)	R	11	9	R
22	R	11 (9)	I	5	0	I
24	R	11 (9)	J	0	0	J
38	R	11 (9)	M	9	0	M
41	R	11 (9)	N	8	0	N
42	N	8 (12)	S	9	9	S
			U	8	17	
44	S	9 (11)	O	6	0	O
			U	8	15	
45	S	9 (11)	P	0	0	P
			U	8	14	U
50	P, S	9 (11)	None			None
52	P	0 (20)	None			None
59	None	0 (20)	Q	6	0	Q
			T	7	2	T
61	T	7 (13)	V	4	0	V
62	V	4 (16)	None			None
64	None	0 (20)	W	7	0	W

Resource Smoothing

In the above example, the peak workforce was limited to 20 workers. The question arises as to what is the minimum peak workforce requirement that does not result in the project being delayed. This can be determined using the above procedure and progressively reducing the limit on the workforce until the time to complete the overall project is increased. Careful examination of Figure 5.8 indicates that a workforce of 20 workers is required from the start of day 39 until the end of day 41. This cannot be reduced without delaying the overall project, so 20 is the minimum workforce to complete the project in minimum time. Of course, if the available workforce is less than this (say 17 workers) the above procedure can be used to schedule activities so that the project finishes in the minimum possible time (which in this case will be more than 71 days). Then a decision will need to be made as to whether it is better to accept this delay in the completion of the project or to hire additional workers.

5.6 SUMMARY

Project planning may be defined as "the process used to implement a plan and hence to achieve a designated objective, through the efficient use of available resources. It involves the regular monitoring of progress and the scheduling of activities as appropriate to achieve the objective in the required time frame." It has existed as a science since the late nineteenth century.

A number of techniques to assist in scheduling the activities of complex engineering projects. One of the most commonly used techniques today is the critical path method (CPM). CPM can be used to estimate the total time to complete a complex engineering project as well as the earliest start time for each activity that forms part of the project. CPM can also be used to identify which are the critical activities (i.e. those whose delay will result in a delay to the overall project). The total float for each activity is defined as the amount of time that the activity can be delayed without affecting the duration of the overall project. The total floats for all activities can be determined using the CPM technique. Free float is another attribute of each activity that can be determined using the CPM. It is defined as the amount of time that the activity can be delayed without affecting the timing of subsequent activities.

An older technique for scheduling activities that is still used today is called a Gantt Chart. It is difficult to depict all of the precedence relationships between activities on a Gantt Chart. A Gantt Chart together with a CPM diagram for a project can be used to reduce the peak workforce requirement of a project.

PROBLEMS

5.1 A swimming pool is to be constructed in the backyard of a house as shown in Figure 5.9. The pool will be 3 m by 1.5 m in plan and range in depth from 2.4 m at one end to 1.0 m at the other. It will be constructed of reinforced concrete and covered in ceramic tiles on the top and inside. Access to the site will be via a driveway, but the carport will need to be dismantled to allow access for earthmoving equipment. Assume that you are the contractor who will construct the pool from a complete set of plans and specifications. Make a list of the activities involved and draw up an organisational network that shows the interdependencies between them.

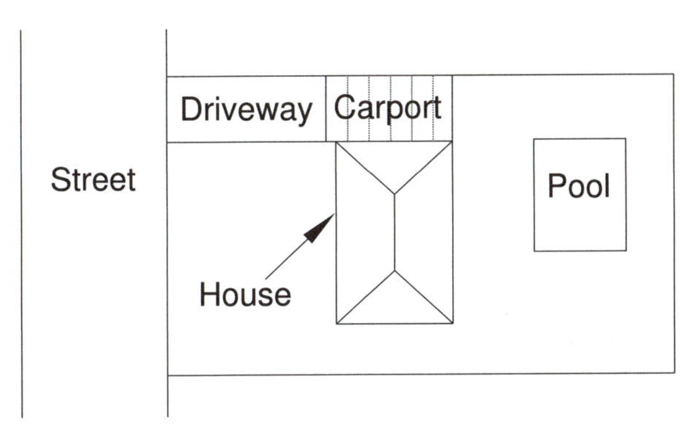

Figure 5.9: Site of a proposed swimming pool

5.2 Table 5.7 contains a list of activities associated with the construction of a steel frame industrial building.

Table 5.7: Activities involved in constructing a steel frame building

Activity Number	Description	Duration (days)	Preceding Activities	Number of People
1	Establish site office	3	None	3
2	Clear topsoil from site	3	1	6
3	Order steelwork	2	None	1
4	Fabricate steelwork offsite	14	3	8
5	Deliver steelwork to the site	2	1,4	4
6	Order steel reinforcement	2	None	1
7	Fabricate steel reinforcement offsite	7	6	6
8	Deliver steel reinforcement to site	1	1,7	2
9	Deliver cladding to site	1	1	2
10	Excavate for foundations and floor slab	4	2	7
11	Place reinforcement for foundations and floor slab	4	8,10	6
12	Pour concrete in foundations and floor slab	2	11	6
13	Cure concrete	7	12	0
14	Erect steelwork	8	5,13	10
15	Place cladding on steelwork	4	9,14	5
16	Complete internal fit out of the building	6	15	8
17	Cleanup site	2	16	4
18	Remove site office	1	17	3

(a) Draw an organisational network for this project.
(b) Determine the earliest start time (EST) and latest finish time (LFT) for each activity.
(c) Determine the critical path and the minimum time to complete the project.
(d) If Activity 4 is delayed by 7 days what will be the delay in the total project?

5.3 The organisational network for an engineering project using arrow notation is shown in Figure 5.10. The duration of each activity in days is indicated in the figure. Compute the following for each activity: earliest start time, latest finish time, latest start time, earliest finish time, total float, free float and interfering float. Also determine the critical path and the minimum time to complete the project.

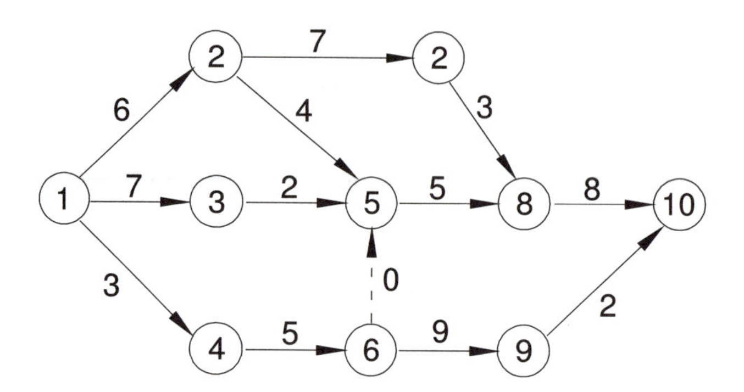

Figure 5.10: Organisational network for an engineering project

5.4 Draw a Gantt Chart for the project described in Question 5.2 assuming that each activity commences at its earliest start time. Determine the total float, free float and interfering float for each activity.

5.5 Table 5.7 contains an estimate of the number of people required to carry out each activity.
a) Using the Gantt Chart developed in Question 5.4, estimate the peak workforce requirement for the project assuming that all activities commence at their earliest start times.
b) If the available workforce is limited to 16 people who can carry out any of the activities associated with the project, reschedule activities so that the project can be completed in the minimum possible time.

REFERENCES

Archibald, R.D. 2003, *Managing High-Technology Programs and Projects.* Third Edition (New York: John Wiley & Sons).

Burke, R. 2001, *Project Management. Planning and Control Techniques* (Chichester, UK: Wiley & Sons Ltd).

Faniran, O. 2005, *Engineering Project Management* (Pearson Education Australia).

Griffis, F.H. and Farr, J.V. 2000, *Construction Planning for Engineers* (Boston USA: McGraw-Hill).

Hyman, B. 1998, *Fundamentals of Engineering Design* (Upper Saddle River, New Jersey: Prentice-Hall Inc.)

Meredith, D.D., Wong, K.M., Woodhead, R.W. and Wortman, R.H. 1985, *Design and Planning of Engineering Systems, Second Edition* (Englewood Cliffs, New Jersey: Prentice-Hall Inc).

Shtub, A., Bard, J.F. and Globerson, S. 2005, *Project Management. Processes, Methodologies and Economics* (Upper Slade River, New Jersey: Pearson Education Inc.).

CHAPTER SIX

Management Processes and Skills

Management is about making the best use of available resources, including money, people, time and materials, in order to achieve designated goals. To complete engineering projects in a satisfactory manner requires teamwork and full use of the skills of individuals. In this chapter skills to manage oneself and also a team are discussed. The importance of personality characteristics of individuals and the roles individuals take in teams are also outlined. Skills that are required in managing day to day activities and provide the basis for undertaking many basic engineering activities such as planning, design and construction are also described. Managing conflict, time management and effective communication are skills for successfully undertaking an engineering project.

6.1 INTRODUCTION

In Chapter 1 it was shown that throughout history many major engineering accomplishments have changed the way societies work. The design and construction of the pyramids, development of railway systems that crossed continents, the planning and construction of major dams such as Hoover Dam and the Three Gorges Dam, the development of the space shuttle and developing of the personal computer are examples of engineering projects that are very different in their nature, but that have changed their societies. Such projects were undertaken using various leadership and management styles, varying from the one extreme of an autocratic and military nature to a *laissez faire* creative atmosphere in which personal computers were developed. While the management structures varied in these examples, the success of the project always rested on individuals with special skills who were able to work together in teams to achieve an effective solution using the resources at hand.

Engineering management as a discipline has its origins in the industrial revolution of the eighteenth and nineteenth centuries. During this period, cottage industries were replaced by large industrial organisations in the mining, minerals processing, manufacturing and construction sectors. In the late nineteenth and early twentieth centuries, a number of engineers and managers attempted to develop a more rigorous approach to managing people, machinery and resources to achieve a defined end.

An engineering manager has to be capable of dealing with the processes related to the planning, design, construction and operation of engineering systems. At the same time the engineer's managerial responsibilities include the allocation of human and financial resources to enable tasks to be performed. Because of the nature of the engineering profession, engineers are members of teams and may take on management roles soon after graduation. Engineers therefore have to develop good communication and interpersonal skills to perform in teams and manage people. This chapter addresses some of the concepts of management and outlines skills required by engineers to manage the resources of people and time.

6.2 MANAGEMENT HISTORY AND PROCESS

The earliest management theories developed from the desire to manage workers and organizations more efficiently. The discipline of Scientific Management commenced in the 1880s with the work of F.W. Taylor and the time and motion studies of Taylor and Gilbreth in the early 1900s. This was followed by Gantt's analysis of project management and the use of charts as outlined in Chapter 5. Henri Fayol (1841–1925), a French mining engineer, described management in terms of planning, organisation, command, coordination and control. Fayol's Administration Industrielle et Generale (published in 1916) was translated into English by Constance Storrs in 1949. From the translation Fayol says "To manage is to forecast and plan, to organise, to command, to coordinate and to control. To foresee and provide means examining the future and drawing up the plan of action. To organise means building up the dual structure, material and human, of the undertaking. To command means maintaining activity among the personnel. To coordinate means binding together, unifying and harmonising all activity and effort. To control means seeing that everything occurs in conformity with established rule and expressed command" (Fayol, 1949).

Henry Ford had adapted time and motion studies to the assembly line in 1913 for the production of cars. However, in the 1400s in the Italian city of Venice, which was known for its naval production facilities, assembly-line techniques were used to fit out galley ships for war. A Spanish traveller in 1436 described the Venetians' process (George, 1968): "And as one enters the gate there is a great street on either hand with the sea in the middle, and on one side are windows opening out of the house of the arsenal, and the same on the other side, and out came a galley towed by a boat, and from the windows they handed out to them, from one the cardage, from another the ballistics and mortars, and so from all sides everything which was required, and when the galley had reached the end of the street all the men required were on board, together with the complement of oars, and she was equipped from end to end. In this manner there came out ten galleys, fully armed, between the hours of three and nine." This was a production line 500 years before Henry Ford (George, 1968).

More recently, Shtub et al. (2005) defined management as "the art of getting things done through people." They identified the following seven functions of management: planning, organising, staffing, directing, motivating, leading and controlling. These activities are described in more detail in Table 6.1. The similarities to Fayol's list of management functions are obvious. The modern emphasis on motivating and leading staff, as distinct from commanding them, should be noted. It undoubtedly reflects a more egalitarian perspective than that which existed in the early part of the twentieth century.

Max Weber described bureaucratic management in the 1920s and emphasized that order, system and rationality in management leads to equitable treatment for employees. Throughout the 1920s and 1930s an effort was made to understand human behaviour in the workplace through studies such as that led by Elton Mayo at the Hawthorne plant of Western Electric Co (US). The workers were tested to see how their work environment affected their production. Many other human relations researchers contributed to the field, including Mary Parker Follett, Abraham Maslow, Kurt Lewin, and Renais Likert. In the 1950s, a further group of researchers including Douglas McGregor, Chris Argyris, Frederick Herzberg, Renais Likert, and Ralph Stogdill proposed behaviour theories. There was a mix of psychologists and managerial academics working on organizational development and restructuring in the 1960s and 70s, including people such as Fred Emery and Peter Drucker (Ullman, 1986; McShane and Travaglione, 2003).

Table 6.1 Functions of management (adapted from Shtub et al. 2005)

Function	Description
Planning	Setting goals for the organisation. Identifying a course of action or plan that will lead to the achievement of these goals.
Organising	Assigning people and resources to activities, delegating appropriate authority and establishing a structure for reporting.
Staffing	Ensuring that appropriate human resources are available for the desired activities. Ensuring that adequate training and reward structures are in place.
Directing	Orientating staff and resources towards achieving the goals of the organisation.
Motivating	Encouraging individuals to achieve their best regardless of the tasks undertaken.
Leading	Setting an example for others to follow. Encouraging the development of group pride and loyalty.
Controlling	Monitoring performance relative to the plans. Taking action when the desired outcomes are not being achieved.

The development of understanding an organization as a system began in the 1950s, with an emphasis in this approach on explaining outputs in terms of transformed inputs, taking into account interaction with the surrounding environment. Ludwig von Bertalanffy, a biologist, developed a general systems model which contributed to this thinking. Other early systems contributors included Kenneth Boulding, Richard Johnson, Fremont Kast, and James Rosenzweig. Following World War II a contemporary School of Management evolved which included Deming and Juran, who were the proponents of Total Quality Management which transformed many industries in both Japan and the United States.

The Contingency School of the 1960s emphasized that management processes were dictated by the unique characteristics of each situation. It was resolved that the complexity of organizations was such that no single management strategy supplied the complete answer.

In summary, it has been seen that ideas have come from many different views of management, with different tools proving to be useful for solving managerial problems. Over time it has been resolved that managerial actions must be decided on a situational basis. It is essential that organizations are considered as open systems and be designed to consider individual needs for harmonious and continuing survival.

Human Needs

All engineers require some management expertise to undertake engineering work. The planning, design and construction of projects, the innovative design of engineering systems, practical problem solving, managing operations of large engineering water and energy utilities and decision-making at all levels of an organisation require the management of the needs of a large number of individuals. Many engineers will work as project managers where the role is to manage people and resources committed to a project to ensure that it is completed on time and within budget. The needs of each individual in a project group affect the operation of the group, so much so, that the needs and tasks of both the individual and the group need to be considered together.

Intrinsically, under an optimistic view of life or the McGregor X style, it is assumed that individuals want to do their best (McShane and Travaglione, 2003). Research into what makes people work harder and achieve greater output was, and still is, of paramount importance to management in organizations. Maslow (1970) developed his Hierarchy of Needs, shown in Table 6.2, upon which many management researchers based their work.

Abraham Harold Maslow

Abraham Harold Maslow (1908–1970) was born in Brooklyn, New York, one of seven children born to his Jewish immigrant parents from Russia. His parents pushed him hard for academic success. He was lonely as a boy, and found refuge in books. Maslow's thinking was surprisingly original because he researched positive mental health. Maslow became the leader of the humanistic school of psychology, that emerged in the 1950s and 1960s, which gave rise to the idea that people possess inner resources for healing and growth (Public Broadcasting Service, 1998; Boeree, 2003).

"I was awfully curious to find out why I didn't go insane" - Abraham Maslow.

Maslow and those researchers who followed developed models of the reasons for an individual's work behaviour. Note that these are only models and, in reality, human beings are very complex creatures who have many different aspirations that drive them. Maslow's theory suggests that, as a person satisfies one level of need, then behaviour is motivated to meet the next level of need. This theory has been built upon to explain people and organizational needs of today with a number of alternative variations of the five levels evolving. These models include Alderfer's ERG (Existence Relatedness Growth) theory, Herzberg's motivator-hygiene theory and McClelland's learned needs, all of which are used to explain employee motivation (McShane and Travaglione, 2003).

Table 6.2 Abraham Maslow's Hierarchy of Needs (adapted from Maslow, 1970)

Basic Needs	Elements
Physiological	Food, air, water, sleep, comfort
Safety/Security	Physical, emotional security, fairness and justice, absence of threat, consistency and predictability
Social/Love and belonging	Love (both giving and receiving; separate from sex), affection, friendship
Esteem	Personal achievement, adequacy, confidence, freedom, independence, and the desire for reputation, prestige, recognition, appreciation from others
Self-Fulfilment (Actualization)	Realising a person's potential, basically satisfied people who expect the fullest creativity

6.3 WORKING IN GROUPS AND TEAMS

A group is a term which is vague in concept and can be considered to be any number of individuals who interact together. A team can be considered to be a group of people who work well together to achieve a common goal. Effective teams must have members who are willing and able to complete the task set, as well as work in a team environment. The size of effective teams has always been somewhat subjective because large tasks need large teams but generally teams should be small enough to maintain efficient communication and coordination of the team members. Larger groups will always break into smaller informal groups to allow effective contributions from all members.

The three circles model of Adair (1983), shown in Figure 6.1, illustrates the three areas of needs in any group or team. Adair does not take credit for the model and its origin is unknown. It has been proposed (Johnson and Johnson, 2000) that one of the keys to human development has been the ability to form and work in small effective groups. Groups are central to much of human life and engineering is no exception.

Whether as a member of tutorial or practical groups as part of undergraduate study or as part of a large consulting firm working on multi-million dollar projects, engineers will generally find themselves part of a group and in many cases will take, or be expected to take, a leadership role.

Effective groups are a force to be reckoned with; however, not all groups are effective and this can lead to significant problems. According to Johnson and Johnson (2000), an effective group:

- achieves its goals (task);
- maintains good working relationships among members; and
- adapts to changing conditions in the surrounding organization, society and world.

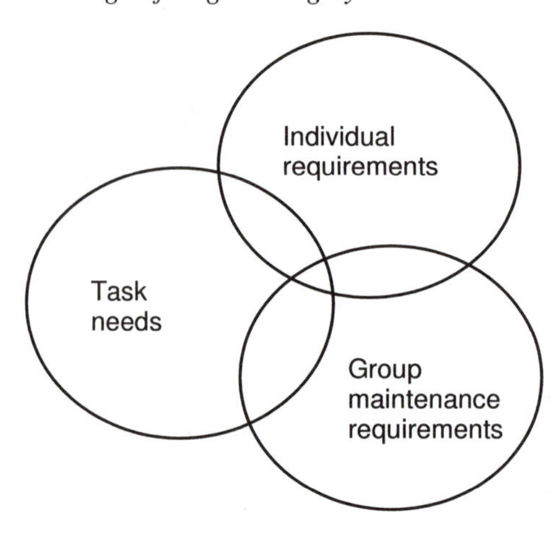

Figure 6.1 Three Circles Model (adapted from Adair, 1983)

What is missing here is the attention to the individuals in the group which is a factor that can make or break a group's activity. There is a trend for more group work rather than less. A Fortune 1000 survey in 1993 found that 91% of companies had implemented some form of team working (up from 70% in 1987), and in Australia in 1991, 47% of manufacturing companies had employees in teams (up from 8% in 1988).

This means that engineers will not only work in groups and teams but in all likelihood take some form of leadership role in them as well, given that engineers are more likely to be in senior positions. For this reason it is important to know how teams work and, more importantly, to know what to do when they start to break down. Groups are often formed just for a particular project and the members may not have worked together before. Members, therefore, can feel a little awkward with one another. Since the group will probably be working within a time limit, it is important for individuals to understand how groups function in order to improve the effectiveness of the team as soon as possible. This will be discussed later in Section 6.6. Figure 6.2 shows some of the actions which are undertaken in both the problem solving process of group activities and the interpersonal process where both group maintenance and self oriented behaviours are outlined.

6.4 LEADERSHIP

The question "What makes a leader?" has been posed many times. In business, sport, industry and academia there are facilitators, coaches, management academics and practitioners trying to explain a leadership concept that is broad enough to include leaders varying from Mohandas (Mahatma) Gandhi, Margaret Thatcher, Elizabeth I, Joseph Stalin, Bill Gates, John F Kennedy, Martin Luther King, Mao Zedong, Nelson Mandela, John Monash and Aung San Suu Kyi. We will see later that different leaders have different personality styles.

Developing leadership skills in the workforce is a strategy of many organizations. Books, materials, training courses and conferences on the subject are plentiful. The challenge is to have the capacity to try individual actions that aim to improve the way things are done through development of an individual's style. Adair (1983) extended the simple three circles model for the functioning of groups and teams to show how leaders of teams need to function by maintaining processes in each of the three areas of task, group and self. The leader has the responsibility to ensure that all areas of need are addressed, the task is achieved, by building and maintaining the team and by ensuring the development of the individual.

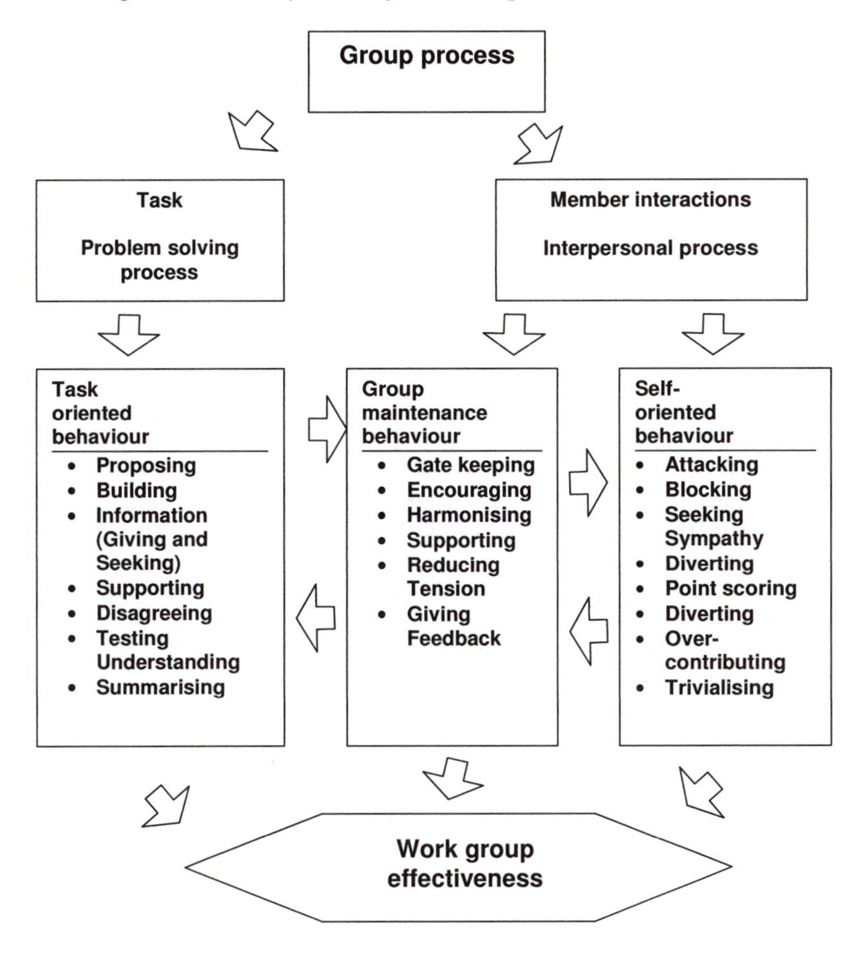

Figure 6.2 Behaviours undertaken for Group Effectiveness (adapted from Adair, 1983).

Teams need a leader to succeed. An effort should be made to designate someone to play that role at or before the first meeting. Sometimes a person is appointed to the leadership role because of their place in the organisation. However, there is no need for that leader to be permanent and omnipotent. The leader needs to ensure that the task and maintenance functions of the three circles model are performed. The roles of the members within the group can be decided by natural aptitudes for different functional roles or different people can be assigned to take on functions akin to their particular styles. Belbin (1981) outlined eight roles that any member of a group prefers to operate in, as shown in Table 6.3. This was later extended to nine roles with a specialist type added.

Table 6.3 The proposed team roles of Belbin (adapted from Belbin, 1996).

Type	Characteristics
Coordinator/Chairperson	respected, mature and good at ensuring that talents are used effectively
Shaper/Driver	dynamic and challenging, usually leads
Plant	very creative, the ideas person
Resource Investigator	extrovert, good at making outside contacts and developing ideas
Monitor Evaluator	shrewd and prudent, analytical
Implementer	practical, loyal and task orientated
Completer/ Finisher	meticulous and with attention to detail, also full of nervous energy
Team Worker/Supporter	caring and very person orientated
Specialist	high technical skill and professional, operates narrowly

The Coordinator /Chairperson takes the responsibility of keeping people on track, and pays attention to group processes. This person with the help of the Team Worker/Supporter, ensures that all members participate and notices when someone is upset. However, it is apparent that when both a Shaper and a Coordinator work in a team, one of them has to adopt a secondary preferred role.

The Resource Investigator can also act as a person who serves as liaison between the team and the rest of the world. In the context of groups at university, this person interacts with the academic supervisor and other groups. We all learn from others but mostly from our own experience. It can be said that someone with a modest amount of natural ability, who works hard at observing the member interactions of group maintenance and self oriented behaviours, as well as the task and problem solving process, will forge ahead of a person of high natural ability who relies on instincts and never addresses his or her faults. To understand the principles of leadership and to work hard at them will ensure success. As Adair (1983) stated; "Good leadership is often so silent, so self-effacing, that you are hardly aware of it, but bad leadership always shouts at you."

Laws of Leadership

A leader needs vision, discipline and wisdom (Newman, 1994). Vision shows that the leader knows what the long term goal is and discipline is used to ensure that energy, time and resources are directed to achieve the goal. Wisdom can be considered to be the ability to apply knowledge and experience to any situation.

The leader also shows courage when different situations demand it. How does one create courage? The answer is as Mark Twain said; "Courage is a resistance to fear, mastery of fear—not absence of fear." To make some difficult decisions requires courage and this decision making ability is an attribute that is needed by a leader. The leader empowers others to make decisions, as the team depends on all members contributing to the process of achieving the designated goal.

Friendships and humility are developed throughout one's career and being a good listener and trusted confidante enables mentoring of the team, avoiding rivalry and assisting in producing the best outcomes. At some time in a career there is a need for having someone we can open up to; no one survives alone. In developing these relationships a leader exercises tact and diplomacy, always showing impartiality to team members. At all times the leader must be prepared to learn from those within the team and those outside the team to produce the best outcome. The leader exudes inspirational power and enthusiasm for tasks, while at the same time ensuring that others want to do their best. This is done by empowering those that he or she works with. A leader's role is but to serve the team so that the team will say, "We did this ourselves".

6.5 BEHAVIOURAL STYLES OF INDIVIDUALS

One important consideration when thinking about groups and teams and how they will work is the melding of individual members as a team which depends on the individual's personality. It has been noted earlier that one view of personality can incorporate the sorts of tasks that people like to take on (leader, innovator, or keeper of the peace among many others). But there are many other ways of classifying it and an over-riding one is how people relate to the challenge of a task. Two personality models will be described here from the many typologies that are available. A search in the literature and the internet can yield many free tests to discover what preferences of behavioural style an individual has. It is important to note here that there is no 'right' personality. The main task is to recognize the differences and work with them, rather than have them work against the team.

Two models will be discussed. Firstly, there is a behavioural model which is nice because of its simplicity called DISC which is an acronym for: Dominance, Influencing, Steadiness, and Compliance. It parallels the writings of the Greek Hippocrates who established some terms in 370 BC for four temperaments: Sanguine, Choleric, Phlegmatic, and Melancholic. Plato in 340 BC also determined 4 categories, labelled: Guardian, Artisan, Scientist and Philosopher (McShane and Travaglioine, 2003). The second model is that based on Jungian theory and adopted by Isabel Myers and Katharine Briggs for their MBTI model but now comes under other guises such as Keirsey's Temperament Sorter (Keirsey, 1998).

DISC behavioural styles

DISC follows a theory developed by Dr W.M. Marston in 1928 and further developed in the 1940s (Cole and Tuzinski, 2003). This theory suggested that a person's preferred behavioural style falls into one of four categories. Although all four styles are used by a single person, one style tends to describe people better than the others. DISC does not measure skills, experience, values, intelligence, beliefs or knowledge. DISC behavioural styles are shown in Table 6.4.

Table 6.4 DISC Behavioural Styles

	Tasks and Results	Ideas and People
Direct Style	D Direct, Dominant, Doer	I Influencer, Inspired, Persuader
Indirect Style	C Conscientious, Cautious, Critical	S Steady, Supporter, Stable

The descriptors help us to clarify differences in people and remove barriers to improve communication. All styles are necessary and valuable. The four styles are determined from whether behaviour is direct or indirect and whether behaviour is oriented towards tasks and results, or people and ideas. By determining whether someone else is direct or indirect, task or people oriented, should assist in developing a better relationship and better communication with this person.

Tests that can be used to assess behavioural style can be found on the internet. Following testing, or more importantly an examination of style, as shown below, an allocation to one of the following groups is made: direct (eagle); influencing (parrot); conscientious (owl); and supportive (dove). The allocation of particular birds in this case makes for easy recognition.

Direct or Indirect Style? Faster or Slower Paced

Some people are direct and work more quickly, take risks, are forceful, talkative and tend to make decisions quickly. The Dominant Director and Influencing Persuasive types fit into the category of direct styles, as shown in Figure 6.3. The indirect styles are quieter, patient, cooperative, more cautious and easy-going. They are good listeners and tend to take their time in making decisions and take fewer risks. Examples of indirect types include the Supporter and Conscientious styles.

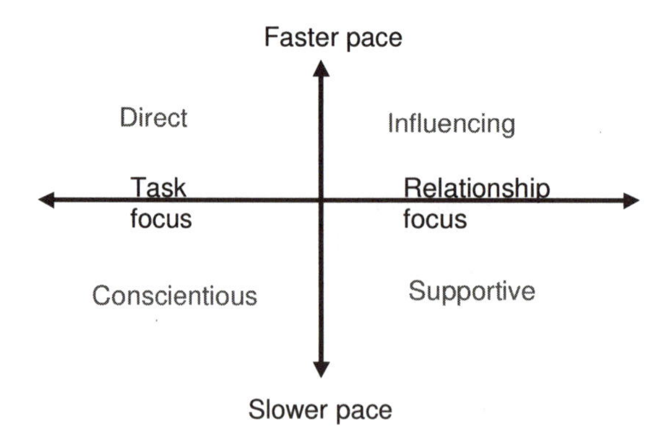

Figure 6.3 DISC Personality types and their simple characteristics

People and ideas or tasks and results?

People who are relationship oriented are generally open, appreciative and supportive, and find it easy to make friends. These people can be enthusiastic and can share their feelings. They tend to go with the flow in a relaxed manner. The Steady Supporter (S) and Influencing (I) styles stress people and ideas. They tend to promote harmony. The other types of people seem to be biased to task and results. This type likes structure, procedures, guidelines and facts. They prefer to get to the point and don't like their time wasted. They usually take a considerable time to show their warm side and tend to keep their feelings to themselves.

Note that no one operates entirely in one quadrant and we all tend to have a mixture of all the styles in different percentages. Each style has weaknesses and it is important to be aware of our own weaknesses so that we can improve in those areas. One style will be the dominant style of an individual, although in different situations an alternative style will be used. Is the dominant style Direct or Indirect? Is there a preference for tasks and results, or people and ideas? Are people who have similar or different characteristics easier to get along with? Why?

Eagle (Direct)

The eagle personality is decisive and strong willed. Eagles are keen to get to the point and quickly. They are adventurous, take risks and are forceful. At work eagles are results orientated, but to be easier to work with, eagles should do more listening and improve their consideration of others. In various other typologies this personality has been labelled as a Bear, Lion, Guardian and Controller.

Parrot (Influencing)

The parrot is enthusiastic and likes to express emotions. Parrots are talkative, optimistic and confident. In groups they are persuasive and gregarious. Parrots need to pay more attention to detail and improve their follow-through on tasks. In various other typologies of four characteristics this personality has been labelled as a Monkey, Otter, Artisan and Promoter.

Dove (Supportive)

The dove is dependable, diplomatic, patient and a team player. Doves like to get results, are stable, good listeners and are sincere and patient. However, they could work at coping with change and improve their decision making ability. In various other typologies of four characteristics this has been labelled as a Dolphin, Golden Retriever, Philosopher and Supporter.

Owl (Conscientious)

The owl personality is orderly, cautious and reserved. Owls have high standards, are careful, analytical, diplomatic and accurate. They could, however, open up more and attempt to move out of their comfort zone if they are to make more impact on their team. In various other typologies of four characteristics this personality has been labelled as a Beaver, Scientist and Analyst.

6.5.3 Myers Briggs type indicator

In 1920 Carl Jung theorised that people were all different, having various degrees of four characteristic functions, Thinking, Feeling, Sensation and Intuition, and two attitudes, Introversion and Extraversion (Jung, 1923). He proposed that people's psychological orientation was formed by a dynamic mix of attitudes and functions. Isabel Myers used Jung's typology to establish a procedure for determining type in individuals and added the dimension of Judging and Perceiving to Jung's original typology. The Myers-Briggs Type Indicator (MBTI) was developed from decades of research by accumulating information on individuals' behaviour and attitudes of people in all spheres of work and life. The MBTI made available the theory of Jung to a much wider audience and was popularised through the 1980s and 1990s and is now embedded in many organisational management programs. A modified version has been developed by Keirsey and Bates (1978) and Keirsey (1998). Boeree (2004) has reviewed the work of Jung and has developed a questionnaire to map the elements of the MBTI as well, and is freely available on the web. The test has four scales. The first is the **Extroversion (E) – Introversion (I).** This scale demonstrates how we interact with the others. Overall about 50% of the population is extroverted and 50% introverted.

The second scale is **Sensing (S) – Intuiting (N)** with approximately 73% of the general population sensing and 27% intuitive. This scale shows our preference for how we deal with understanding the world either through step by step approaches, or through using vision and insight. A sensing person tends to assimilate a series of facts in a linear fashion, while the intuiting person absorbs the same information through conceptual jumps and development of patterns from abstractions. The S types dislike solving new problems without prior experience on how to solve them. On the other hand the N types prefer to solve new problems and dislike carrying out the same thing over and over again. Of course, people tend to share both sets of qualities to some extent.

Thinking (T) – Feeling (F) is the third scale and these functions are distributed in a proportion 40% thinking and 60% feeling, with 60% of men being thinking types, while over 70% of women are feeling types. This scale is related to how we make decisions. This is the only scale that has a gender bias.

The last scale is **Judging–Perceiving** (J–P) and was included by Myers and Briggs to help determine the superior function of an individual. Judging people tend to be more cautious whereas Perceiving people tend to be more spontaneous.

The Judging – Perceiving continuum also determines the superior function. Those people who have a "J" and are extrovert have the superior function on the thinking feeling continuum. Alternatively, extroverted and "P" means that the superior function is on the sensing feeling scale. An introvert deemed to be judging will have the superior function of senser or intuiter, while an introvert with perceiving will be superior on the thinking feeling scale. It has been found that J and P types are approximately evenly distributed throughout the population.

One function from each continuum is combined to identify a type represented by four letters such as ESTJ. Table 6.5 shows the various types for the general population and Table 6.6 the various types for engineers. The tables could be rolled to make a cylinder which puts ISTP and INTP side by side and ESTJ and ENTJ side by side. We are more likely to favour the adjacent styles when not operating out of the main style. Each of the functions is a continuum, so situations will determine how much along that continuum we are likely to be.

From Tables 6.5 and 6.6 it can be seen that engineers do have a different personality profile from the general population and we find that most engineers are driven by thoughts and ideas, make decisions based on logic and facts, and tend to be organised and punctual (STJs). On the other side of the table there are the more creative NTs who like to solve new problems, have an insight into the future and do not like routine. As with all these groupings, it is necessary to appreciate that different people see things differently, make decisions differently, interact with others differently and have different preferences.

Table 6.5 Myers Briggs Types Summarised (% for general Population).

Type Preferences	Sensing Thinking	Feeling	Intuiting Feeling	Thinking
Introversion Judging	ISTJ 11.6% Archivist	ISFJ 13.8% Devoted Carer	INFJ 1.5% Counsellor	INTJ 2.1% Builder
Introversion Perceiving	ISTP 5.4% Artisan	ISFP 8.8% Reticent artists	INFP 4.4% Idealist	INTP 3.3% Analyst
Extraversion Perceiving	ESTP 4.3% Negotiator	ESFP 8.5% Performer	ENFP 8.1% Enthusiast	ENTP 3.2% Pragmatic Politician
Extraversion Judging	ESTJ 8.7% Executive	ESFJ 12.3% Loyalist	ENFJ 2.5% Empathic leader	ENTJ 1.8% Visionary Commander

The % of each type come from Myers et al (1998)

Table 6.6 Myers Briggs Types Summarised (% for Engineers) after (Culp and Smith,2001).

Type Preferences	Sensing Thinking	Feeling	Intuiting Feeling	Thinking
Introversion Judging	ISTJ 23%	ISFJ 5%	INFJ 2%	INTJ 14%
Introversion Perceiving	ISTP 6%	ISFP 2%	INFP 5%	INTP 6%
Extraversion Perceiving	ESTP 5%	ESFP 1%	ENFP 4%	ENTP 5%
Extraversion Judging	ESTJ 8%	ESFJ 4%	ENFJ 2%	ENTJ 7%

6.5.4 Diversity – an important dimension of a team

> *We know that a group of intelligent, motivated men and women of many different backgrounds and experiences makes an ideal engineering team. These individuals will see the world through different eyes and bring unique perspectives to the engineering task at hand. Their diversity will yield a*

diversity of solutions, which ultimately leads to the best solution. – Dean of Engineering, University of Illinois at Urbana-Champaign (quoted in Daniel, 2002).

When people talk of diversity it is, common to classify differences into a number of categories. These are:

- professional - level of expertise, career stage;
- demographic - gender, age, nationality, language;
- psychodynamic - attitudes, personality type, sexual orientation;
- physiological - energy level, health; and
- world view - values, tendency to prejudice, tendency to bias, ethics stance.

As an example, the Adelaide Car Component Company had a workforce with the following statistics:

- 1160 employees;
- (58 managers, 161 professionals, 80 sales and service, 861 on shop floor);
- 71% male, 29% female;
- 52 ethnic groups.

One would expect diversity based on professional, demographic, psychodynamic, physiological and world view – in fact over all categories. There is a tremendous potential here for great things (and for disasters!). A characteristic of teams with diversity is that they are susceptible to splitting into subgroups along gender, ethnic or other dimensions (Lau and Murnighan, 1998), but they do have the capability of having a great synergy to get better solutions to problems.

6.6 GROUP AND TEAM DEVELOPMENT

Many common sense strategies for working in groups are now discussed and, if used in conjunction with knowledge of the stages of team development, can assist in moving the team forward.

Getting to know other group members

This appears to be elementary, but many groups never get to know the other team members. Generally, different team members have very different values, motivations, abilities and personalities. The first thing to do as a group is to make sure everybody introduces themselves. Make sure everyone has written down all other members' names. Suggest that everyone uses each person's name once at the initial meeting. Many people will immediately forget names if heard once or never quite hear them the first time. Ensure that everyone gets each other's name (and do not be afraid to ask for the spelling). Following this get to know what they like doing, where they are from, show interest and generally find out about them. Some members will be totally involved and others will be apathetic if they are allowed to. Try not to let people remove themselves from the group because they then become dead weight, leading to frustration and resentment in some group members.

Something which is hard to accept is that when team members do not participate it is not entirely their fault. It is also the fault of the leader and all other team members. One cannot be responsible for all team members all the time but some simple skills will enable fuller participation by all.

Our style and approach is adapted to the situation, depending on whom we are dealing with; from being patient to being very direct. The key to our understanding, is knowing those we are dealing with, and learning to change our individual style so that better understanding is achieved. While each person has a dominant style of interaction it can be advantageous to understand each others' styles and adapt when necessary. Keep a watch on the process and practise this skill.

From the MBTI we know that approximately 25% of people are introverted. These people find it against their nature to immediately talk in a group situation as they find it stressful. They tend to think things through before talking. Some people consider others' feelings and will not tell them when they have made mistakes. In a group with many extroverts, the introverted person will find it difficult to participate. To speak up to gain attention is not in his or her nature. It is the extrovert's responsibility to make an effort to include the introvert, to not dominate the conversation with them, and to not take the floor away from them. If the team wants to succeed, the team must actively manage the process of inclusion of others.

Stages in team development

When teams come together to undertake a specific task it has been found that they tend to go through a series of quite well defined stages. In 1965, Tuckman developed a model that had four stages of forming, norming, storming and performing. Some stages are very productive, some less so. It is important to realise that this is standard behaviour and that if one is part of a group these are the different stages that should be expected to happen and need to be worked through. Two extra stages of pre-group and adjourning can be added, as shown in Figure 6.4 and described below.

Stage 1 – Pre-group

At this stage the task is usually broadly defined, the group is undefined, the resources are undefined and there might appear to be little that anyone can do. However, there is. Members can decide commitment at this early stage, and show it by, for example, arriving on time, or behaving in a way that demonstrates their willingness to be part of the group.

Stage 2 – Forming

As the groups are formed, inclusion or exclusion is of paramount importance as members try to work out their place in the group. There may be superficial conversation and people orientate to each other and the task. It is suggested that even at this early stage there will be a search for direction and people will be looking for someone to provide strong leadership. The culture of the group is established during this stage. When teams are beginning, each member considers their identity within the group. Identity can be considered to be a combination of personality, behaviour, competencies, and position in the social structure of the group (Borgatti, 2002). Certain members will fight for dominance, others will like to be seen as being smart, others will play a comedian role, and some just want to be liked. Knowing Belbin's roles helps a team through this stage by allowing the preferred styles of individuals to be incorporated in planning roles for the group by the leader.

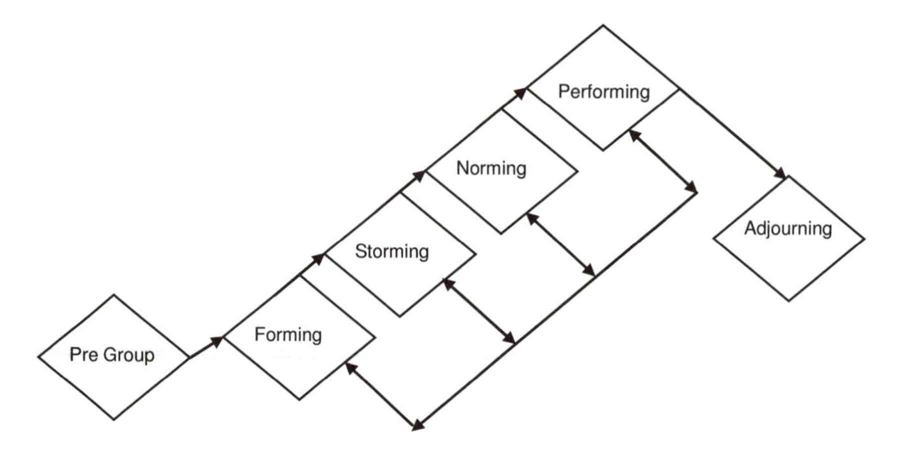

Figure 6.4 Stages of Group Development

Stage 3 – Storming

Once the initial nerves are overcome, it is time for people to be confident enough to cause trouble! There may be the first conflict with personal agendas revealed. The interactions between members may be quite uneven and there is the tendency to rebel against the leader. The interactions allow a pecking order to be established. It is important to resolve conflict, otherwise apathy can set in, and doubts may surface about the ability to cope with the task.

Stage 4 – Norming

Following the early disruptions, there may emerge a sense of renewed hope. Members can become more tolerant of each other and cohesion increases, and harmony becomes important. The roles of the members are established and they can become less dependent on the leader, although the leader is still necessary.

Stage 5 – Performing

The group then enters its most productive phase. Members "get on with the job", although minor problems may still occur. The role of the leader has changed and he or she is now seen as a peer and resource person. The roles have all been determined and there is consolidation of the status hierarchy.

Stage 6 – Adjourning

The final stage of the team is its disbandment. As things wind down there is time to check that the goals have been achieved and to try and cope with the end of the group. Many will make plans for future meetings and there may even be some sadness. Most importantly, there will often be a level of excitement based on what was achieved.

As teams evolve, if new members join the group, there is a tendency to revert to an earlier stage of development. This can be seen in the Figure 6.4 where a number of cycles are generally undertaken, especially if all members do not attend meetings or carry out the work allocated to them. It is important that teams assess their performance from time to time. Most teams start out well, and then drift away from their original goals and eventually fall apart. This is less likely to happen if, from time to time, the team facilitator or leader asks everyone how they are feeling about the team, and does a public check of the performance of the team against the mission/vision statement.

6.7 TEAM MEETING SKILLS

The skills required by an engineer are many and varied. These include the management of time and people. Managing people requires the manager to know the inner workings of his or herself. Once an individual knows how others see them and how they interact with others, the better those individuals will become in leading groups and teams. Managing the resources that are available is paramount for any engineer carrying out projects. The ability to facilitate meetings and to manage one's own time are two skills that are part of being an efficient manager and leader. Meetings can be useless if there is no control of the discussion.

Effective Meetings

To have effective meetings, it is important to set an agenda, start and finishing times, the location for the meeting and ensure the chairperson keeps control of the process and time schedule. One simple thing that helps a lot is having an agenda. Brainstorming sessions have their own format. Meetings should have a designated outcome but if the meeting is just for a one way distribution of information this can be done prior to the meeting and the agenda set for discussion. Some simple processes are:

A **Attendees to assign a chairperson** who should encourage active participation to have the best possible decision making process. Remember the meeting is for the benefit of all attendees. Listen to other points of view and do not be afraid to offer your own opinions.

G **Group dynamics** are important to allow all members to have a say. Ensure quieter members contribute, as they often have excellent ideas. Be firm with dominators who talk all the time, not allowing others to contribute. Remember the goals of the meeting.

E Expect outcomes by explaining the purpose of the meeting, exploring ideas and allowing all members to contribute to the discussion. Each member should feel that a contribution is expected and will be valued.

N Note-taking for an accurate record of the meeting is essential. This is a skill to practise and apply in meetings. The more that one does note taking the more adept one becomes at doing it. Rotating the position of the official recorder of the meeting is important for all members to get experience.

D Designate an action against each item. Decide "who is doing what" by what date so that everyone is clear on what is required. Create an action list.

A Announce and Advise what is on the designated **Action** list, by circulating the minutes of the meeting as soon after the meeting is completed and set the next agenda, including time and place! The first item on any agenda should be a "status check" which is where the facilitator asks each person how things are going and whether actions have been undertaken and, more importantly, whether there is anything that needs to be discussed.

As an individual attending a meeting, check the contribution that can be made by asking: Should this be a brainstorming session? See guidelines on brainstorming and creativity in Chapter 4. What preparation is required for the meeting? Has the work that has been allocated on the action list been completed? Are there other members who were not at the meeting who need to be involved?

Some goals that successful chairpersons try to undertake are: to establish the vision in conjunction with the group; ensure the team focus on the task; actively pursue participation from all members of the team; protect individuals from direct personal attack and establish conflict resolution strategies; suggest alternative processes when the team is stalled; and summarize and clarify the team's decisions. The chairperson accomplishes these goals by doing the following:

- Stays neutral and ensures that good seating arrangements are used (e.g. best in a circle);
- Keeps the meeting on time, even if it's going well (or people will try to avoid coming next time);
- Expresses out loud what seems to be happening (e.g. "George and Paul (in a side conversation) can you please give your opinion on the issue at hand"; "nobody seems to be saying much since Belinda suggested");
- Ensure conflicts, snide remarks and put downs are addressed immediately to help foster a team spirit even referring the perpetrators to reading material on group behaviours; and
- after a person has not contributed for a while, ask for his or her opinion.

Most importantly, before it is decided that a meeting needs to take place, examine if there is a better method to achieve a similar outcome. Email, telephone and/or video-conferencing may be more efficient alternatives.

Handling Conflict in Teams

Too much conflict may be a bad thing, but some is considered good because it may lead to more ideas being considered. The important issue in conflict is to ensure it is controlled and does not get out of hand. This is not as easy as it sounds because there are many sources of conflict, many different situations in which it occurs and therefore many different ways of handling it. The ways of handling conflict are shown in Figure 6.5. It is seen that there are five basic strategies for resolving conflict and that these depend on the behaviour and attitude of the person causing the conflict, and the managerial style of the person attempting to resolve it. These will now be discussed in a little detail.

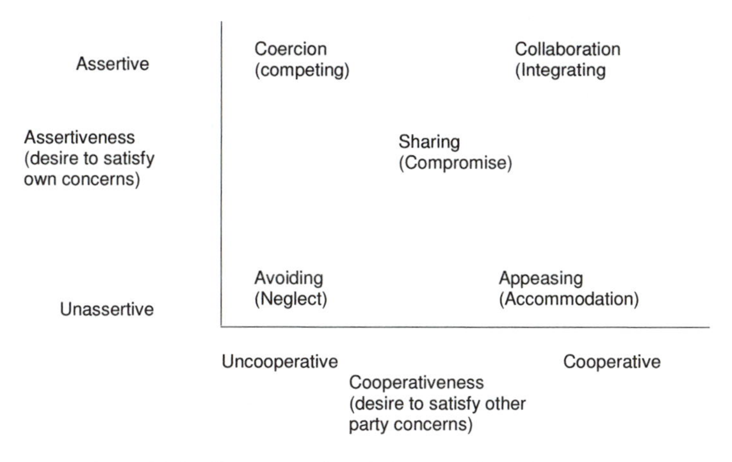

Figure 6.5 Conflict types and resolution strategies. (adapted from Thomas, 1976)

Coercion

With coercion the leader essentially imposes his or her will on the person who has caused the conflict. It is a power-orientated approach and is useful if quick action is needed, or for unpopular decisions. While it works well in these situations it may cause problems later because, although quashed at the moment, the conflict may not be resolved and the loser may become angry because there is competition for their own concerns. At this point, a negotiated resolution becomes less likely.

Accommodation

In accommodation, an unassertive leader works with a cooperative team member. Both neglect their own needs to suit others (to a certain extent). It is a useful conflict strategy if a team needs to preserve harmony, or when one of the two realise they are wrong. It has the advantage in that it allows others to learn by their mistakes, and is useful in building credit for later. On the down side, because they have given way the feeling of self worth may be diminished, and winners may press for further domination.

Avoidance

Conflict from an uncooperative team member can also be handled by avoiding the issues altogether. Avoidance does not address the conflict directly but sidesteps and attempts to postpone any resolution. As a strategy, avoidance is useful when the issue is trivial, or the leader's power is low. It may also be applicable when the potential damage is greater than the benefits of resolution, or if there is a need to allow people to cool off. On the other hand, it generally gives low satisfaction and may even increase the opponent's frustration.

Compromise

Compromise seeks a mutually acceptable solution to the problem, or at least a temporary solution. It can be used where two opponents have equal power, or to achieve a temporary settlement in complex matters, or when goals are only moderately important. The main disadvantages with it are that the conflict is thought likely to re-emerge, and both sides may be dissatisfied with the outcome.

Collaboration

If there is time, working together using successful negotiating behaviours can bring an excellent result to conflict. Collaboration can be used to find a solution when both parties' concerns are too important to ignore, to maximise commitment, and when it is important to learn from others. It can, however, take time and energy to work through it.

Elephants and Ants analogy for choosing priorities

If ants were on the menu for dinner, how many would be needed for a feed? A great many! A lot of hunting would be required. On the other hand if an elephant was on the menu for dinner how many would be needed? One would satisfy anyone's appetite.

The activities that we perform in our daily lives can be related to the elephants and ants analogy. When work is about killing ants there is a confusion of activity with accomplishment. What kind of sustenance would be obtained from a large number of ants? Choosing to carry out easy small tasks is done because they can be done quickly, so as to achieve the illusion that a lot has been accomplished, when in reality productivity has been poor. One falls further and further behind because the elephant hunting has been overlooked.

Elephant hunting or focussing on high pay off activities leads to productivity in the longer term. If one is busy stomping ants all day long, then the elephants will be totally ignored.

If the situation exists of constantly being busy, spending time answering unimportant emails or dealing with minor interruptions then there is a need to change from these "ant stomping tasks" to work on higher payoff activities. The ants still need to be dealt with but at a lower priority and with less time allocation. Try to spend less time with ants each day. **Don't Confuse Activity with Accomplishment** (Adapted from Vance, 1993).

6.8 PERSONAL TIME MANAGEMENT SKILLS

Vilfredo Pareto (1848–1923), an economist, made the well-known observation in 1906 that 20% of the population owned 80% of the property in Italy which was later generalised into the so-called Pareto Principle (for many phenomena, 80% of consequences stem from 20% of the causes). This principle was further generalised in 1937 by Joseph Juran (1904–), the quality management guru of the 20th century. Applied to time management this translates to 20% of the time being expended on "the vital few" situations or problems to account for 80% of the results.

People can waste much time if they do not have direction or know how to recognise what is important. Time is irreversible and can never be regained. Engineering managers and leaders have many demands in any one day and therefore time is required to be managed to ensure effective use of it.

Vance (1993) emphasises the need to focus on important activities rather than activities which just keep every-one busy and being reactive to distractions or so-called urgent needs. This is also the basic premise of Covey (1989) in "The Seven Habits of Highly Effective People". The quick way to improve work productivity, family and social activities is to identify the 20% of the activities that will reap 80% of the benefits. It is necessary to reduce the time allocated to activities which give us small benefits, the so called "ants". Identify time wasters from the list in Table 6.7.

Table 6.7 List of Time Wasters.

Waiting	Travel
Attempting too much at once	Unrealistic Time objectives
Lacking objective/priorities and planning	Telephone interruptions/conversations
Inability to say "No"	Indecision
Jumping in	Bureaucratic processes and form filling
Lack of self discipline	Television watching/DVD
Internet/computer games	Idle conversations
Failure to listen	Poor organisation/filing system
Duplicating effort from not finishing	Unanticipated interruptions or visitors
Communication problems with others	Ineffective meetings
Micromanagement/too much supervision	Environment with visual/noise distractions
Confused responsibility	Doing urgent rather than important tasks
Other......	Other......

(adapted from *"The Executive's Guide to Modern Management Skills"* by Richards, 1982)

Procrastination

One must overcome the state of mind that is procrastination, which is the delaying of planned activities. What we have to do in most cases is just get started—make a plan and do it. Do not give any excuses for putting off doing activities or even not planning to do them at all. Analyse how time is spent, for example, by jotting down what is done for a week for each quarter of an hour, and then implement a few methods to eliminate the identified bad habits that will gain the most time. Rank these according to level of importance. Do this so that there are four groups of three activities. Once this has been done, undertake actions to operate on the top three as a priority, to help manage time.

First of all, we need to consider allocating time to doing important rather than urgent unimportant tasks. This can be achieved by developing a plan with priorities which is updated along with a daily to do list and weekly and monthly goals.

To accomplish good time management needs self discipline. This will require changing old habits and developing new ones. One has to be persistent in using the time-saving hints developed in the time management plan. Once plans are put in place and gains in time have been achieved, attend to the next areas for review.

A Simple Time Management Plan

Effective time management is crucial to accomplishing tasks and goals, as well as having time for personal enjoyment. Organization and Task Focus are the primary requirements. There needs to be persistence and self discipline, an awareness of overcoming procrastination and a long term vision.

- **Plan activities** – Take time to plan activities. It is important to allow time to plan wisely. Establish priorities for the day, the week, short-term, mid-term, and for the long-term. **Dividing large tasks** into a series of small manageable tasks will let the large task be easily accomplished. Each small task should have a deadline.
- Set time aside to do high priority tasks. Maintain accurate calendars; abide by them and adjust priorities as a result of new tasks. This can be done by using checklists and to-do lists and in some cases doing the most difficult task first.
- Do one task at a time, if possible. When starting a task, try to complete it before starting another task-this can be done by allocating enough time for the task. **Set start and stop times for activities.** This will need estimates, but these will improve with practice. Challenge the theory, "Work expands to fill the allotted time." Therefore, establish deadlines for all tasks. Ensure meetings have a specified purpose, have a time limit, and include only essential people. Do not waste other people's time.
- Do it, delegate it, and dump it (Vance, 1993) Do **not put unneeded effort into tasks and activities which do not require perfection.** Handle correspondence quickly with short letters and memos. Save time for other activities. Learn to know when to stop a task, using the Pareto Principle. Delegate as much as possible and empower subordinates. Throw unneeded things away.
- **Learn to say no.** By making the mistake of saying yes to too many things, priorities are then decided by others.
- **Avoid committing to unimportant activities, no matter when they are.** Ask what can be planned for this time slot, such as a holiday, a weekend hiking or camping or a fun weekend with family.
- **Get Started** – The classic time waster is avoiding starting a project. The most important item is to do it now. “A journey of a thousand miles starts with one

step" – Lao Tzu. The ability to start work quickly results in achievement and satisfaction. Five minutes now can achieve the start and the journey is begun. Give rewards for finishing tasks, as this tends to stop procrastination and helps starting.
- **Develop a routine** to do certain tasks like answering emails when there is time but not the energy to do other tasks. Set aside time for reflection.

Time Management – Long Term Goals

Keep long term goals in sight. Have checklists with items such as: "department meeting at 2:00" and "ring so and so, write letter/memo on…" but ensure time is put aside for the relationship tasks. Other examples include:

- Meet with staff on a regular basis both formally (interviews) and informally (morning coffee at least once a week)
- Develop a plan for the organisation to use only recycled paper
- Enrol to study Italian because in 4 years I want to be fluent for (insert your goal).
- Exercise each day (e.g. at least 15 minutes walk no matter what the weather).

Planning

As a planner, memorising this poem of Rudyard Kipling (1865–1936) will assist in developing a focus on the crucial facets of any plan. In planning, responses need to be made to each of these questions (Adair, 1988).

I keep six honest serving-men
 (They taught me all I knew);
Their names are What and Why and When
 And How and Where and Who.
I send them over land and sea,
 I send them east and west;
But after they have worked for me,
 I give them all a rest. (From *The Elephant's Child*)

Try not to get caught up in short-term demands which put pressure on things that are more important that one should be doing but cannot find the time. Approximately 30% of items on the "to do list" should be long range items that would normally be put aside for when there is enough time! These long term goals are very much embedded in students earning a degree but also goals for fitness, exercise and sports, relaxation and enjoyment need to be developed. Spontaneity can be fun, but make sure that there is time available for the things which will enable long term goals to be accomplished.

6.9 SUMMARY

The resources of time, money, materials and people are used in making things happen. In this chapter the need to manage people and time has been addressed.

First of all, in learning how to manage oneself it makes it easier to manage others. To manage others it is imperative that one manages oneself in a way that promotes the desire in others to do well. In this way teams and groups will achieve desired outcomes and some fun will be had along the way.

As an engineer you will not only work in groups and teams but in all likelihood take some form of leadership role in them as well, as many engineers rise to senior positions. The importance of knowing how these roles work and, more importantly, to know what to do when they start to break down cannot be stressed enough.

It has been shown that effective time management is crucial to accomplishing tasks and goals as well as providing time for personal activities.

PROBLEMS

6.1 Write down five groups/teams that you have been part of in the last year and determine if they were really a group or a team. Is it possible to classify them easily or are there elements of both in some of them?

6.2 Break into groups, as designated by lecturer. These can be groups of different sizes ranging from 3 to more than 10. Discuss what are the important elements of having a successful meeting. Prepare a set of hints for running meetings.

6.3 Use the results of the personality test (http://keirsey.com/drummers.html) to assess the attributes of the people in the engineering project that you are in. What are the advantages of doing this in a formal way?

6.4 Working again with the engineering project group of which you are/were a member, list the range of diversity that the members exhibit. Take some care with this because, due to cultural differences, some may not be happy to discuss various aspects of themselves. You may find sexual orientation is one area that, quite reasonably, many are keen to avoid discussing, and this should be taken into account.

6.5 In groups where you are involved through the year, make a careful observation of conflict and how it is resolved. Are there people who seem to always want to use the same resolution strategy no matter what the situation? Does this work?

6.6 Irrespective of the nature of the conflict or the feelings of the leader, what strategies of conflict resolution might by appropriate for a problem that arises on the day before a report is due?

6.7 Choose three of the leaders listed in Section 7.4.1 and find out what their main attributes of leadership were against the 10 leadership laws.

6.8 Activity for group review: assess what might be going wrong in your team and think how to remedy it.

How is the group functioning?	**Score from 1 to 7 1 agree, 4 is sometimes, 7 is disagree**
Group clarifying what the task or objective is	
Group continuously checking on progress against timeline and seeks reasons for non compliance	
Group clarifying or recording what has been decided	
Group clarifying who is going to do what	
Group clarifying what has to be done by when	
Group established procedures for handling meetings	
Group keeping to agreed procedures	
All members listening to each other	
Not allowing individuals to dominate and others to withdraw	
Not compromising individuals needs for the sake of the team	
Group recognising the feelings of members of the team	
Members contributing equally to the progress of the team	

Sum the 12 scores. A score of 12 would be perfect, 12 to 29 shows team is working well, 24 to 50 requires more effort, 50 to 84 would be disastrous and requires a meeting to change attitudes in the group

REFERENCES

Adair, J., 1983, *Effective Leadership*, Aldershot: Gower, 228pp.

Adair, J., 1986, *Effective Team Building*, Aldershot: Gower, 212pp.

Ancona, D.G., Kochan, T.A., Scully, M., Van Maanen, J. and Westney, D.E., 1999, *Managing for the Future: Organizational Behavior and Processes.* 2nd ed., Southwestern.

Belbin, M., 1981, *Management Teams, Why They Succeed or Fail*, Heinemann: London.

Belbin, M., 1996, *Team Roles at Work*, Butterworth.

Berens, L.V., Ernst, L.K. and Smith, M.A., 2004, *Quick Guide to the 16 Personality Types and Teams: Applying Team Essentials to Create Effective Teams,* Telos Publications.

Boeree, C.G., 2004, *Personality Theories-* Abraham Maslow, http://www.ship.edu / ~cgboeree/maslow.html, Retrieved 16/8/2005

Boeree, C.G., 2004, Personality Theories- Carl Jung, Retrieved 16/8/2005, http://www.ship.edu/%7Ecgboeree/jung.html.

Borgatti S.P., 2002, Introduction to Organizational Behaviour, last revised April 2002, http://www.analytictech.com/mb021/mbtidim.htm, Retrieved August 2005.

Bynner, W., 1944, Tao Te Ching, Translation, The Way of Life, http://home.switchboard.com/ TaoTeChing

Cleland, D.I. and Kocaoglu, D.F., 1980, *Engineering Management*, McGraw Hill.

Cole, P. and Tuzinski, K., 2003, *DiSC* Indra Research Report, Inscape publishing, 31pp.

Covey, S.R., 1989, *The Seven Habits of Highly Effective People*, Simon and Schuster NY.

Culp, G. and Smith, A., 2001, Understanding psychological type to improve project team performance. *Journal of Management in Engineering, ASCE, 17*(1), 24-33.

Daniel, D.E., 2002, Letter from the Dean. *Engineering Outlook.* 42(1&2), p. 2.

Duch, B., 2000, *Working in Groups,* University of Delaware, USA, http://www.physics.udel.edu/~watson/scen103/colloq2000/workingingroups.html, Retrieved Aug 2005.

Fayol, H., 1949, *General and Industrial Management*, translated from the French edition (Dunod) by Constance Storrs, Pitman, London,

George, C.S., 1968, *The History of Management Thought,* Englewood Cliffs, N.J.: Prentice-Hall.

Hofstadter, D., 1979, *Gödel, Escher, Bach: an Eternal Golden Braid* (commonly *GEB*), Basic Books.

Johnson, D.W. and Johnson, F.P., 2000, *Joining Together. Group Theory and Group Skills.* Seventh Edition, Allyn and Bacon, 643pp.

Jung, C.G., 1923, Psychological Types, Routeledge and Kegan Paul, London.

Keirsey, D. and Bates, M., 1984, *Please Understand Me, An Essay on Temperament Styles*, Prometheus Nemesis Book Company.

Keirsey, D., 1998, *Please Understand Me II*, Prometheus Nemesis Book Company. http://keirsey.com/drummers.html

Lau, D. C. and Murnighan, J. K., 1998, "Demographic Diversity and Faultlines: The compositional Dynamics of Organisational Groups", *Academy of Management Review*, 23, April, pp. 325–340.

Maslow, A.H., 1970, *Motivation and Personality*, Harper and Row.

McShane, S. and Travaglione, T., 2003, *Organisational Behaviour on the Pacific Rim*, McGraw Hill.

Myers, I.B., McCaulley, M.H., Quenk, N.L. and Hammer, A.L., 1998 *MBTI Manual*. Palo Alto: Consulting Psychologists Press.

Newman, B., 1994, *The Ten Laws of Leadership*, BNC Publications.

Pinker, S., 2002, *The Blank Slate*. Penguin, 509pp.

Public Broadcasting Service, 1998, A Science Odyssey, People and Discoveries- Abraham Maslow http://www.pbs.org/wgbh/aso/databank/entries/bhmasl.html. Retrieved 31 August 2005.

Richards, C., 1982, *The Executive's Guide to Modern Management Skills* Rydge Publications, 197pp.

Samson, D. and Daft, R.L., 2003, *Management*, Pacific Rim Edition, Thomson.

Schein, E.H., 1970, *Organisational Psychology*, 2nd Edition, Prentice Hall.

Shtub, A., Bard, J.F. and Globerson, S., 2005, *Project Management. Processes, Methodologies and Economics,* Upper Slade River, New Jersey: Pearson Education Inc.

Thomas, K., 1976, "Conflict and Conflict Management", Chapter 21 in Dunnette, M.D. *Handbook of Industrial and Organizational Psychology*, pp. 889–935.

Tuckman, B.W., 1965, Developmental Sequence in Small Groups, *Psychological Bulletin*, 63, pp. 384–399.

Turla, P. (no date) Time Management Tips for Getting Results by Peter "The Time Man" Turla, www.TimeMan.com, Retrieved 16/08/2005.

Ullmann, J. E. (ed), 1986, *Handbook of Engineering Management*, John Wiley & Sons.

APPENDIX 6A THE MBTI DESCRIPTIONS

The following list comes from a number of sources including Keirsey and Bates (1978), Keirsey (1998) and Boeree(2004).

ENFJ (Extroverted feeling with intuiting): Mikhail Gorbachev, Mao. These people are the conversationalists of the world. They tend to idealize their friends. They make good parents, but have a tendency to allow themselves to be used. They make good therapists, teachers, executives, and salespeople.
ENFP (Extroverted intuiting with feeling): Leon Trotsky. These types of people love novelty and surprises and tend to be imaginative. They are big on emotions and expression. They tend to find reasons to do whatever they want. They tend to improvise rather than spend time preparing. They are good at sales, advertising, politics, and acting.
ENTJ (Extroverted thinking with intuiting): Bill Gates, Margaret Thatcher, Napoleon Bonaparte. In charge at home, they expect a lot from spouses and kids. They like organization and structure and tend to make good executives and administrators.
ENTP (Extroverted intuiting with thinking): Richard Feynman, Walt Disney and Nikola Tesla. These are lively people and tend to be outspoken, not humdrum or orderly. As mates, they are a little dangerous, especially economically. They are good at analysis and make good entrepreneurs. They do tend to play at one-upmanship.
ESFJ (Extroverted feeling with sensing): Michael Pallin. These people like harmony. They tend to be very active committee members and work best with encouragement and praise. They may be dependent, first on parents and later on spouses. They wear their hearts on their sleeves and excel in service occupations involving personal contact.
ESFP (Extroverted sensing with feeling): Picasso. Very generous and impulsive, they have a low tolerance for anxiety and are sometimes labelled as performers or artisans. These people like public relations, and they love the phone. They tend to know what is going on and join in eagerly.
ESTJ (Extroverted thinking with sensing): Joseph Stalin, Harry S Truman. These are responsible mates and parents and are loyal in the workplace. They are realistic, down-to-earth, orderly, and love tradition. The majority of engineers fall in this category. They often find themselves joining civic clubs!
ESTP (Extroverted sensing with thinking): Theodore Roosevelt, Henry Ford. These are action-oriented people, often sophisticated, sometimes ruthless - our "James Bonds." As mates, they are exciting and charming, but they have trouble with commitment. They make good promoters, entrepreneurs, and con artists.
INFJ (Introverted intuiting with feeling): Carl Gustav Jung, Mahatma Gandhi. These are serious students and workers who really want to contribute. They are private and easily hurt. They make good spouses, but tend to be physically reserved. People often think they are psychic. They make good therapists, general practitioners and ministers.
INFP (Introverted feeling with intuiting): Albert Schweitzer, Audrey Hepburn.

These people are idealistic, self-sacrificing, and somewhat cool or reserved. They are very family and home oriented, but do not like to relax. These people are found in psychology, architecture, and religion, but never in business.

INTJ (Introverted intuiting with thinking): Stephen Hawking, Dwight Eisenhower, Isaac Asimov These are the most independent of all types. They love logic and ideas and are drawn to scientific research. They can be rather single-minded.

INTP (Introverted thinking with intuiting): Albert Einstein, Marie Curie. Faithful, preoccupied, and forgetful, these are the bookworms. They tend to be very precise in their use of language. They are good at logic and mathematics and make good philosophers and theoretical scientists, but not writers or salespeople.

ISFJ (Introverted sensing with feeling): Mother Therese. These people are service and work oriented. They may suffer from fatigue and tend to be attracted to troublemakers. They are good nurses, teachers, secretaries, general practitioners, librarians, middle managers, and housekeepers.

ISFP (Introverted feeling with sensing): Mozart, Auguste Rodin. They tend to be shy and retiring, not talkative, but like sensuous action. They can be good at painting, drawing, sculpting, composing, dancing and they like nature. They are not big on commitment.

ISTJ (Introverted sensing with thinking): John D Rockefeller. These people are dependable pillars of strength. They often try to reform their mates and other people. They make good bank examiners, auditors, accountants, tax examiners, supervisors in libraries and hospitals, business and boy or girl scouts! Engineers are a greater proportion than average in this category.

ISTP (Introverted thinking with sensing): Michael Jordan, Lance Armstrong. These people are action-oriented and fearless, and crave excitement. They are impulsive and dangerous to stop. They often like tools, instruments, and weapons, and often become technical experts. They are not interested in communication and are often incorrectly diagnosed as dyslexic or hyperactive.

APPENDIX 6B THE MBTI DIMENSIONS

(Compiled from Keirsey (1998); Ancona, Deborah, Kochan, Scully, Van Maanen, and Westney (1999); Borgatti (2002))

Interacting with Others (E/I)

Extraverts	**Introverts**
Prefer variety and action Communicate freely Often impatient with long, complicated jobs Like having people around Are interested in the activities of their work and in how other people do it Are often impulsive Develop ideas by discussion Like greeting people Learn new tasks by talking and doing Enjoy meeting new people Seek out social gatherings When speaking publicly will often improvise Can be impulsive	Prefer working alone without interruptions Prefer quiet for concentration Tend not to mind working on one project for a long time uninterruptedly Are interested in the ideas behind their work Tend to think before they act Develop ideas by reflection Dislike intrusions and interruptions No strong need to meet regularly with others When speaking publicly will prepare in depth and speak from a plan Consider consequences before acting socially Sometimes have problems communicating

Understanding the World (S/N)

Sensing	**Intuition**
Prefer practical problems Prefer established systems and methods Like using experience and standard ways to solve problems Enjoy applying what they have already learned: like to work with tested ideas May distrust and ignore their inspirations Seldom make errors of fact Like to do things with a practical bent Like to present the details of their work first Prefer continuation of what is, with fine tuning Usually proceed step-by-step Patient with routine detail Like to have schedule for working Searches for standard problem solving approach	Dislike doing the same thing over and over Like solving new complex ambiguous problems Are impatient with routine details Like to float new ideas Enjoy learning a new skill more than using it See possibilities and implications Tend to follow their inspirations May ignore or overlook facts Like to do things with an innovative bent Have creative vision and insight Like to present an overview of work first Prefer change, sometimes radical, to continuation of 'what is' Usually proceed in bursts of energy Likes innovative approaches

Making Decisions (T/F)

Thinking	**Feeling**
Try to establish objective decision criteria Measure decisions against payoffs Can be seen as hard hearted, detached and cold Decide according to situation Tend to relate well only to other thinking types Negotiate on the evidence Concern for fairness based on rules Like analysis and clarity Situation oriented Use logical analysis to reach conclusions Want mutual respect among colleagues May hurt people's feelings without knowing it Tend to decide impersonally, sometimes paying insufficient attention to people's wishes Tend to be firm-minded and can give criticism when appropriate Look at the principles involved in the situation Feel rewarded when job is done well	Personal subjective decision criteria Measure decisions against beliefs Can appear overcommitted to a point of view Believe in deciding on personal considerations Nostalgic Negotiate on rights and wrongs of issues Fairness comes from values and beliefs Like harmony based on common values Objectives emerge from beliefs Principles oriented Use values to reach conclusions Want harmony and support among colleagues Enjoy pleasing people Often let decisions be influenced by their own and other people's likes and dislikes Tend to be sympathetic Dislike telling people unpleasant things Look at the underlying values in the situation Feel rewarded when people's needs are met

Allocating Time (J/P)

Judging	**Perceiving**
Like clarity and order Work best when they can plan their work and follow their plan May decide things too quickly Concerned with resolving matters Dislike ambiguity Can be inflexible once decision is made Emphasize decision taking over information getting Like to get things settled and finished May not notice new things that need to be done Tend to be satisfied once they reach a decision on a thing, situation, or person Reach closure by deciding quickly Feel supported by structure/schedules Focus on completion of a project	Enjoy searching and finding Procrastinate decisions while searching for options Can tolerate ambiguity Concerned to know, not resolving problems Open minded and curious Emphasis on diagnosing over concluding Concern is to know Enjoy flexibility in their work Like things open for last-minute changes Tend to procrastinate Tend to be curious and welcome a new light on a thing, situation. or person Adapt well to changing situations and feel restricted without variety Focus on the process of a project

CHAPTER SEVEN

Communication

Effective communication is needed in all facets of engineering work. In undertaking a project, engineers have to listen to and understand others, question and discuss ideas, give feedback, write reports, take an active part in meetings, engage in conflict resolution, and develop and manage teams. In order to communicate effectively with others, it is necessary to be able to present information clearly and simply and to interpret the words and emotions and non-verbal messages of others. This chapter discusses the nature of human communications and strategies for achieving effective and clear communication.

7.1 INTRODUCTION

Communication is the process by which humans interact with each other. Language, both written and spoken, enables us to develop ideas and plans, and then to communicate them to others. Language and communication together make social interaction possible. Without them, there would no society and it would be impossible to engage in engineering work. It is the basic ability of humans to make and interpret vocal utterances to make known their needs, wants, ideas and feelings, and to change these spoken words into written records for posterity, that make us different from other animals on earth.

Human beings have many **channels of communication** available to them. Like other animals, they can employ all of their five senses to communicate; however, social communication occurs predominantly through the use of eyes, ears and voice. For example, in face-to-face inter-personal communication between two people, one listens and watches while the other speaks, until the roles change. However, the message sent out by the speaker is not contained solely in the words used. Non-verbal communication also takes place, both intentionally and unintentionally, through the speaker's hand movements and facial expressions. The tone of voice, independently of the words used, can also transmit information. Likewise, non-verbal messages are sent by the listener back to the speaker, intentionally and unintentionally, through facial expressions which may contain a variety of signals such as agreement or disagreement, confusion, disbelief and non-comprehension. This is an instance of feedback. Various channels of communication are thus used to send information back and forth, even in this most common of situations.

Interpersonal communication can also take place through other channels that employ sight, with or without sound. For example, written notes and letters can be transmitted by post or by email, or even by carrier pigeon. Devices such as telephone, radio and television provide other potential channels. Throughout history, humans have devised a wide range of ingenious ways to communicate with each other using sight and sound to overcome the barriers of both space and time.

In the discussion to date we have focused on a specific form of communication involving two people. There are obviously other possibilities and various **forms of communication** to consider. Perhaps the simplest is **self communication**, which occurs, for example, when we talk to ourselves or write notes and messages to ourselves as reminders of things we want to remember and to record information we need to keep. The case of **person-to-person communication** has already been discussed. In **group communication** more than two people talk and listen to each other. It may be that everyone in the group speaks; alternatively, several may speak while the others listen. Meetings allow direct communication to occur in a group of people. Group communication can also occur through indirect channels such as written messages. In **mass communication** one or several people send messages to large numbers of people. In modern society, mass communication typically occurs through television, radio and the internet. The communication in such circumstances is often one-way, in that the receivers do not have any opportunity to send messages back to the senders. There is often no **feedback** in such situations. Feedback can play an important role in communication: it provides one means for checking whether the message received is the same as the one sent. Mass communication can also take place through books and other printed documents. An instance of mass communication occurs when a lecturer makes a presentation to an audience, with or without visual aids, in a hall. A play, staged in a theatre, is yet another form of mass communication. In the last-mentioned instances it is interesting to reflect on the extent to which feedback can occur.

In human communication, the message that is sent is unfortunately not always the same as the one received. Poor communication and miscommunication can occur in many ways. The fault may be that the original message is ambiguous or misleading. Alternatively, the error may be in a misreading of the received message. Messages can also be distorted and corrupted during the process of transmission, perhaps as a result of the channels used to send the messages. Non-verbal messages are susceptible to misinterpretation. Some people seem to be naturally good and effective communicators and almost always get their message across. Others do not communicate clearly and effectively, even in the simplest circumstances. Some people are good listeners, others less so. Depending on the channel of communication used, the message can become distorted during transmission.

In all forms of successful human communication, irrespective of the channels employed, there must be at least one person sending information (the sender) and at least one person receiving the information (the receiver). Some of the problems of communication are illustrated in the model shown in Figure 7.1, where there are two parties who send and receive information in turn. Each message is encoded and sent, but is susceptible to **noise** which may result in loss of clarity and distortion. The noise depends on the channel used for communicating. When received, the message is decoded. The presence of noise, coupled with the possibility of faulty encoding and decoding, means that the message received may not be the same as the message sent. The model stems from Shannon's original model of communication in a telephone network (Shannon and Weaver, 1949).

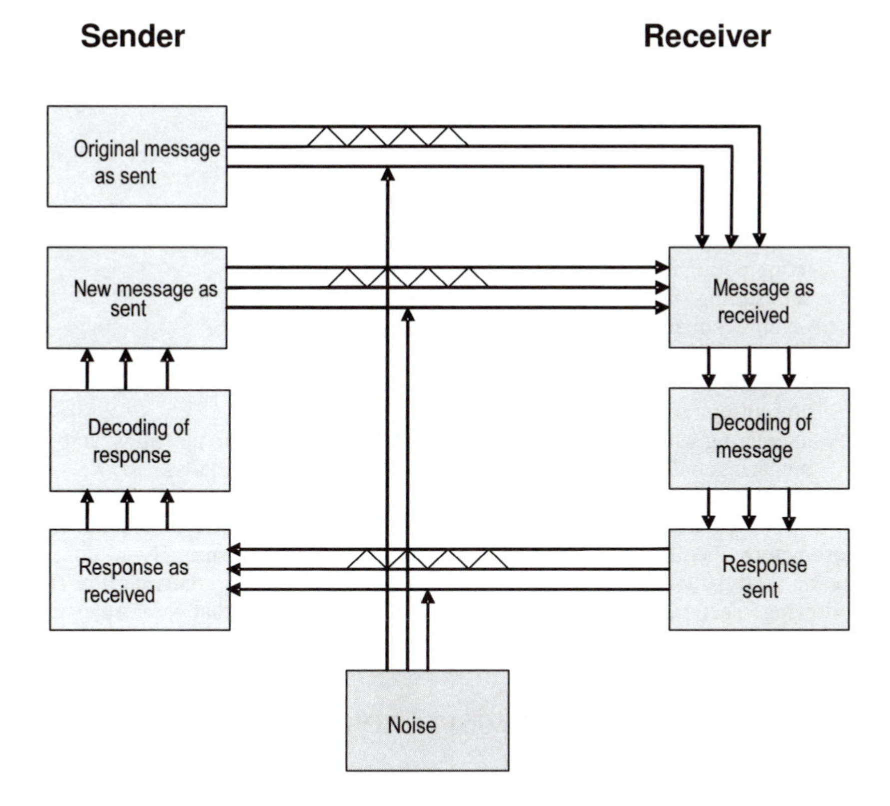

Figure 7.1 Model of the communication process

Communication plays a particularly important role in all engineering work. It dictates how problems are formulated and how they are solved. The quality of the communication among team members and between client and engineer determines the quality of engineering plans and design solutions. Good communication is necessary in order to carry good designs and plans through to successful implementation. Frequent and effective communication must occur among the members of any engineering team if it is to be successful. Lack of communication, poor communication and miscommunication in an engineering project can lead to catastrophe, as was the case in the Space Shuttle Challenger disaster where different companies' management and NASA were told of critical safety indicators by engineers but allowed the Shuttle to be launched against many people's objections.

In engineering communication, whether direct or indirect, a great deal of information can be transmitted clearly and unambiguously through the use of sketches and drawings. These provide an added communication channel. Experienced engineers, when talking to each other, usually resort to simple sketches to get their ideas across clearly. The ability to sketch and the conventions used in sketching are therefore of great importance to engineers. It is very important to learn these skills. Sketches and figures are also used in more formal channels of communications, such as reports, design drawings, plans, and even books, and lecture notes.

Engineers must be able to communicate effectively in activities such as the following:

- managing teams;
- participating in team work;
- counselling staff under their supervision;
- giving feedback to team members;
- resolving conflicts among team members;
- presenting concepts to other engineers;
- presenting proposals to clients;
- interacting with the community in consultations and public meetings; and
- preparing and describing clearly documents and design details.

Good engineering is of necessity based on good and effective communication. Fortunately, the ability to communicate effectively can be learnt. Even those who are naturally good communicators should not rely on their native gifts. In the following sections various aspects of communication that are relevant to engineering work are discussed.

7.2 PREPARING FOR COMMUNICATION

One of the first and most important keys to successful communication is adequate preparation. Before an effective and successful engineering message can be sent, the content of the message has to be clearly worked out and fully understood by the sender. This is particularly important if the message is complex, which is often the case for engineers.

Even when the content of the message has been worked out, it is necessary for the potential sender to put time and effort into choosing the best way in which to present the information so that it will be most easily understood by persons receiving the message. In preparing for sending, it is important to evaluate the potential receivers, their backgrounds and expertise, and their level of understanding of the technical matters involved in the engineering message. For example, the communication of design information to non-specialist clients and the interested public will require a different form of presentation from that used for engineering colleagues who have a good technical understanding of the field. It may well be that a broad overview is better for non-specialists than detailed data which will not be understood.

The sequencing of information also needs careful attention. While a logical sequence is usually preferable, it will not always be the best. A logical but long and complex argument can rapidly become boring and hence lead to inattention and misunderstanding.

Methods for organising thoughts and for presenting ideas and concepts.

Before an idea can be communicated it must be developed. There are a number of simple tools which can assist in the development of ideas, ranging from hierarchal lists, using word outline for ease of developing hierarchy, mind mapping, and picture and concept mapping (Checkland, 1981; Buzan, 1988; Horan, 2002). These methods advocate a systems approach to thinking by using images to link key ideas and hence enable transference of ideas between thinking and communicating. The mind mapping terminology was popularised by Buzan (1988) to be used for problem solving but also for learning, communicating and remembering. It appears that the future will include more tools for these techniques, as Bill Gates (co-founder of Microsoft) stated that "a new generation of 'mind-mapping' software can also be used as a digital 'blank slate' to help connect and synthesize ideas and data – and ultimately create new knowledge" (Gates, 2006). The basics of developing a mind map will now be discussed.

Mind mapping

A mind map is a graphical representation linking related ideas and concepts. It is created by starting with a central image, word or concept under consideration. Around this central concept up to 10 main ideas which relate to this concept can be drawn or placed. Ideas can be generated by using Kipling's six question words: What? Why? When? How? Where? and Who? Each of these images, words or concepts are considered and a further 10 main ideas or concepts which relate to each of them are generated. Without too much effort a large number of related ideas can be produced. With this simple technique it is possible to get a clear understanding of almost any problem.

As an example of what is meant, a mind map has been created using Free Mind (2006) of a mind map. This is shown in Figure 7.2. Despite the fact that mind maps follow the way the brain is believed to work, mind mapping may seem an unnatural way to record or process information. Experience has shown that a little persistence pays off and people who use mind mapping find it a valuable tool that can be employed to assist in communication. These techniques can be an instrument in developing a systems approach to problem solving and most web pages store their information in a similar way, with links to relevant information being the cornerstone of the world wide web.

One of the key reasons why mind mapping is so successful is that it forces the listener to not just record the points a speaker or article is making, but to engage with them and to consider not just their importance but also their relationship to the general theme or argument. This results in a deliberate sense of active listening or thinking that works the brain harder tending to produce greater understanding. Lazslo (2001) also noted that "The full potential of human communication unfolds only when the communicators understand the strands of connection through which they communicate". To communicate at a high level requires people to make use of the different paths of communication.

Preparation for communication is not purely the task of the sender. The receiver also has preparatory work to do in order to ensure good communication. Even when attending a meeting purely to listen and to obtain information, it is worthwhile planning for the event by thinking about what can be achieved and what information can be obtained. It is rare that the attendees at a meeting all have the same aims and agendas. With minimal preparation it is easier to question speakers to obtain clarification and to ensure they provide the required information.

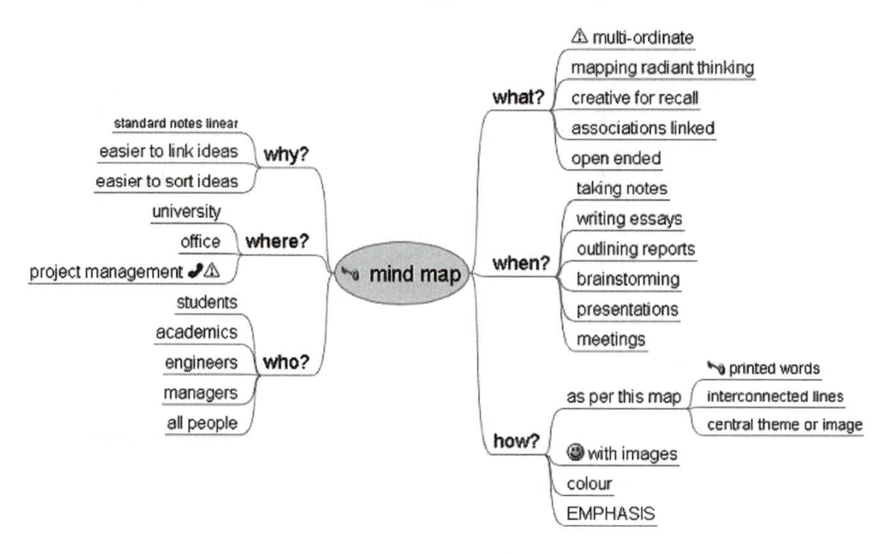

Figure 7.2 A mind map of the concept of mind maps.

Preparation can even be worthwhile prior to self communication. Probably the most important thing is to take sufficient time to organise yourself properly and develop techniques for storing information so that it is can easily be found and retrieved. Being systematic is something that is well worth learning and developing.

7.3 ACTIVE LISTENING

People who are good listeners have a high success in getting the information they want from others. It has been suggested that they follow a process called active listening (Steinmetz, 1979). Active listening includes being encouraging, being reflective, probing for information, and using a summary technique.

One way of being encouraging is to use open rather than closed questions. A closed question is one where the answer is either a single word or a short phrase. Closed question words include: Is, have, has, does, could, can, will, are and shall. For example: "How old are you?" and "Where do you live?" are closed questions. Closed questions have the following characteristics:

- they request factual information;
- they are quick and easy to answer; and
- they enable the questioner to control the conversation.

On the other hand an open question is one that generally receives an extended answer. Open question words include: describe; what; why; and how. Thus

"Describe what you think you will be doing in five years time?" is an open question. Open questions have the following characteristics:

- they ask the respondent to think and reflect;
- they try to elicit opinions, knowledge and feelings; and
- they deliver control of the conversation to the respondent.

The open question gets the respondent thinking and giving useful information about themselves or the subject being discussed. Note: some question words can elicit either a short answer or a long answer depending on how they are phrased. Some examples of closed and open questions are given in Table 7.1.

Table 7.1 Samples of Closed and Open questions

Closed question	Open question
Are you happy with the job?	What do you like or dislike about the job?
Are you sad about the decision?	What did you think about the decision?
Do you like your manager?	How do you feel about your manager?

Once the open ended question has been asked, an active listener will give free rein to the person answering and so encourage an uninhibited response. The aim is to get the responder to talk freely, and part of this freedom to talk involves the listener not talking. Silence can promote talking but so can non-verbal communication. Appropriate body motion should not to be distracting, artificial or forced. Movement, through the head, eyes, hands or any other body part which signals interest and attention, improves communication. Awareness of other people's non-verbal communication is paramount for clear understanding.

Reflective skills intend to help the listener keep track of the message. Reflecting skills improve the dialogue, by providing, as well as asking for, information. The most effective reflecting device is "paraphrasing". It serves to confirm that the person has been listening, but also requests more information. Paraphrasing helps to clarify a message or attempts to reflect the feelings of the speaker by summarising what has been said in fewer words.

Applying active listening skills takes practice and people should beware of jumping to an incomplete picture of what others are talking about and then offering a master plan to solve the problems, without recognition of the thoughts, feelings and emotions of the person they are supposedly listening to. Rogers and Farson (1979) stress the importance of listening as "an important way to bring about changes in people". They suggest it involves three activities: listening for the full meaning including both the content and the underlying emotions; responding to feelings (if the real message is an emotional one then the response should address that) and noting all the cues by observing the non-verbal messages.

7.4 NON-VERBAL COMMUNICATION

Non-verbal communication can be thought of as all the information transmitted other than the message contained in the actual words used. It includes both the visual element and the vocal tones (Samson and Daft, 2003). The face is used a great deal to transmit information and this is usually the first visual element noticed. People often notice the hand signals, shrugs, head movements and body movements but do not necessarily interpret the whole message. Non-verbal communication can be used for:

- expressing emotion (e.g. smiling to show happiness);
- conveying attitudes (e.g. staring, glaring to show aggression); and
- demonstrating personality traits (e.g. open palms to show accepting qualities).

Non-verbal behaviour varies across cultures, although the six emotions of anger, fear, disgust, sadness, happiness and surprise seem to prevail around the world. The ability to understand and use non-verbal communication is valuable for any workplace environment. Some non-verbal communication strategies are easier to use than others. For example, when going to a meeting or interview, the way a person dresses can portray a certain image. Other non-verbal communication strategies such as being punctual, animated and demonstrating friendliness by a smile are able to communicate an image about a person in a much more effective way than the spoken word.

Facial expressions, eye movements, and head movements constitute a significant percentage of recognised non-verbal communication. This is because people tend to look at a person's face and eyes when speaking with them. It is estimated that the human face can show more than 50,000 different expressions (Hamilton and Parker, 1990) and most people are able to interpret the meanings or feelings associated with many of the common ones, such as surprise, joy, suspicion, anger or menace. The skill to read small facial changes and eye movements requires greater observation and awareness but is useful to ensure that one obtains all the information being transmitted in a conversation.

Being able to interpret body language and gestures is vital in establishing and maintaining good working relations with work colleagues and clients. If clients exhibit discomfort or unwillingness to participate, then it is possible to alter the environment and the communication strategy to help them feel more comfortable. It is doubly important to be aware what messages your own body language is sending, to avoid sending an incorrect or ambivalent message, and, more importantly, to ensure that you send the correct message. A useful strategy is to watch for reactions when speaking and then to try and spot contradictory messages when, for example, the voice says one thing and the body says another. This can indicate deception but one should always check this out by follow up questions in the context of the conversation.

A simple test can be performed to illustrate the importance of non-verbal communication: listen to someone with your eyes closed, then listen and watch. It should be evident that there are differences and that it is much easier to understand someone when you can see them.

Non-verbal behaviour patterns for communication

To deliver a greater impact when a person delivers a message, the use of non-verbal behaviours will raise the level of interpersonal communication. The following material has been compiled from a variety of sources (Eunson, 1994; Hunt, 1979; James, 1995; Quilliam,1995; Tidwell, 2005; Pease and Pease, 2004).

One of the most important behaviour patterns involves eye contact. Eye contact signals interest in others and increases the speaker's credibility and helps to regulate communication flow. However, people should be aware of cultural differences. In Western countries people who make eye contact open the flow of communication and give a message which shows interest and credibility. On the other hand people from Japan, China, Africa, Latin American, and the Caribbean avoid eye contact to show respect. Western cultures see direct eye to eye contact as positive but there are peculiarities even within a country. For example, in the United States African-Americans use more eye contact when talking and less when listening, with the reverse being true for Anglo-Americans. While a prolonged gaze is often interpreted as a sign of sexual interest, Arabic cultures use prolonged eye contact because they believe it indicates interest, and hence assists them in gauging the truthfulness of others. A person who doesn't reciprocate is seen as untrustworthy.

Facial expressions, such as smiling, transmit friendliness, happiness, warmth, and a connotation of welcoming. People who smile frequently are perceived as more likable, friendly, warm and approachable. Smiling can be contagious and people tend to mirror smiling to gain rapport. They will be more comfortable and will want to listen more. It appears that facial expressions have similar meanings world wide with respect to smiling, crying, or showing anger, sorrow, or disgust. Many Asian cultures suppress facial expression as much as possible. Many Mediterranean (Latino/Arabic) cultures openly express grief, whereas it has been a trait of men of English speaking cultures to hide grief or sorrow. Overall, there is a tendency for women to smile more than men but in some groups such as African-Americans there is little difference in the degree of smiling between genders.

Gestures are a natural part of human communications and failing to make gestures while speaking may lead to a perception of the speaker being uninterested or boring. Congruent gestures with speaking capture the listener's attention, and facilitate understanding. It is impossible to list all gestures but it is important to remember that what is acceptable in one's own culture may be offensive in another. Some cultures are restrained in gestures while others are animated, with the consequence that the restrained cultures feel the animated cultures lack manners. On the other hand, animated cultures often feel restrained cultures lack emotion. The use of hands for pointing or counting differs from country to country. For example, in Australia and the United States pointing is done with the index finger, whereas in Germany the little finger is used. The Japanese prefer to use the entire hand (in fact, most Asian cultures consider pointing with the index finger to be rude).

Posture and body orientation can also be used for communication. Standing erect and leaning forward communicates to listeners that the speaker is approachable, receptive and friendly. Interpersonal closeness results when the speaker and listener face each other. Speaking with the back turned or looking at the floor or ceiling should be avoided as it communicates lack of interest. There are also cultural differences for a range of what might appear to be quite common actions. In Japan and Korea bowing shows rank and is a sign of respect. Slouching is considered rude in most Northern European countries. Standing with hands in pockets is disrespectful in Turkey. Sitting with legs crossed is offensive in Ghana and Turkey, but traditional in Korea. Showing the soles of the feet is offensive in Thailand, Malaysia, and Saudi Arabia.

Proximity, the distance between speaker and listener, is a key area where problems can occur. Cultural norms dictate a comfortable distance for interaction with others and people should look for signals of discomfort caused by invading other people's space. Some of these signs are rocking, leg swinging, tapping, and gaze aversion. The context is also important: what is perfectly acceptable on a crowded train or bus is not acceptable in a normal business meeting. In Arab cultures, invasion of space and to stare closely into the eyes is considered normal.

Although speaking is the essence of verbal communication, it can also have non-verbal aspects to it. The way the words are spoken, through such aspects as tone, pitch, rhythm, volume and inflection can have an important effect on the message being transmitted. One of the major criticisms of many speakers is that they speak in a monotonic voice. Listeners perceive this type of speaker as boring and dull.

Touch is generally culturally determined, with each culture having a clear concept of what parts of the body one may touch or not touch. Touch generally shows emotions or control and varies between genders and cultures. In Western cultures, particularly in the United States, Australia, Canada and the United Kingdom, handshaking is common (even for strangers). People of Islamic and Hindu backgrounds avoid touching with the left hand because to do so is a social insult, as the left hand is used for toilet functions. It is the custom in India to break bread only using the right hand (a difficult task for non-Indians). Islamic cultures don't approve of touching between genders in public (even hand shakes), but consider such touching (including hand holding and hugs) between same-sex people to be appropriate. Many Asians avoid touching the head because the head houses the soul and a touch puts it in jeopardy. Cultures such as the English, German, Scandinavian, Chinese, and Japanese which have high emotional restraint have little public contact, whereas those which encourage emotion, such as the Latino, Middle-Eastern and Jewish, have more frequent touching.

Judgements, based on looks and dress, seem to be a trait of all cultures. Europeans and Americans appear to be almost obsessed with dress and personal attractiveness, partly induced by marketing and wealth. Around the globe there are differing cultural standards on what is appropriate, attractive in dress and what constitutes modesty. Many companies have different dress codes which can be used in the corporate world as a sign of status.

It appears that as verbal codes are used within a particular cultural context they evolve over time (Underwood, 2003). This also happens to non-verbal codes. Through travelling and working in different countries, an increased awareness of how specific gestures are different from one country to another is developed. For example, total confusion in a conversation can result from Indians shaking their head in response to matters being discussed. One needs to be very cautious and rephrase conversations to adjust to body language in this environment.

People tend to define a space bubble or proximity by their own culture and when dealing with some Middle-Eastern and Asian cultures one finds that space limits are vastly different from those in Western cultures. In some situations people stand so close, it is possible for Westerners to feel quite uncomfortable with a constant desire to back away. Other cultures find Europeans cold and aloof because of this tendency to back off.

Checklist: Tips for Improved Nonverbal Communication

► Make yourself at ease with the person you are communicating with. Avoid being too close or too far away. (Within 600–700 mm is a comfortable range for city dwellers of Anglo Saxon heritage.)
► Be attentive and try to relax, but avoid slouching or sitting rigidly. Show interest by leaning slightly toward the other person.
► Avoid staring or glaring, but try to maintain frequent eye contact.
► Respond with non-verbal communication while the other person is talking by simple nods for approval or agreement.
► All non-verbal gestures should be natural, smooth and unobtrusive. Do not allow gestures to dominate your words. Be aware of gestures that reveal negative emotions and frustration.
► Slow down your rate of speech to a little slower than normal, to avoid indicating impatience. Use the tone of your voice to give a feeling of warmth and acceptance.
► Do not mumble but maintain a clearly audible voice, not too loud nor too soft.
► Avoid using your limbs, hands and feet as barriers.
► Appropriate genuine smiling assists in gaining rapport.
► Closing eyes and yawning can block communication, so attempt to be alert when interacting with others.

Adapted from Messina (1999)

7.5 ORAL PRESENTATIONS

Woodrow Wilson, the 28th president of the United States said, "If I am to speak ten minutes, I need a week for preparation; if fifteen minutes, three days; if half an hour, two days; if an hour, I am ready now". This clearly indicates that good presentations do not just happen, and if someone makes it look easy then it is likely that much more effort and practice has gone into the talk than appearances would indicate.

Oral communication skills are required by all engineers, as a consultant to present a company's case to win work, to present results to senior management and clients, or to present details of projects to the community. Technical presentations can involve both oral and written communication. In all communications, an awareness of the audience is essential to ensure a consistent style, flow and clarity of detail. There is no easy way to ensure a presentation goes well. However, there are some key points that should be addressed well before the time of the presentation.

A common failing with presentations is running short or running out of time. Speakers should consider their timing very carefully and practise their talk to ensure that it will run to time. It is at the end of the talk where it is likely that the key points are to be made and where, if the ending is rushed, much impact will be lost.

Checklist: Oral Presentations

Preliminaries

► What is the objective of the talk?
► What do you want your audience to think or do as a result of this talk?
► Analyse the audience: what do you know about them, their backgrounds, age, needs, motivation?
► Consider issues like the time of day, number in audience, and size of room.

Opening

► How will you hook the audience?
► What will your first words be?
► Ensure that you thank the chairperson and address the audience!
► What is the purpose of the talk?
► What benefit will the talk be to the audience?

Signposts/Material Content

► Can you create a mind map showing the content of the talk?
► Point the way through your talk – use headings.
► Research your talk thoroughly. Preparation is the key to a good talk.

Notes

► Short memory aids only.
► Key words or ideographs or even a mind map on a small card.
► **DO NOT READ**

Visual aids

► Ensure visual aids are clear, easy and appropriate for ideas being expressed.
► Present them effectively - do not have colours that make the text unintelligible.
► Use a sans serif font (e.g. Arial) and a minimum font size of 18pt.
► Have a maximum of 30 to 35 words per slide : fewer is better.
► Describing a picture or graph is better than a slide full of words.
► Reinforce, explain and illustrate your ideas.
► Aim for approximately one minute per slide (as a rough guide)

Delivery

► Transmit energy and enthusiasm – use positive gestures and eye contact.
► Ensure voice is clear and well modulated.
► Body behaviours have to be congruent – generally be yourself.

Closing

► What are the main points to be reinforced in the summary?
► What is the final impression I wish to make?
► What will be my final words?
► Summarise and finish with a well prepared strong close.

Question time

► Remember, it is the chairperson's responsibility to handle question time.

Rather than just attempting to say everything more quickly, it is better to leave out a whole section so that time can be spent on giving a good summary of the rest of the work and leaving the audience with a positive feeling about the presentation. If there are people in the audience who are particularly interested in the details of the work, they can always seek the speaker out after the talk.

7.6 WRITTEN COMMUNICATION

Written communication, including reports, letters, memoranda, sketches and drawings, is perhaps the form of communication where the most effort is expended. In talking and presenting material orally, instant feedback is generally received. An audience that is slowly drifting off to sleep is letting you know your presentation is not going well and that something needs to be done. A written report receives no such instant feedback, but can generate the same consequences, and there is no chance to make immediate amendments. During the normal course of events, engineers will be required to write resumés, business letters, memoranda and reports for many different reasons. In the following sections suggestions are given to make writing as easy as is possible and to ensure that it conforms to some basic standards. The idea that writing should follow a standard may seem somewhat constraining but it should be remembered that one of the main reasons for communication is to present information clearly and to make the job of the reader as easy as possible. If this means a little extra effort for the writer, so be it.

Checklist: Writing

- ► Avoid jargon.
- ► Have a structure with an introduction, main content and a conclusion or recommendations.
- ► Ensure all written material is checked for spelling, grammar and punctuation.
- ► Read the whole piece from top to bottom.
- ► Have colleagues read at least key sections of the report
- ► Leave sufficient time so that it is not rushed. Rushed sections are likely to be at the end and this is where the key results are found and impressions made.

Memoranda

The purpose of a memorandum is to give information to one or more people at once within an organisation. It is, at the same time, a useful and formal way of having your views or advice on a particular topic entered onto the official record. Although emails have largely replaced the written memorandum in the modern engineering office, are still used for more formal or more important transmissions, and for this reason it is important to be able to write and structure one in an appropriate format.

Memos are usually restricted in length and contain the important information in a clear and easy to find format. Much is made to whom the memo is supposed to go to and, in some ways, this highlights just how important it is to the receiver.

Checklist: Memorandum

To:	[Name of person to whom the memo is being sent]
Cc:	[Name of other people who need information]
From:	[Name of sender]
Re:	[What the memo is about]
Date:	[Date memo sent]

§ Make the message as short and clear as possible
§ Use bold text to make the most important information clear
§ Make sure that all memoranda are proof read and spell checked.

The email

A form of modern communication used every day in business now is the email. A few simple rules and a little care can eliminate the drudgery of having to respond to numerous emails every day. In many ways the email is similar to the memorandum. However, there is one important difference. The email has, by its nature, an urgency and an informality about it.

Checklist: emails

- ► Put enough details in the subject line so that meaning of the email is conveyed, such as “Remember Salinity Project meeting Tuesday 25 July”.
- ► To reduce number of emails received, when appropriate insert “No Reply required” in the “Subject” line or the opening line of your e-mail.
- ► Avoid overuse of CAPITAL LETTERS as this conveys anger and shouting or “punctuation marks” or bold you'll appear to be putting a lot of stress on the message!!
- ► Cut and paste standard replies for frequently asked questions or requests.
- ► Although emails tend to be less formal, do a spell check before sending.
- ► Only reply to the sender when this is necessary, rather than to everyone on the sender's list.
- ► Only have one subject per email. People often respond to your first and last questions, but overlook or forget the others, so keep things simple.
- ► When replying to a series of questions, embed your answer after the question or other remarks at appropriate places.
- ► Any email you send could be forwarded to others so make sure you are aware of that and hence make the receiver aware of the fact that you do not want it to be forwarded on by marking it as “Confidential”.
- ► Use plain text, simple language, short paragraphs and keep messages under 25 lines long, when possible, as this is the length that can be read on one screen.
- ► Alert recipients when sending large attachments.
- ► When forwarding a message, put comments at the top rather than at the end.
- ► Always obtain permission from an author before sending on personal emails.
- ► Send information using a text format, if at all possible, rather than HTML.

A former NASA engineer who now runs time management seminars, asked a group of time management participants for ideas on improving time used on writing and sending emails. The email checklist was developed from some of those responses (Turla, 2005) and combined with those in McShane and Travaglione (2003).

Emails can be quick to send and almost demand a quick response. Engineers should ensure that their responses are well considered because, once again, the reply is on the public record.

Business letters

When reports are written, they will generally be sent with a covering letter. That letter then is the first thing that the client will read and therefore is important in creating the first impression, similar to the cover of a report. Most companies have quite strict standards for letters and it is better to know what they are, rather than risk the task of many re-drafts, or worse, upsetting a client or boss. A typical letter is shown to bring out some of the key elements that are generally required.

XYZ Environmental Solutions
PO Box 2222
ADELAIDE 5001
1 January 2006

Ms Caroline Chong
Manager
Engineering Constructions
PO Box 888
SYDNEY 2003

Dear Caroline,

RE: Contract for Sediment Ponds AW2005/89

The body of the letter should be clear, concise and courteous. The way you organise the body of your letter will depend on the reason for writing it but many of the other features will be common to all formal business letters.

Yours sincerely,

W Smith

William Smith

The company letterhead will generally give contact details of the company. The date is important both for sequencing of communications and filing but it also can have legal ramifications in situations where the time advice given or received is crucial to subsequent actions. In a business letter, the date is located under the company letterhead or the address. Formality generally applies to the details of the person to whom the letter is written, including his or her title and company details. As an aid to understanding it is helpful to state clearly what the contents of the letter are in regard to (RE:). Once again, this is to make the receiver's task as easy as possible. A business letter should be structured; that is, it should have a beginning, a middle and an end. In the main body of the letter the main information is presented. A key aim is to ensure that the letter is not misunderstood and for this reason it is essential that the language is clear and simple to avoid misunderstandings. There are no awards for showing a large vocabulary in a business letter. The most important thing is to ensure that you communicate your message as effectively as possible.

The final words should tell the person what actions are required or perhaps give thanks for any help given. It is customary to sign off; "Yours faithfully" in a formal business letter and when the name of the person is unknown; and 'Yours sincerely' if the person is known personally. Nowadays it is acceptable to use less formal closings such as "Kind regards" and "Best regards" if there is less formality in your letter. The closing signature can be located on either side of the letter. Most companies have a template for writing business letters in their own style and therefore this should be used.

Report Writing

Writing reports takes more time and is harder work than most people think. A poorly presented report, where the author is unaware or uninterested in the most basic rules, will be read with this in mind and will most likely lead to any information or recommendations being discounted or totally ignored. The engineering report conforms to quite a standard format. The aim is to ensure that the reader, no matter what their interest in the work, will be able to get the appropriate level of detail which they require: that level of detail is determined by the reader, and not the writer. For this reason it is quite common for the same material to appear a number of times in the same report so that it can be found in an efficient manner. This will be discussed further in the coming sections.

Main Components of a Technical Report

The checklist for formal reports shows the structure that should be developed for formal reports.

The title page should include the title of the report, the names and affiliations of the authors, the date (month and year) of publication and perhaps some company identifiers.

The abstract or executive summary should take only a few pages at most and should summarise the whole report in a highly condensed form. It should allow anyone who just wants to know what the report is about to find what work or research was undertaken and what conclusions were reached, without having to look at the body of the report at all. It should not read as a teaser to the main report and generally does not include references, or extended descriptions.

Most word processors will generate tables of contents, lists of figures and lists of tables automatically and this feature should be used to develop these pages where possible. The information required is taken from the actual report so any changes can be reflected in an updated table of contents at the touch of a button.

Checklist: A formal report

► Title Page includes title, authors and date

► Executive Summary includes:
reason for study;
summary of what work was carried out ; and
summary of key findings and/or recommendations of study.

► Table of contents (may be automatically generated)

► Introduction
starts on page 1;
sets out reason for report / study; and
refers to copy of design brief (if available) in Appendix.

► Other Chapters
develop an argument that leads logically to the conclusion;
lead to a solution or outcome;
have sections and sub-sections that are numbered consistently; and
ensure all 'facts' from literature are referenced.

► Illustrations and Tables
ensure all have labels and captions in the proper place (below or above);
captions allow the figure or table to be understood independently of text;
all appear after first reference in text; and
ensure all figures are referenced in text.

► References
listed in alphabetical order in reference section; and
listed in a standard format (e.g. Harvard system).

► Appendices should only contain information that distracts from the flow of the argument in the main report and the reader of the report is not likely to want to see. This might include extended mathematical derivations, long data sets, the full design brief, or other similar material.

► Miscellaneous
spell check entire document;
write in the third person i.e. "This was carried out" rather than "we did this";
search for and eliminate 1st person text (I, me, my, we, us, our);
read the whole report before submission;
one sentence paragraphs should be avoided;
single paragraph sections or subsections should be avoided;
avoid vague terms such as 'very large', 'really expensive'; and
avoid use of capital letters in middle of sentences unless for proper names.

Most reports start with a chapter entitled "Introduction". It contains a brief description of the need for the report, gives general background information and may even give a brief summary of the contents. A report should be written in the form of an argument that takes the reader from the problem to a solution and a clear introduction is necessary to show how this argument will be made.

The body of the report may contain a number of different chapters that set out in logical order (rather than the order in which the work may have been carried out) the argument the author or authors are making.

The final chapters of a report tend to be a summary and set of conclusions and recommendations. In this chapter or chapters the authors summarise and integrate any conclusions that have been made in report. Recommendations may be to do with how information should be used or what further work needs to be carried out. Following the report, there will generally be a need to list the sources of information which were referred to in the report. There are a number of standard ways of presenting this information. Most will contain, as a bare minimum, the author(s), the date of publication, the title of the work, the journal or report or book or conference proceedings that it appeared in and/or the publisher. Also given are page numbers if the work was in a journal, or the total number of pages if the work is a book. The general rules also apply to web references. Here one additional piece of information is generally required: the download date. This is because web references change quickly and it is useful to know when information was sourced.

The last element of a report may be a single appendix or a series of appendices. Information and results which are useful background material but are not needed for understanding the main arguments in the report, should be included in an appendix. Bulky test results or mathematical derivations can also be included but note that not all test results need to be included. Generally within an engineering report there will be many figures and drawings to enhance the communication and to enable the reader to fully grasp the subject material. In design, the main form of communication is the engineering drawing.

Drawings

Much engineering communication is completed through drawings. The ideas evolve in sketches and doing calculations to formalise a design. The design of a system, whether it is an engine, a building, a water treatment plant or a space shuttle will be communicated through the various phases of a project from preliminary investigation to the construction phase of a project, as described in Chapter 3, by a very large number of drawings. With computer aided drafting techniques, many different forms of drawings have evolved, including 3D details, animations and traditional plan and cross sectional views. An important part of any drawing is the Title box because this has boxes for various signatories such as project manager, designer, drawer and checker. Signing off on these indicates the people responsible for the design.

7.7 COMMUNICATION IN GROUPS

The success or failure of large engineering projects inevitably depends on communication within and between groups. Engineers must therefore know the basics of communicating within groups and facilitating groups and group meetings.

Communication is an integral part of developing synergy in a group to develop trust and cooperation. Covey (1989) describes the synergy that comes from good communication as an understanding that the whole is greater than the sum of the parts. It comes from respecting differences between individuals and relies on building on strengths to compensate for weaknesses. Most importantly it requires both trust and cooperation, as seen in Figure 7.3. Note that if there is high trust and high cooperation then a win/win situation can result.

Many of the activities for good group performance were outlined in Chapter 6 with the leader of a group having to be aware of the styles of the members of the group and the need to ensure participation from all. Simple processes for conducting meetings were also outlined to ensure effective meetings.

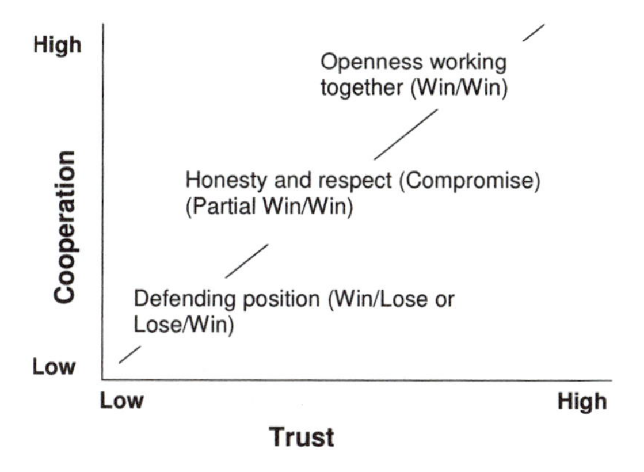

Figure 7.3 Levels of Communication (adapted from Covey, 1989)

7.8 PLAGIARISM

Plagiarism is the unacknowledged use of the material written by others with or without the intention of passing it off as your own work. With the development of electronic information systems, plagiarism is on the increase, as it is becoming much easier to source material from a range of different areas.

Plagiarism is wrong and dangerous for a number of reasons. If you have read something it is actually in your interest to cite the source and to quote directly if it is something that is going to help your argument. That is what is referred to as scholarship. Plagiarised material is often quite easy to detect, either by simply reading the material and reflecting on the changes in language, or by using one of the many plagiarism detectors that are available nowadays that can check submissions against a vast range of publicly available material from around the world. It is essential, therefore, to reference all the sources of information which are used in any report or design. When using material, ensure that the source is credited somewhere in the material, e.g., according to Smithers and Smith, 1999. When exact phrases from the original source are used, be sure to use quotation marks to set off any exact text. Paraphrasing can also be deemed to be plagiarism when the same ideas are used as in the original source, so make sure all ideas are referenced.

Contrasting Styles of Communication

Bennis (1997), the author of 26 books on leadership, has a favourite example of two contrasting styles of communication, those of the British Prime Ministers Gladstone and Disraeli.

It was said of Gladstone, a 19th century Liberal Prime Minister, that if you had dinner with him, you came away believing that he was the world's brightest, wittiest, most charming man. However, if you had dinner with Disraeli, a peer who became a Conservative Prime Minister, you came away believing that you were the brightest, wittiest, most charming person in the country.

7.9 SUMMARY

Good communication skills are required by all engineers, whether in managerial positions or technical specialist areas. Communication involves more than simply sending a message; it necessitates that the correct message is received by using feedback and active listening. Managing personnel in project teams also requires an understanding of how the individuals within the team can facilitate or block communication between members. In all communications, an awareness of who is receiving the message is essential to ensure a consistent style, flow and clarity of detail. There is no easy way to ensure an oral presentation or a written report is well received other than by putting a lot of effort into the structure and flow of the material, enquiring and that it is relevant to the topic being discussed, and looking for feedback.

In larger organisations the manager of groups must be aware that there are some people who listen and others who do not. A strategy that assumes everyone understands what has been said can lead to disaster, therefore asking listeners and readers if clarification of details are needed can avoid miscommunication. A good rule is to attempt to keep communication channels open and be receptive to communication at all levels. It is imperative that other people's body language is observed when involved in discussions and when there are conflicts. It is important that as a manager and leader of engineering teams you have congruent communications, body language and actions, thus imparting a feeling of trust amongst the team. This example to fellow team members will then be the catalyst to open communication within the group and across the boundaries of other groups within the organisation.

Use simple language to convey your message in all reports, memoranda and letters. Remember to read and check the spelling of all outgoing correspondence. Always reference all sources of ideas and information.

PROBLEMS

7.1 What processes are important in giving a good presentation. What checks of the room where you are going to present would you make?

7.2 Give guidelines for reading body language by the following visual clues:

- (a) facial expression;
- (b) posture;
- (c) gestures;
- (d) clothing, grooming and environment.

7.3 Give guidelines for reading body language by the following auditory clues:

- (a) Specific words that are spoken;
- (b) Sound of the voice;
- (c) Rapidity of speech, frequency and length of pauses.

7.4 Every one will benefit from improving listening skills. Search the internet to find information on becoming a better listener. Select one or two areas that you want to improve on and try those over the next 48 hours. Write a one paragraph report on why you think you have improved or not.

7.5 Write an essay of no more than 1000 words explaining the importance of good communication skills for engineers.

REFERENCES

Bennis, W., 1997, *Organizing Genius, The Secrets of Creative Collaboration,* Transcript of Interview, March 26, http://www.pbs.org/newshour/gergen/march97/ bennis_3-26.html, Retrieved Feb 2006.

Buzan, T., 1988, *Make the Most of Your Mind.* Pan.

Checkland P., 1981, *Systems Thinking, Systems Practice*. John Wiley & Sons, Chichester.

Covey, S.R., 1989, *The Seven Habits of Highly Effective People*, Simon and Schuster NY.

Eunson, B., 1994, *Communicating for Team Building*, Jacaranda Wiley Ltd.

Free Mind 2006, http://freemind.sourceforge.net/wiki/index.php/Main_Page, Retrieved Feb 2006.

Gates, B., 2006, The Road Ahead - How 'intelligent agents' and mind-mappers are taking our information democracy to the next stage. *Newsweek,* Jan. 25, 2006 http://msnbc.msn.com/id/11020787/site/newsweek/, Retrieved Mar 2006.

Haith-Cooper, M., 2003, An Exploration of Tutors' Experiences of Facilitating Problem-Based Learning. Part 1. An Educational Research Methodology Combining Innovation and Philosophical Tradition. *Nurse Education Today*, 23, 58–64.

Hamilton, C., Parker, C., 1990, *Communicating for Results : A Guide for Business and the Professions*, Wadsworth Publishing Company, Belmont, California.

Horan, P., 2002, A new and flexible graphic organizer for IS learning: The rich picture, in E and A Zaliwska (eds), *IS 2002 Proceedings of Informing Science and IT Education Conference,* Informing Science: Informing Science Institute: http://ecommerce.lebow.drexel.edu/eli/2002Proceedings/2002.

Hunt, J., 1979, *Managing People at Work - A Manager's Guide to Behaviour in Organizations*, McGraw Hill UK.

James, D.L., 1995, *The Executive Guide to Asia-Pacific Communications: Doing Business Across the Pacific*, Kodansha Press, New York.

Laszlo, E., 2001, Human Evolution in the Third Millenium. *Futures*, 33, 649-658.

Manage Effective Workplace Communication, 2004, http://ntcci.harvestroad.com.au/wppuser/ntcci/training/courses2004/work_communication.html

McShane S. and Travaglione T., 2003, *Organisational Behaviour on the Pacific Rim*, McGraw Hill. www.mhhe.com/au/mcshane

Messina, James, J., 1999, *Tools for Communication: A Model of Effective Communication,* http://www.coping.org/communi/model.htm, Retrieved Aug 2005.

Myers, D.G., 2004, *Psychology*, 7th ed., Worth Publishers, Inc.

Pease, B. and Pease, A., 1999, Why men don't listen & women can't read maps : how we're different and what to do about it, Pease Training International, Mona Vale, NSW, Australia.

Rogers, C. and Farson, R., 1979, Active Listening, in Kolb, D., Rubin, I., and MacIntyre, J, *Organizational Psychology* (3rd ed.), New Jersey: Prentice Hall.

Samson, D. and Daft, R., 2003, *Management,* Pacific Rim Edition, Thomson.

Shannon, C.E. and Weaver, W., 1949, *The Mathematical Theory of Communication.* The University of Illinois Press, Urbana, 91pp.

Steinmetz, Lawrence L., 1979, *Human Relations: People and Work*, Harper and Row.

Tidwell C., 2005, *Intercultural Business Relations Course*, Class Notes 18, Non-verbal Communication modes, http://www2.andrews.edu/~tidwell/bsad560/NonVerbal.html, Retrieved Aug 2005.

Underwood, M., 2003, *Communication, Cultural and Media Studies Info-base*, http://www.cultsock.ndirect.co.uk/MUHome/cshtml/index.html, Retrieved 30 August 2005.

CHAPTER EIGHT

Economic Evaluation

This chapter deals with the economic evaluation of engineering projects. The purpose of economic evaluation is to identify the benefits and costs of a project and hence to determine whether it is justified on economic grounds. Alternative solutions to an engineering problem can also be compared based on their respective costs and benefits. It is usually necessary to discount the future benefits and costs of a project in order to make comparisons in terms of present value. Various economic criteria have been proposed for the comparison of projects. Each of these has its merits, although the best on theoretical grounds is net present value.

8.1 INTRODUCTION

An engineering project involves the transformation of limited resources into valuable final products or outputs. For example, the construction of a dam involves the use of resources such as concrete, steel, human effort, and machine time. The dam is used to produce outputs such as water for domestic and industrial purposes, flood control, and recreation.

The engineer must be concerned not only with the technical feasibility of the project (i.e. will it work?) but also the economic feasibility. Economic evaluation is aimed at assessing whether the value of the final products exceeds the value of the resources used by the project. The values of the outputs when measured in economic terms are called the benefits of the project. Similarly the values of the resources used in its construction and maintenance when measured in economic terms are called the costs of the project.

In the private sector, benefits and costs are measured by cash flows into and out of the firm, respectively. That is, a conceptual boundary is drawn around the firm and benefits, and costs are represented by cash flows across this boundary.

For public sector projects, benefits and costs must be considered for society as a whole. In this case, benefits and costs are not necessarily associated with cash flows. For example, the benefits of a public transport system are not necessarily measured by the revenue which it generates. The government may choose to run the system at a loss in order to promote a more sustainable future. In this case the benefits to the users of the system would probably exceed the revenue generated by it. If, for example, public transport were offered free of charge, few people would argue that the benefits to society were zero.

Sometimes there is a different perception of benefits and costs at various levels of government. For example, the federal government may provide a subsidy of 50% of the capital costs of new sewage treatment plants. When a state government is evaluating a particular plant it would consider this subsidy as a reduction in the cost of the project. On the other hand, the federal government would consider the subsidy as merely a transfer payment from one branch of government to another and therefore not a true benefit or cost to society as a whole. In fact, the federal government may require the state to provide economic justification for the project in its (i.e. the federal government's) terms without including the subsidy effect.

8.2 THE TIME VALUE OF MONEY

The costs and benefits (or revenue) of an engineering project usually occur over a long time period. For example, consider the construction of a new freeway. A typical time stream of benefits and costs is illustrated in Figure 8.1. Costs will be high during the construction phase which may last for three or four years. Annual maintenance and repair costs will be low initially, but will increase due to ageing of the pavement, bridges, and other components. The benefits to road users will be primarily due to savings in vehicle operating costs, savings in travel time and a reduction in the number of accidents in relation to the pre-existing road network. These benefits would normally increase with time due to increasing volumes of traffic using the freeway.

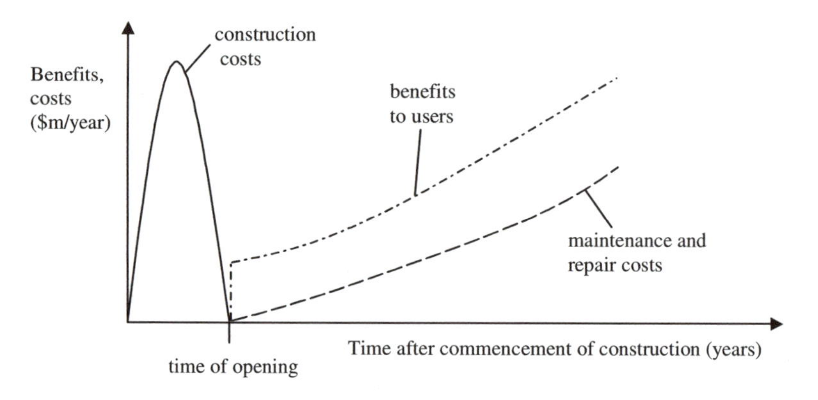

Figure 8.1 The time stream of benefits and costs for a freeway.

In carrying out an economic evaluation of such a project it must be recognized that benefits or costs incurred in ten years' time cannot be directly compared with those incurred in the current year. Given the choice between $1000 now and $1000 in 10 years' time, very few people would choose the latter. In the first place, the effects of inflation are such that fewer goods and services could be purchased with the future sum than at present. However, even in the absence of inflation, human nature is such that there is a preference to consume goods now rather than later, and to postpone costs if possible. This is clearly evidenced by the fact that many individuals are willing to borrow money at an interest rate which exceeds the rate of inflation. Therefore in carrying out economic evaluation it is necessary to discount future benefits and costs in order to make them directly comparable with benefits and costs incurred now. This is carried out using discounting formulae derived from considerations of compound interest.

8.3 DISCOUNTING FORMULAE

In all the economic calculations that follow there are two basic assumptions:

- there is a single rate of interest (discount rate) that applies into the future to both borrowing and lending;
- interest is paid in a compound fashion; that is, if interest is earned then it is added to the capital and taken into account for future calculations of interest.

If one looks at interest rates and how they vary in time (see Figure 8.2) it may be thought that the first assumption is questionable. However, despite the fact that rates do vary, it is usual for calculations to be carried out assuming an unvarying rate. If the rate does change at a later time, adjustments are made, again on the assumption of a constant rate into the future.

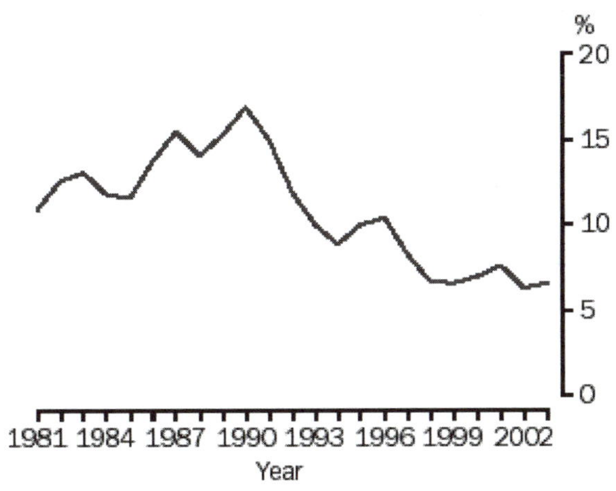

Figure 8.2 Interest rates for housing loans in Australia. Source: Australian Bureau of Statistics (2006)

Present Worth - Lump Sum

If a sum of money P is invested for one year at an interest rate of i, then after one year the value will be:

$$F = P(1+i) \tag{8.1}$$

where F is the future sum and P the present sum. For example, \$100 invested at a rate of 5% (i = 0.05) for one year would yield \$105 at the end of the year. If that money is invested for a second year then the value becomes:

$$F = P(1+i)(1+i) = P(1+i)^2 \tag{8.2}$$

After two years the initial \$100 yields \$110.25. After a period of *n* years it can be shown that the basic formula to calculate the future worth of a sum of money can be written:

$$F = P(1+i)^n \tag{8.3}$$

where *F* is the future sum, *P* the present sum, *i* the interest rate and *n* the number of years (or to be more precise, the number of time periods over which the interest is paid). Some results for a simple situation are shown in Table 8.1.

Table 8.1 Future sums after a variable number of years with an interest rate of *i* and an initial sum of \$1000.

years	Future return from \$1,000 invested for *n* years at various interest rates			
	2%	5%	8%	10%
1	1020	1050	1080	1100
2	1040	1103	1166	1210
5	1104	1276	1469	1611
10	1219	1629	2159	2594
100	7,245	131,501	2,199,761	13,780,612

It is convenient to show the process on a timeline (Figure 8.3) where the horizontal axis represents time and the vertical axis the units of money. While this appears trivial for the present example, once the situation becomes more complicated, the benefit of the time plot will become more apparent.

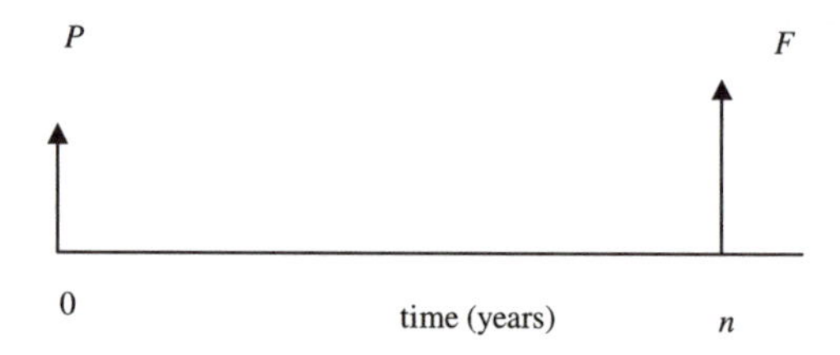

Figure 8.3 Timeline showing future value based on present value.

By rearranging Equation (8.3) it is possible to see how much must be deposited today to yield a certain sum of money after a given period.

$$P = \frac{F}{(1+i)^n} \tag{8.4}$$

where P is the present value of a lump sum of F, to be paid in n years time.

Worked Example

What is the present worth of $1000 that will be received in 10 years' time assuming a 10% compound interest rate ?

Solution The timeline for the problem is shown in Figure 8.4. The present sum can be calculated from:

$$P = \frac{1000}{(1+0.10)^{10}} = \$385.54$$

The important point to grasp is that under the interest rate specified, $385.54 today and $1,000 in 10 years time have equal worth. Alternatively, if $385.54 were placed in a bank account paying 10% p.a. interest, the sum would accumulate to $1,000 in 10 years time.

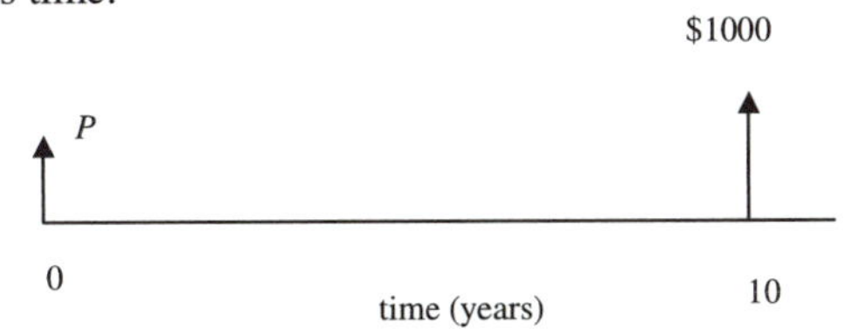

Figure 8.4 Timeline for the worked example.

Doubling your money at 20%

A problem that was written on a stone tablet dating from 1700 BC asked: how long will it take for a sum of money to double if invested at 20% interest rate compounded annually? If the future sum is double the present sum then it is possible to write:

$F/P = (1 + i)^n = 2$

With i = 0.20 (20%) the equation can be solved by taking logarithms of both sides:

$n \log(1.2) = \log(2)$ so that $n = \log(2)/\log(1.2) = 3.80$

Hence the solution is 3.8 years to double your money at 20% interest rate, compounded annually.

Present Worth - Uniform Series

It is often necessary to find the present worth of a uniform series of annual payments. These may be regular mortgage repayments for a home owner or the cost of leasing a particular piece of equipment for a manufacturer. The timeline for this is shown in Figure 8.5 where $\$A$ are paid per year up to n years and these are to be made equivalent to a single sum $\$P$ at the present time.

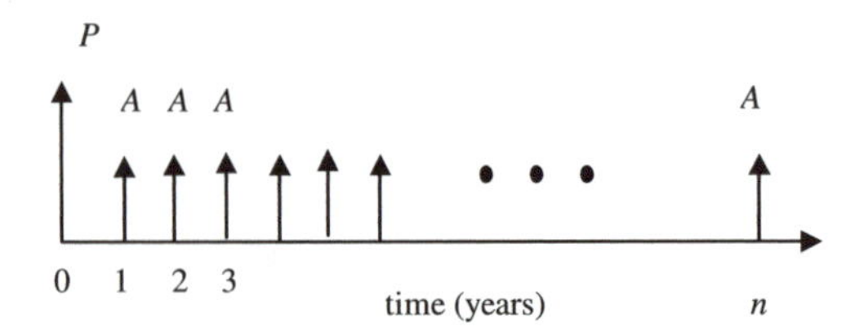

Figure 8.5 Timeline showing the equivalence between a present value, P, and a series of n annual payments of A.

Note that the first payment of $\$A$ is at the end of the first year. It is possible to discount the annual payments one by one. In this case:

$$P = \frac{A}{(1+i)} + \frac{A}{(1+i)^2} + \ldots \frac{A}{(1+i)^n} \tag{8.5}$$

However, by treating this as the sum of a geometric series, it is possible to write this in a more convenient way:

$$P = A\left[\frac{1-(1+i)^{-n}}{i}\right] \tag{8.6}$$

It is also possible to transpose the equation to determine what annual series is equivalent to have a present value of $\$P$:

$$A = \left[P\frac{i}{1-(1+i)^{-n}}\right] \tag{8.7}$$

where P is the present worth of a uniform series of payments of $\$A$ over a period of n years.

Worked Example
A project is expected to yield \$3m a year in benefits for 30 years. What is the present value of the benefits assuming a discount rate of 8% ?
Solution Applying Equation (8.6):

$$P = 3\left[\frac{1-(1+0.08)^{-30}}{0.08}\right] = \$33.77 \text{ m}$$

Loan Repayments and Outstanding Balances

When a loan is taken out, it is normal to repay the total debt over a number of years with repayments that are maintained at the same level over the whole period. This then raises the problem of knowing how much is left to repay at any particular time. For example, in the case of a loan of \$100,000 taken out at a 7.5% annual interest rate over a 25 year period, how much is still owed at the end of 1 year, or 5 years or 24 years? In the following calculations it is assumed that repayments are made annually. The same procedure may be applied to monthly or weekly repayments. The annual repayment on the loan can be calculated from Equation (8.7):

$$A = 100{,}000\left[\frac{0.075}{1-(1+0.075)^{-25}}\right] = \$8971.07$$

To determine the outstanding debt after x years, it is necessary to consider the present value of the annual repayments that remain to be paid over a period of $(n\text{-}x)$ years. Therefore, the outstanding debt after 1 year is a series of annual payments over 24 years. In this case:

$$P = 8971.07\frac{1-(1+0.075)^{-24}}{0.075} = \$98{,}528.96$$

Therefore, after a single annual repayment of \$8971.07 there is still \$98528.96 left to repay, a debt reduction of only \$1,471.04. The other \$7,500.34 has gone in interest repayments. As time goes by, the total repayment remains the same, but the sum being paid as interest gradually reduces while that going to reduce the outstanding debt gradually increases. This is shown in Figure 8.6 for the current example.

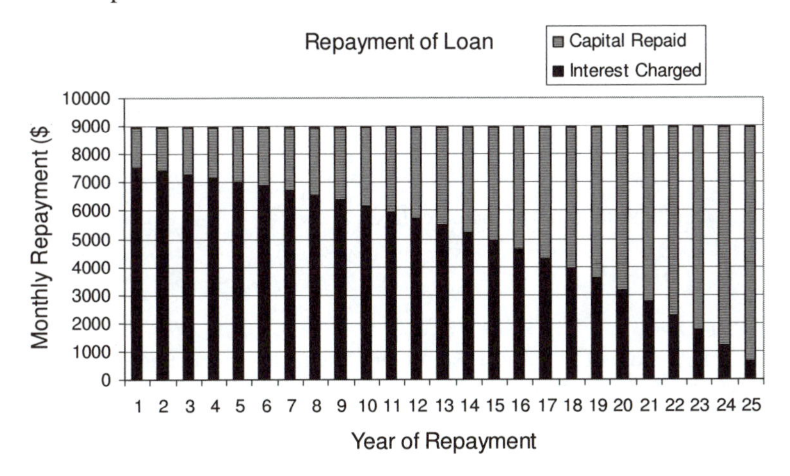

Figure 8.6 The split between interest (solid) and capital repayment (hatched) of each annual payment is shown. Note how the interest component gradually reduces over time.

To illustrate the effect that more frequent payments has on the calculations the same problem is repeated for monthly repayments for the same annual interest rate. An annual interest rate of 7.5% equates to 0.625% per month and the monthly repayment (*M*) can be calculated as:

$$M = 100{,}000 \frac{0.00625}{1-(1+0.00625)^{-300}} = \$738.99$$

Over one year the total annual repayment is calculated from 12 monthly payments and is equal to \$8867.89 (a slight increase on the previous figure). After one year there will be 288 repayments left to make leaving a total of \$98,584.09 still owed (again, slightly more than for the annual repayments).

The subject of different repayment intervals will be dealt more formally in the next section where it will be seen that it is possible to determine a relationship between the same interest rate charged annually or monthly.

Effective Interest Rate

Interest rates are generally quoted as a percentage per year, but the compounding period (the frequency with which the interest is charged or paid) is often less than a year. As we have just seen, this makes a difference to the calculations, and in order to allow a valid comparison the concepts of a nominal interest rate and an effective interest rate have been developed. If the nominal interest rate is r per annum and it is charged in m periods over a year then the interest rate applied at each period is (r/m). For example, an annual rate of 18% charged monthly means that the interest charged each month is 1.5%. Therefore if \$$P$ is invested at the start of the year the balance at the end of m periods can be calculated as:

$$F = P(1 + r/m)^m \tag{8.8}$$

If this monthly rate is to be quoted as an equivalent annual rate then it must produce the same future sum at that equivalent annual rate:

$$F = P(1 + r/m)^m = P(1+i) \tag{8.9}$$

so the equivalent annual rate can be calculated as:

$$i = (1 + r/m)^m - 1 \tag{8.10}$$

Worked Example
A credit agency claims to charge 18% p.a. but interest is calculated monthly. What is the effective rate of interest ?
Solution By applying Equation (8.10) it is possible to determine the effective annual rate of interest. Hence, 18% per year charged monthly is equivalent to 19.56% per year charged annually.

$$i = (1 + r/m)^m - 1 = (1 + 0.18/12)^{12} - 1 = 0.1956$$

Note that in a number of countries, all financial institutions must publish their interest rates as an equivalent annual rate so that consumers can make informed choices.

Interest and the Natural Logarithm

If the interest rate is 100% and $1 is invested for 1 year then the way interest is charged (annually, monthly, etc.) can be seen to change the outcome considerably. If charged annually the amount invested at the end of the year will be $2. If it is paid six monthly the amount will be $2.25. These, and other possibilities, climaxing in continuous payment of interest are:

Interest is paid ...	Final Sum
Annually	$2.00
Six Monthly	$2.25
Quarterly	$2.44
Monthly	$2.61
Weekly	$2.69
Daily	$2.714567
Hourly	$2.718127
Continuously	$2.71828182845 (*e*, the base of the natural logarithm)

The effects of inflation on economic evaluation

Inflation is represented by a rising general level of prices. It is important to note that during an inflationary period not all prices are necessarily rising nor are all prices necessarily rising at the same rate. The effect of inflation is that a dollar next year will buy less than a dollar today. There have also been periods in history when the general level of prices in some countries dropped. This is called deflation.

Changes in the general level of prices may be measured by a price index. This is defined as 'the average price of a mixture of goods and services in a given year' divided by 'the average price of the same mixture of goods and services in a base year' (expressed as a percentage). A common price index is the consumer price index or CPI. The CPI measures changes in the prices of goods and services purchased by moderate-income families. It is based on the prices of a large number of representative items each weighted according to its relative importance in the typical family budget. Movements in the consumer price indexes for a number of countries are shown in Table 8.2.

Of particular relevance to engineering projects are the price indexes for construction and building costs. These indexes for a number of North American cities are published in the quarterly journal Engineering News Record. In 1986, the average construction cost index for 20 cities in the United States was 300.8, relative to a value of 100.0 in 1970. When this is compared with a 1986 CPI value of 283.8 for the USA (relative to the same base year) it is apparent that construction costs rose faster than consumer prices over those 16 years. In Australia the Australian Bureau of Statistics publishes price indexes for materials used in house building and other building in its Monthly Summary of Statistics.

It is most important to distinguish clearly between the concepts of inflation and the time value of money. Even in the absence of inflation, very few people would be prepared to lend money at zero interest. As most people would prefer to consume goods now rather than later, there is clearly a real time value of money. In inflationary times, lenders expect a real rate of return on their money which exceeds the inflation rate. For example, anyone who receives 8% p.a. interest on an investment when the inflation rate is 10% p.a. is clearly losing money. Before we consider how inflation is taken into account in economic evaluation some definitions are required.

Table 8.2 Consumer price index for a selection of world countries. Base year (100.0) = 1989–90. Source: Australian Bureau of Statistics (2006)

Country	1994	1995	1996	1997	1998	1999	2000
Australia	116.5	121.1	123.9	125.4	126.9	129.4	136.4
NZ	110.5	111.9	113.7	114.9	116.9	118.7	123.5
HK	151.4	160.3	167.6	173.9	172.0	166.6	164.8
Indonesia	150.3	163.7	174.1	232.7	368.3	367.1	402.6
Japan	107.8	107.3	108.2	112.4	112.4	111.6	111.0
Korea	138.0	144.4	151.3	162.1	169.0	172.1	179.2
Singapore	114.5	116.0	118.1	119.4	118.5	120.7	122.9
Taiwan	119.1	122.5	125.7	127.2	128.2	129.3	130.9
Canada	113.4	116.0	118.8	120.6	122.0	125.0	128.1
USA	118.0	120.9	124.3	125.8	127.2	130.9	135.3
Germany	115.8	117.0	118.2	120.3	120.7	121.8	124.2
UK	124.8	128.3	131.5	134.6	137.2	139.3	141.4

Actual dollars (or then-current dollars) are dollars which are current in the particular years that the benefits and costs occur. For example, if a man bought a block of land in 1950 for $5000 (in then-current dollars) and sold it in 1980 for $20000 (in then-current dollars) has he made a profit on the investment? The answer is yes if $20000 in 1980 will buy more goods and services than $5000 in 1950. Otherwise, he has lost on the investment.

Constant-worth dollars are dollars which have the same purchasing power at a defined point in time e.g. 1988 dollars. For example, a new power station is designed to provide 8000 GWh of electrical energy per year under certain operating conditions. The benefits to consumers of this energy are estimated to be $600m in 1990 (the first year of operation). Thereafter the station will continue to provide 8000 GWh of energy per year for its estimated operating life of 40 years. The annual benefits expressed in actual dollars will continue to increase throughout the life of the station due to inflationary increases in the price of electricity and the prices of goods produced with it. However, when expressed in constant-worth dollars (e.g. 1990 dollars) the annual benefits are likely to be constant over the life of the station as the same amount of energy is produced each year.

The nominal interest rate is the rate received on invested money (or paid on borrowed money) when calculated in terms of actual (or inflated) dollars. The real rate of interest on an investment is the rate received net of inflation. For example, if money is invested at a nominal interest rate of 18% p.a. when the annual inflation rate is 10%, the real interest rate (i.e. above inflation) is approximately 8% p.a.

The effect of inflation on the economic evaluation of engineering projects can be taken into account in one of two equivalent ways:

- by using actual dollars and the nominal interest rate throughout;
- by using constant-worth dollars and the real rate of interest throughout.

The first method is commonly used in evaluating private sector investments in which the future cash flows are estimated in actual dollars and the nominal interest rate is usually known. The second method is more commonly used for evaluating public sector investments, as it is often easier to estimate future benefits and costs in constant-worth dollars. In this way there is no need to estimate an inflation rate. The discount rate used is the real rate above inflation. As will be discussed in Section 8.10, the choice of the discount rate in the public sector is the source of some controversy. In this book, constant-worth dollars and real interest or discount rates will be used unless otherwise stated.

Relationship between nominal and real interest rates

Consider the annual maintenance costs of a bridge which requires a constant number of person-hours of maintenance each year during its life of n years. (This is a gross simplification, as it would be normal for the amount of maintenance to increase as the bridge deteriorates with age.) When expressed in constant-worth dollars the annual maintenance cost of the bridge is a constant amount (say $\$A$) per year for the life of the bridge, as shown in Figure 8.7.

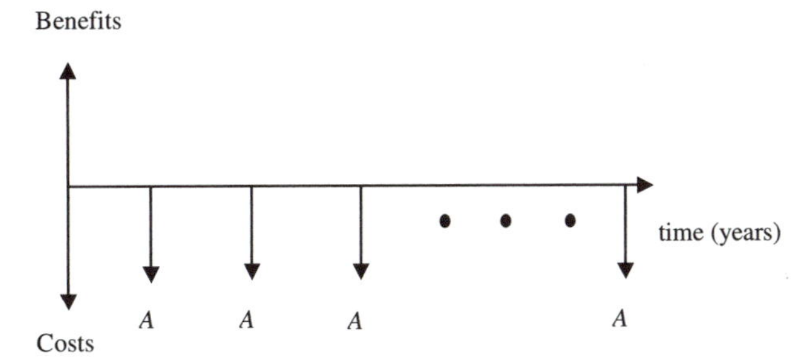

Figure 8.7 Bridge maintenance costs in constant-worth dollars.

If the anticipated inflation rate for the life of the project is f % p.a. (compound), then the maintenance costs in actual dollars are shown in Figure 8.8. The nominal rate of interest is assumed to be known and is denoted by i_n. The objective is to find a relationship between the real interest rate, i, the nominal interest rate, i_n, and the inflation rate, f. The present worth of the series shown in Figure 8.8 is denoted by P. It can be determined by applying Equation (8.4) to each annual payment using actual dollars and the nominal interest rate:

$$P = \frac{A(1+f)}{(1+i_n)} + \frac{A(1+f)^2}{(1+i_n)^2} + \cdots \frac{A(1+f)^n}{(1+i_n)^n} \tag{8.11}$$

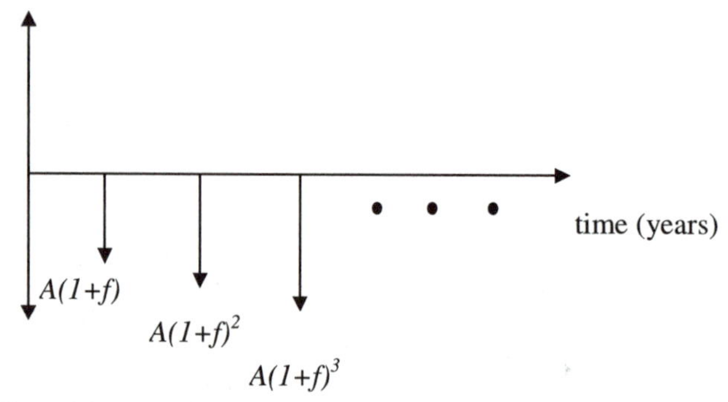

Figure 8.8 Bridge maintenance costs in actual dollars (f = annual inflation rate).

The right-hand side of Equation (8.11) can be simplified, as it is the sum of a geometric series:

$$P = \frac{A(1+f)}{(1+i_n)}\left\{\frac{1-(1+f)^n/(1+i_n)^n}{1-(1+f)/(1+i_n)}\right\} \tag{8.12}$$

Now, the present worth of the series shown in Figure 8.7 is found using constant-worth dollars and the real interest rate, i. Rewriting Equation (8.5):

$$P = \frac{A}{(1+i)} + \frac{A}{(1+i)^2} + \cdots \frac{A}{(1+i)^n} \tag{8.13}$$

It should be apparent that the two equations (Equations (8.11) and (8.13)) are equivalent if:

$$(1+i) = \frac{(1+i_n)}{(1+f)} \tag{8.14}$$

Therefore, the real rate can be calculated as:

$$i = \frac{(1+i_n)}{(1+f)} - 1 \tag{8.15}$$

Worked Example
In Kenya in 1998 the nominal rate of interest for borrowing was 21% and the annual inflation rate 15% (Adeoti et al., 2000). What was the real rate of interest under these conditions?
Solution By applying Equation (8.15) it is possible to show:

$$i = \frac{(1+i_n)}{(1+f)} - 1 = \frac{1+0.21}{1+0.15} - 1 = 0.052 \text{ or } 5.2\%$$

8.4 EVALUATION CRITERIA

Various economic criteria have been proposed for use in comparing engineering projects. The following criteria are described briefly and compared:

- payback period;
- net present value;
- equivalent annual worth;
- benefit–cost ratio; and
- internal rate of return.

Payback Period

The payback period is defined as the time it takes for the project to generate sufficient net benefits to cover its initial cost of construction and implementation. For example, if a project has an initial capital cost of \$10m, annual benefits of \$3m and annual operating costs of \$1m the payback period can be calculated as:

Net annual benefits = \$3 m – \$1 m = \$2 m
Payback period = \$10 m / \$2 m = 5 years

Payback period case studies

Best (2000) quotes a payback period of 4 years for the lighting and heating of a building for Lockheed in the 1980s where natural light was used in place of electric lights, and where this also reduced the need for air conditioning. It was further stated that productivity gains due to the healthier environment in the building gave the company a competitive advantage in its bid for a large contract, the profit from which would have paid for the entire building.

Morgan and Elliot (2002) used payback period to justify the implementation of a mechanical mixing system for a water supply reservoir. The motors used only 8.5% of the power required for a traditional aerator plant and had an estimated payback period of 4 years based on direct savings in energy costs.

The price of solar cells has been reported (The Australian - 8/8/01) as falling to 20% of that of 25 years ago. Rooftop systems that can meet half a home's electricity needs for more than 20 years now cost as little as US\$10,000. This was quoted as having a five to six year payback period in California compared with 20 years a few years ago.

Net Present Value

The net present value (NPV) project is defined as the present value of all benefits minus the present value of all costs. It can be written as:

$$NPV = PVB - PVC \tag{8.16}$$

where *PVB* is the Present Value of Benefits and *PVC* is the Present Value of Costs. By writing the formulae for PVB and PVC we can write:

$$NPV = \sum_{t=0}^{n} \frac{B_t}{(1+i)^t} - \sum_{t=0}^{n} \frac{C_t}{(1+i)^t} \tag{8.17}$$

where B_t is the benefit in year t, C_t is the cost in year t, i is the discount (or interest) rate and n is the life of the project (years). The calculation of NPV therefore brings all costs and benefits back to the present time to allow them to be compared at the same point in time. It is important to note that while Equation (8.17) is mathematically correct it is unlikely that it would be the way the actual calculation was carried out. For example, if the costs are a series of annual payments, Equation (8.6) would most likely be used to bring them back to the present.

Equivalent Annual Worth

The Net Present Value brings all benefits and costs to the present time and gives the answer as a single present day sum of money. There may be an argument for representing the project in terms of its net annual earnings over the period of the project. In this case the Equivalent Annual Worth may be more appropriate. The equivalent annual worth (EAW) is found by converting all project benefits into a series of constant annual benefits for n years, where n is the life of the project. Similarly all project costs are converted into a series of constant annual costs for n years. The equivalent annual worth is then equal to the equivalent annual benefit minus the equivalent annual cost. A reliable method of calculating EAW is to determine it from the NPV applying Equation (8.7) which gives:

$$EAW = NPV\left[\frac{i}{1-(1+i)^{-n}}\right] \tag{8.18}$$

Benefit-Cost Ratio

The benefit-cost ratio (B/C) is widely used for evaluating projects in the public sector. As normally defined it is the present value of all benefits divided by the present value of all costs:

$$B/C = \frac{PVB}{PVC} \tag{8.19}$$

where *PVB* is the Present Value of Benefits and *PVC* is the Present Value of Costs. It can be calculated as:

$$B/C = \frac{\sum_{t=0}^{n} \frac{B_t}{(1+i)^t}}{\sum_{t=0}^{n} \frac{C_t}{(1+i)^t}} \tag{8.20}$$

where the terms are as defined previously. Again, it is worth noting that while Equation (8.20) is mathematically correct it is unlikely that it would be the way the actual calculation was carried out. It is much more likely that the benefits and costs would be brought back to present worth using appropriate formulae.

Salt interception scheme

Murray Darling Association Inc. (2001) reviewed the cost of a salt interception scheme for Chowilla, near Renmark in South Australia. The scheme, which was designed to reduce the flow of saline groundwater into the river by drawing down the groundwater level, was estimated to reduce salinity in the river by 14 EC (electrical conductivity) units which was said to be worth $1.4 million per year. Using these figures the benefit cost ratio was 1.27 based on salinity benefits only. The article also estimated the benefits to recreation and the conservation of the environment. With both of these included the B/C ratio rose to 1.92. On that basis the project was deemed to be economically viable. The article did not give the discount rate or the project life but a rate of 6% and a life of 100 years gives the required B/C ratio. Other rates and lives give different values, highlighting the need for a proper choice of these values.

Internal Rate of Return

The economic criteria discussed so far (with the exception of the payback period) involve the use of a discount rate to convert all future benefits and costs to present value. The internal rate of return (also called yield) is a different concept in that it is the implied interest rate of the investment. The internal rate of return is the discount rate at which the present value of benefits just equals the present value of costs for the life of the project. The internal rate of return (IRR) of a project, r, can be determined by finding the rate at which the present value of the benefits equals the present value of the costs:

$$PVB = PVC \tag{8.21}$$

It can also be written as:

$$\sum_{t=0}^{n} \frac{B_t}{(1+r)^t} = \sum_{t=0}^{n} \frac{C_t}{(1+r)^t} \tag{8.22}$$

where the terms are as defined previously. In general this will involve the solution of a polynomial in order to determine r.

Internal rate of return case studies

In late 2000 General Motors announced an investment in Victoria in the form of a new engine plant. According to Alan Wood (Australian, 19/12/00) the Australian and Victorian governments offered to contribute $160 m in inducements. General Motors requested a further $25 m so that the project could achieve an IRR of 25%. According to Wood this was remarkably high and double the rate that the Australian Competition and Consumer Commission would normally use in its calculations.

Campus Review (2002) reported results from an OECD evaluation of the benefits of a university education. It was found that the private rate of return to the individual was about 11% and the return to society ranged between 6% and 15%. This was based on higher average earnings and lower risks of unemployment. It was suggested that the rates were higher than conventional investments and therefore money spent on an education was money well spent.

8.5 A COMPARISON OF THE EVALUATION CRITERIA

There are three basic types of decisions requiring economic evaluation. They are:

1. Go or no-go decisions, for example, whether or not to build a new airport;
2. the choice of a single project from a list of mutually exclusive projects, for example, the choice between a bridge and a tunnel for crossing a waterway; and
3. the choice of a number of feasible projects when the total funds available are limited, for example, how to allocate an annual capital works budget between feasible projects.

8.5.1 GO or NO-GO Decisions

In deciding whether or not an individual project is justified on economic grounds any one of the following criteria may be applied:

$$\text{NPV} > 0 \quad \text{or} \quad \text{EAW} > 0 \quad \text{or} \quad \text{B/C} > 1 \quad \text{or} \quad \text{IRR} > i$$

Only one of these needs to be looked at since if, for example, the present value of benefits exceeds the present value of costs then NPV will be positive, EAW will be positive, B/C will exceed 1 and the IRR will exceed the discount rate.

8.5.2 Choice from a List of Mutually Exclusive Options

Often economic decisions are used to choose between a number of different options where only one will be chosen. For example, what sort of car to buy, or which computer to purchase. In each case, one will be chosen in preference to the others. Rawlinsons (2004) gives an example of two air-conditioning systems that are identical in performance but differ in their costs. A summary of relevant information is given in Table 8.3. For this example, an equivalent annual benefit has been added to make the comparison more interesting. An interest rate of 12% and a project life of 30 years are used in the comparison. The question then is: which is better on economic grounds?

As a first step a timeline is drawn for the whole 30 year investment period (see Figure 8.9) for System X. Note that since the plant only lasts 10 years it will have to be replaced twice. In the absence of better information it is assumed that the original prices are the best estimate for the values to be used in subsequent purchases. Note: It is important to compare products over the same time period as different periods can lead to unreliable results.

Table 8.3 Costings for two air-conditioning systems. Source: adapted from Rawlinsons (2004). Note: salvage value is not available at end of project.

Characteristic	System X	System Y
Capital cost	\$115,600	\$158,800
Life of plant	10 years	15 years
Annual costs	\$37,800	\$28,200
Annual benefits	\$80,000	\$80,000
Salvage value	\$3,000	\$7,000

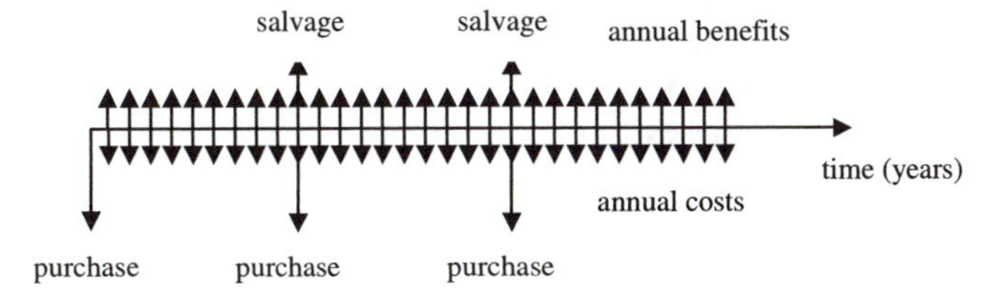

Figure 8.9 The timeline for System X showing the purchase and salvage of three units, each lasting 10 years together with the annual costs.

The Present Value of the Costs can be calculated as:

$$PVC = 115600 + \frac{115600}{1.12^{10}} + \frac{115600}{1.12^{20}} + 37800\frac{1-1.12^{-30}}{0.12} = \$469,289.94$$

The Present Value of the Benefits can be calculated as:

$$PVB = \frac{3000}{1.12^{10}} + \frac{3000}{1.12^{20}} + 80000\frac{1-1.12^{-30}}{0.12} = \$645,691.64$$

Therefore $NPV = PVB - PVC = \$176,401.70$

The Internal Rate of Return is calculated by solving for the rate, r, such that:

$$115600 + \frac{115600}{(1+r)^{10}} + \frac{115600}{(1+r)^{20}} + 37800\frac{1-(1+r)^{-30}}{r} =$$

$$\frac{3000}{(1+r)^{10}} + \frac{3000}{(1+r)^{20}} + 80000\frac{1-(1+r)^{-30}}{r}$$

The solution is $r = 0.347$, that is 34.7%.

The same calculations are now carried out for System Y over the whole 30 year investment period (see Figure 8.10). Since System Y has a longer life it will only have to be replaced once as is evident from Figure 8.10.

$$PVC = 158800 + \frac{158800}{1.12^{15}} + 28200\frac{1-1.12^{-30}}{0.12} = \$414,968.35$$

$$PVB = \frac{3000}{1.12^{15}} + 80000\frac{1-1.12^{-30}}{0.12} = \$644,962.81$$

$$NPV = \$229,994.45$$

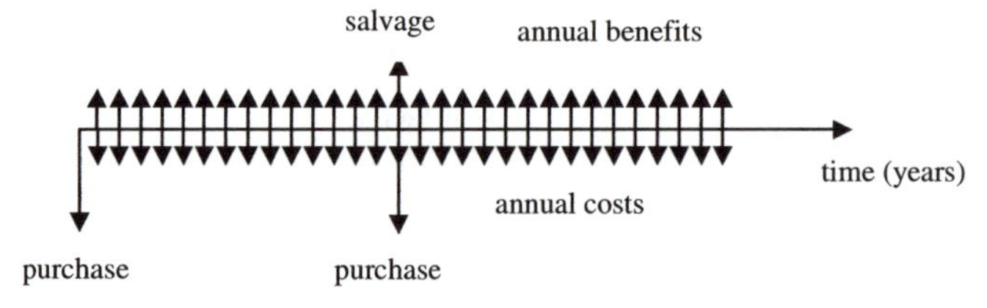

Figure 8.10 The timeline for System y showing the purchase of two units, each lasting 15 years together with the annual costs and salvage benefits.

The Internal Rate of Return is solved in a similar fashion to the previous value. The solution is r = 0.321 or 32.1%. Based on the calculations it is possible to list the various economic indicators as shown in Table 8.4 for the two systems.

Table 8.4 Comparison between System X and System Y

Economic Criteria	System X	System Y
NPV	\$176,401	**\$229,994**
B/C	1.38	**1.55**
IRR	**0.347**	0.321

It is evident that not all indicators agree on the better system, but, in general, it is recommended that the NPV criterion is used for this type of decision and System Y would be selected despite having an inferior Internal Rate of Return.

Least cost decisions

A variation on the above decision-making situation is when one design must be chosen from among a number of alternatives, each of which performs substantially the same function. For example, the choice between two designs for a road bridge, one of which has prestressed concrete girders and the other steel girders. In both cases, the girders act compositely with a reinforced concrete deck. The benefits of the bridge may be assumed to be the same for both designs and may therefore be ignored in the decision. In this case the NPV criterion is equivalent to finding the design which minimises the present value of costs (assuming that the present value of benefits exceeds the present value of costs for the chosen design). Alternatively the choice may be made by minimising the equivalent annual cost. The B/C and IRR criteria cannot be used in this situation.

8.5.3 Capital budgeting problems

It will be the case for many organisations that the cost of the full list of desirable projects exceeds the available funds and a choice has to be made on which projects will go ahead and which will not. In the previous section it was found that the different methods of economic comparison (B/C ratio, NPV and IRR) can give different results when used to choose one from a list of options. This problem is taken into account in capital budgeting.

Capital Budgeting with B/C or IRR

When making a capital budgeting decision based on the use of the benefit cost ratio or the internal rate of return the general procedure is to rank the projects from best to worst in terms of the criterion. Projects are then chosen from the top down until the budget is exhausted. The method is demonstrated with an example.

Worked example
A city council is attempting to plan its capital works for the coming year and must decide which projects should be selected using B/C ratio if the available budget is \$35 m. A list of possible projects is given in Table 8.5. It is assumed that all of the costs are capital costs for this example.

Table 8.5 Potential projects, their benefits and costs.

Project	PVB(\$m)	PVC(\$m)	NPV(\$m)	B/C
A	40	15	25	2.67
B	18	10	8	1.80
C	18	12	6	1.50
D	50	20	30	2.50
E	28	10	18	2.80
F	26	20	6	1.30

Solution Using B/C the projects are ranked in order as shown in Table 8.6 and selected off the top until the budget is exhausted. Selecting from the top gives Project E with a cost of \$10 m. This leaves \$25 m to spend. Therefore Project A can also be selected as it has the next highest B/C ratio and a cost of \$15 m. This leaves \$10 m in the budget. Project D is next on the list in terms of B/C but is too expensive so the search continues. Project B is next and costs \$10 m which exactly exhausts the budget. Therefore the projects selected are E, A and B with a total PVC of \$35 m and a NPV of \$51 m.

Table 8.6 Potential projects, their benefits and costs sorted in order of B/C ratio.

Project	PVB(\$m)	PVC(\$m)	NPV(\$m)	B/C
E	28	10	18	2.80
A	40	15	25	2.67
D	50	20	30	2.50
B	18	10	8	1.80
C	18	12	6	1.50
F	26	20	6	1.30

Capital Budgeting with NPV

When using the NPV the procedure is a little more complicated in that a set of projects must be chosen which maximise NPV, subject to the available capital budget. Therefore it is necessary to set up an optimisation model which is, in fact, an integer linear programming problem. This will be demonstrated using the same example above.

To set up the problem in a standard format the first step is to define variables X_i where i = A,B,C,D,E,F such that if $X_i = 1$ the project is selected, and $X_i = 0$ if not. The aim is to maximise the NPV so a value Z is defined which will be the total NPV over all the projects:

$$\text{Maximise } Z = 25X_A + 8X_B + 6X_C + 30X_D + 18X_E + 6X_F$$

The cost of all projects selected must not exceed the available budget so, it is possible to write:

$$15X_A + 10X_B + 12X_C + 20X_D + 10X_E + 20X_F \leq 35$$

and $X_i = 1$ or 0 (a binary variable).

The solution of integer linear programming problems is beyond the scope of this book. The solution of linear programming problems with continuous variables is discussed in Chapter 13. For simple problems the solution can be found by inspection or a limited trial and error procedure as:

$X_D = X_A = 1$ with all other $X_i = 0$.

That is, undertake Projects A and D and not the others. In this case the projects have a PVC of $35m and NPV of $55m. This NPV is $4m better than the projects chosen using B/C. In general it can be shown that NPV is a better criterion than B/C for capital budgeting projects. Also it can be demonstrated that IRR is inferior to NPV in capital budgeting.

World bank programs and projects

According to Talvitie (2000), when the World Bank lends money for infrastructure programs (e.g. road development) in the form of individual projects in developing countries it does so with the aim of reducing poverty, but is also mindful of ensuring that projects are well engineered and technically feasible. For major road projects, steps are taken to safeguard the environment, cultural heritage, indigenous people, and endangered species, and in fact approximately 50% of project preparation costs are spent on these issues while the design and traffic engineering component may only be 20–25%.

In terms of selecting actual projects, Talvitie states that the project priority setting and program development and evaluation steps occur together rather than simply picking projects off the top of the list until the funds are exhausted; an approach which may lead to a less than optimal mix of projects.

8.6 ADVANTAGES AND LIMITATIONS OF EACH CRITERION

From the previous section it should be clear that each economic criterion may lead to a different choice of project(s) when choosing one or more from a list of feasible alternatives. It is important therefore to consider the advantages and limitations of each criterion in turn.

Payback Period

Although the payback period is often quoted in the press, there are various problems with the method. Firstly, the criterion does not deal with the time value of benefits and costs. Secondly, and equally importantly, the method may ignore significant benefits or costs that might occur following the payback period. On the other hand the advantage of the method is that it is an intuitively appealing concept and can make a compelling argument for implementing a project.

Payback period has also been used in other aspects of economic decision making. For example, in an assessment of the economic viability of a biogas project for Nigeria, Adeoti et al. (2000) promoted the payback period as a measure of project riskiness.

However, because of its limitations, payback period is not recommended for use in the evaluation of major engineering projects.

The Internal Rate of Return

This criterion has the advantage of not requiring a prescribed value of discount rate for its determination. As the determination of a discount rate may be the source of considerable controversy, particularly in the public sector, this is an advantage in project selection. However, this advantage is outweighed by the disadvantages of the criterion, namely:

- to determine the IRR involves the solution of a polynomial equation;
- sometimes more than one value of IRR may be obtained for a particular project.

Worked Example
Figure 8.11 shows the time stream of expected revenue and costs for a proposed open-cut mining development. There is an initial cost to establish the mine and associated works and a high cost at the end of the project due to clean-up and re-vegetation activities. What is the IRR for the project?
Solution Let i = the discount rate of the project. Then the Net Present Value can be calculated as:

$$NPV = 200\frac{1-(1+i)^{-5}}{i} - \frac{620}{(1+i)^{6}} - 400$$

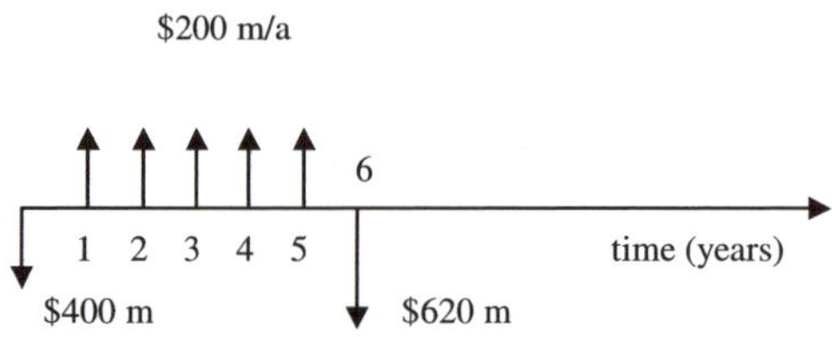

$200 m/a
1 2 3 4 5 6
time (years)
$400 m
$620 m

Figure 8.11 Timeline showing costs and benefits for mine.

Figure 8.12 shows a plot of NPV as a function of i. It can be seen that there are two values of i for which the NPV is zero, namely 3.8% and 16%. Both of these values may be interpreted as the internal rate of return of the project. If the project were being ranked in a capital budgeting exercise it could receive a high or low place depending on the value of IRR chosen. Although this is a somewhat unusual case it does demonstrate one difficulty with using the IRR criterion, i.e. it is the solution of a polynomial equation for which more than one root may exist.

Equivalent annual worth

To determine the EAW of a general time series of benefits and costs it is necessary first to compute the NPV of the series and convert the NPV into a uniform annual payment throughout the life of the project. Considering that EAW gives the same ranking as NPV for projects with equal lives, a strong case can be made for using NPV directly.

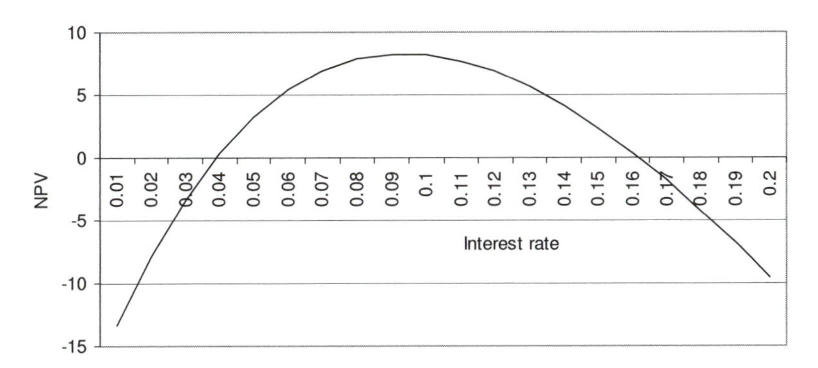

Figure 8.12 Plot of NPV versus interest or discount rate. The IRR is where NPV = 0.

Benefit cost ratio

The benefit-cost ratio has been used extensively in a number of countries around the world for comparison of projects of various sizes. It has some degree of appeal when used in the context of allocating scarce capital funds, i.e. capital budgeting. However, there is some ambiguity in its definition for projects with high recurrent costs (e.g. operating, maintenance, and repair costs). The basic issue here is whether the present value of recurrent costs should be treated as negative benefits and deducted from the numerator of the B/C ratio, or treated as costs and added to the denominator. A different value of B/C ratio will result in each case. For example, consider the economic comparison of two projects with the relevant economic data summarized in Table 8.6.

Table 8.6 Potential projects, their benefits and costs.

Economic Parameter	Project 1	Project 2
PV of benefits B ($m)	1.5	4.0
Construction costs C ($m)	1.0	1.0
PV of OMR costs K ($m)	0.0	2.0
B/C No. 1 = B/(C + K)	1.50	1.33
B/C No. 2 = (B - K)/C	1.50	2.00
NPV = B - K - C	0.5	1.0

OMR cost = operating, maintenance and repair costs

For B/C Ratio No.1 in Table 8.6 the present values of all costs are included in the denominator. For B/C Ratio No.2 the present values of recurrent costs are subtracted from the present value of benefits in the numerator. The rationale for this latter definition is that in many cases it is really the capital costs which are limited; the recurrent costs may be met out of the annual benefits.

If the projects are compared using the first definition of B/C ratio, Project 1 is preferable. Conversely, if the second definition of B/C ratio is used, Project 2 is preferable. On the other hand, NPV always gives an unambiguous ranking of projects as it does not matter whether OMR costs are first subtracted from the benefits or added to the construction costs. In this case, the NPV criterion favours Project 2.

It might be argued that this is not a major problem; why not just define the B/C ratio in one way or the other and proceed? However, in many situations the project benefits are composed entirely of savings in cost. For example, the benefits of most road projects consist of savings in travel time, savings in operating cost, and savings in cost due to a reduction in accidents. In such cases the B/C ratio obtained depends on the arbitrary definitions of benefits and costs, whereas NPV is unambiguous.

Net present value

Most economists advocate NPV as the most appropriate criterion for comparing projects. It measures the net gain to society (or the company) of the proposed project. It can be used in any of the three decision-making contexts discussed above. In the case of a go or no-go decision the choice is independent of the criterion used. In choosing a single project from a list of mutually exclusive projects, NPV appears to favour large projects.

Worked Example
An engineering firm is considering the purchase of a microcomputer since it currently pays for the use of a mainframe computer on a time-sharing basis. Two models of microcomputer are being considered. The relevant economic data are given in Table 8.7. Which option should be chosen?

Table 8.7 Potential projects, their benefits and costs.

Economic data	System A	System B
Purchase Price ($)	10 000	20 000
Annual benefits	4 000	8 000
Life (years)	5	5
PV of benefits ($)	18 914	37 828
NPV ($)	8 914	17 828
EAW ($/year)	1 226	2 452
B/C	1.89	1.89
IRR (%)	28.65	28.65

Solution The benefits represent the estimated savings in computing cost due to reduced usage of the mainframe computer. From Table 8.7 it can be seen that System B has the same life as System A, but twice the cost and twice the benefits.

Its NPV and EAW are therefore twice those of System A while its B/C ratio and IRR are the same. The tendency for NPV (and EAW) to favour large projects is apparent. If there is no constraint on the amount of money that can be spent, the increased NPV of System B is a net gain to the firm and therefore it is the better option. If there are limited funds available for investment due to other equipment needs, the problem should be treated as a capital budgeting exercise.

8.7 ECONOMIC BENEFITS

Unlike the costs, the benefits that accrue from engineering projects are often difficult to quantify. This is because they can occur in the form of quantities that have no direct monetary value. For example, the construction of a new highway may lead to reduced travel time and a reduction in accidents, but how these are compared to the actual costs of building the road is a question that challenges economists and for which there are a variety of approaches.

Desalination benefits

In 2006 the Israeli Government was planning the construction of desalination plants to supplement the water supply for its people (Dreizin, 2006). While the costs were relatively easy to quantify, based on the cost of new plant, running costs due to electricity and the replacement of reverse-osmosis membranes, the benefits were more nebulous. However, by considering effects such as reduced scaling in pipes, extended lifetimes of electric and solar heaters, savings in soap in washing clothes and dishes it was possible to determine a benefit of approximately US\$0.11 /m^3. When savings in pumping costs were taken into account the total benefit rose to approximately US \$0.15 /m^3.

One method for quantifying highway benefits is given by Forkenbrook and Foster (1990) who were able to quantify benefits due to improvements brought on by a new highway based on travel efficiency. Total benefits were of three types:

1. vehicle operating cost savings based on total travel distance, travel speed, curvature and gradient of the road and other factors that can affect the vehicle;
2. accident cost savings (fatal accident \$1.965 m, injury accidents \$16,700, property damage \$3,300); and
3. travel time savings (\$8 per hour for cars, \$15 per hour for trucks).

They applied this method to an evaluation of the ‘Avenue of the Saints’, a new four lane highway connecting St. Louis in Missouri and St. Paul in Minnesota. It is worth noting that they did not take account of other benefits that might accrue such as the likelihood of the area attracting new business since they argued that one person’s gain was another’s loss and the value of these benefits was hard to identify. These benefits will depend to a large extent on the location of the project and the relative level of development of the country. As Talvitie (2000) has noted, travel time savings are rarely important in developing countries while accident rates may be more than 10 times those of developed countries.

Similar issues arise with improvements to airport infrastructure. Jorge and de Rus (2004) suggest that the main benefits that flow from improvements are in passenger comfort, mainly through reduced delays, and a reduced need to divert to other airports. Diversions were costed as being worth €25 per hour for an estimated two hours per passenger. Reduced congestion within the terminal was costed at 10 minutes per passenger at the same €25 per hour.

The determination of benefits is particularly relevant in projects that are designed to have environmental benefits. The salinity diversion project mentioned in an earlier interest box is just one example, but there are few modern infrastructure projects which do not have a significant environmental component. A range of measures have been developed to quantify environmental benefits and costs but a treatment of them is beyond the scope of this text.

Manned mission to Mars

Ehlmann et al. (2005), in a paper promoting a manned mission to Mars estimate the costs of such a venture at between US$20 billion and US$450 billion where the latter included using the moon as an intermediate staging post. With the costs firmly established the challenge was to detail the benefits for such a mission. These, it was suggested, were to be found in the promotion of industry, engineering, technology and science. The authors put forward the US$85 billion satellite industry as a quantifiable benefit of previous space exploration and suggested that future work would contribute in similar ways. They also argued that the work would create new markets, allow for more efficient use of resources and create high-wage jobs.

8.8 ECONOMIC COSTS

The concept of cost is a subtle one which may cause confusion in some engineering studies. Essentially, the same concept of cost is applicable to public sector and private sector projects, although the question of cost to whom must be addressed. For the private sector, only costs incurred by the firm are of relevance, whereas for public sector projects the total cost to all members of society must be evaluated. The correct concept of cost is the incremental opportunity cost of resources used in the project.

The opportunity cost of a resource is its value when used in the best available alternative use. This is a measure of the value which must be forgone by using the resource in the project under consideration. The opportunity cost may be quite different from what was actually paid for the resource.

For example, a company has a vacant block of land which is valued at $500,000 on the open market. The company paid only $100,000 for the land some 10 years ago. If the company is considering building an office block on the land, the opportunity cost of the land is at least $500,000 because the land could be sold at this price. This is the cost which should be used in an economic evaluation of the proposed office block. The opportunity cost may be higher if a better alternative use is available.

As a further example of this concept, consider a contractor who has a crane which costs $500 per day to operate and which can be hired out for $1,000 per day. The opportunity cost to the contractor of using the crane on a particular construction project is $1,000 per day as this is the benefit which he must forgo in order to use the crane.

The opportunity cost concept has particular significance in public sector evaluation. If, for example, a project in the public sector employs labour which was previously unemployed, the opportunity cost of doing so is considerably less than the actual wage rate. As the alternative use of the labour is, in fact, unemployment, the opportunity cost is the value which the workers place on their additional recreation time when unemployed. The question of unemployment benefits paid to the individual by the government does not enter the evaluation as these are a transfer payment between one section of the community and another, i.e. the benefit of this payment to the unemployed equals the cost to the taxpayers, so its net effect on the total community is zero.

The incremental or marginal costs of a project must be carefully assessed. In project evaluation the marginal costs are the additional costs of undertaking the project compared with not undertaking it. There is sometimes a tendency to confuse marginal cost with average cost. The distinction between the two is illustrated by the following example.

Example

An engineer purchases a new car for $20,000. She uses the vehicle to commute to work and for recreational travel on weekends. The estimated annual costs for the vehicle are divided into standing costs (loss of interest on capital, depreciation, registration and insurance) which are estimated to be $8,200 p.a. and running costs (maintenance and repairs, petrol and oil, tyres) which are estimated to be $1,800 p.a. If she travels 20,000 km per year, the average cost of travel is $0.50/km. Now if her employer asks her to use her own vehicle to make a single trip of 20 km to a construction site, what is the cost of this trip? One possible answer would be to use the average cost of $0.50/km and thereby estimate a trip cost of $10. However, the extra trip is incremental to her private travel and therefore the incremental or marginal cost should be used. The registration and insurance costs and loss of interest on capital are unaffected by the extra trip and hence are irrelevant in this situation. Likewise depreciation is more strongly related to the ageing of the vehicle rather than distance travelled so it does not enter the calculation. The only incremental costs are petrol and oil, tyres, and maintenance and repairs (assuming these depend on distance travelled). Therefore, the marginal cost for a short trip is given by as $0.09/km. This gives a cost of $1.80 for the 20km trip. If the engineer is reimbursed at the average cost of $0.50/km, she makes a nice profit on the trip.

Now consider a different situation in which the engineer uses her car for 15,000 km of business travel and 5,000 km of personal travel per year. It could be argued in this case that the vehicle is primarily for business purposes and the private travel is the marginal component. It would then be reasonable for the company to reimburse all standing costs of the vehicle as well as the running costs associated with the business travel. Alternatively the company may provide a vehicle and ask the engineer to pay the running costs associated with private travel.

Sunk and recoverable costs

A further distinction needs to be made between sunk and recoverable costs. Sunk costs are costs which have been incurred in the past and are no longer recoverable. As such they are irrelevant to decisions about future actions.

Worked Example
A firm has a photocopying machine which originally cost $12,000 to purchase and now costs $2,500 per year to operate and maintain. It can be sold for $5,000. An equivalent new photocopying machine costs $15,000 but it is estimated that it costs only $1,500 per year to run because of an attractive maintenance contract being offered by the manufacturer. What are the benefits and costs which should be considered in deciding whether or not to purchase the new machine? A time-frame of 5 years should be used in analysing the decision. At the end of this time the new machine is expected to be worth $4,000 and the old machine nothing.
Solution We need to consider the incremental benefits and costs of the new machine relative to the old. These are shown in Figure 8.13. The incremental costs of purchasing the new machine is $10,000 ($15,000 minus the $5,000 sale value of the old machine). The incremental benefit is the $1,000 per year savings in operating cost plus the $4,000 value of the new machine at the end of 5 years. Note that the original purchase price of the old machine (i.e. the $12,000) is a sunk cost and does not enter the decision. Clearly the decision would be no different if the old machine had cost $200,000 or had been obtained for nothing.

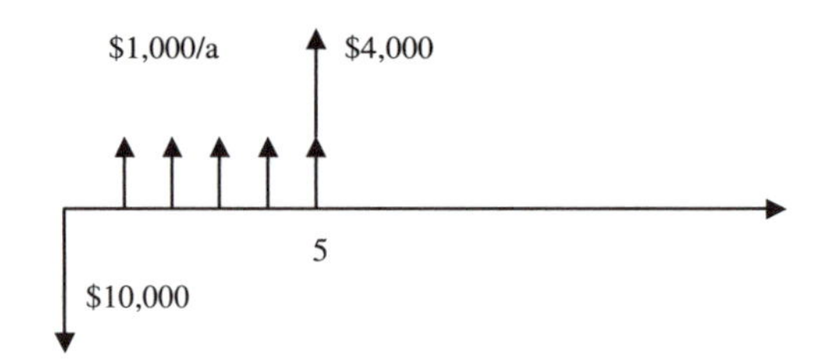

Figure 8.13 Benefits and costs of purchasing a new photocopying machine.

8.9 COST ESTIMATION

In the preparation of tenders, where engineering firms bid to undertake jobs, it is important to be able to estimate costs accurately. For example, the cost of a complete building, or the installation of air-conditioning, or in fact any of the many possible tasks that ultimately make up the total job. To assist in this process there are publications available that contain this sort of information, taking account of the current year, the location of the proposed work and many other variables. In Australia the standard reference for this work is Rawlinsons (2004) Australian Construction Handbook. It contains (among other things) estimates and detailed prices on:

- whole buildings (based on area or usage);
- electrical services (lighting, power, heating, closed circuit television installations, uninterruptable power supply);

- mechanical services (air-conditioners, hot water convectors, water boilers, heat exchangers, pipe work, duct work, natural ventilation); and
- civil services (excavation, laying concrete, installing structural steel).

The costs given are based on average values of actual contracts let in the preceding 12 months in the respective cities. It is difficult to summarise the full scope of the publication but some examples (current for 2004 prices) may assist in the process.

- The price of a 300 to 500 seat cinema in suburban Adelaide that includes air conditioning, ancillary facilities, seats and projectors, etc. is between $4575 and $4975 per seat. A 300 seat theatre would cost around $1.5 m.
- A visitors main door camera unit and directory panel, to call over 12 and up to 36 apartments in Melbourne, would cost between $5,000 and $8,000.
- To install an evaporative air conditioner that meets a recommended 30 air changes per hour in a suburban house with 2.6 metre (nominal) ceiling height in Perth would cost between $30 and $40 per square metre.
- The supply and placing of a strip footing in 25 MPa concrete in Adelaide costs $173 per cubic metre.
- The installation of a set of traffic lights, including fully activated detectors, electrical services between signals, control box, pedestrian push buttons and indication signs to intersection of a dual carriageway access road would be between $110,000 and $135,000 in Sydney.

8.10 SELECTION OF PROJECT LIFE AND DISCOUNT RATE

The choice of project life and discount rate may have an important influence on the economic viability of a particular project. Many engineering projects, for example, roads, water supply, and electricity supply involve high initial costs and benefits which grow steadily with time over many years. The NPV of such projects falls with increasing values of the discount rate, and may become negative for high values. For example, a proposed new highway will cost $10 m to construct. The benefits due to savings in operating costs and travel time are estimated to be $1.2 m in the first year of operation but will grow at 2% p.a. (compound) due to growth in the volume of traffic using the road. Maintenance costs are expected to be constant at $200,000 per year of operation. The NPV ($m) of the road for various values of life and discount rate can be calculated from:

$$NPV = \frac{1.2}{(1+i)}\left\{\frac{1-(1.02)^n/(1+i)^n}{1-(1.02)/(1+i)}\right\} - \frac{0.2\{1-(1+i)^{-n}\}}{i} - 10$$

and the results are shown in Figure 8.14.

Clearly for any given value of project life, the NPV decreases with increasing discount rate. For any given discount rate the NPV increases with increasing project life. It should also be noted that the economic desirability of the project depends strongly on the discount rate. For example, if i = 5%, NPV is positive provided the project life exceeds 12 years. On the other hand, for a value of discount rate of 15% the NPV is negative for all values of project life. For high values of discount rate, the NPV is relatively insensitive to changes in project life. For example, with a discount rate of 10% the NPV is $2.0 m if the project life is 35 years and $2.76m if the project life is 50 years.

There has been considerable controversy associated with the selection of discount rate (and project life), particularly for public sector projects. For example, consider the case of a public authority which is required to justify all projects on an economic basis and only undertake projects which have a positive NPV. Clearly the list of viable projects will be much greater if the authority uses a low value of discount rate than if a high value is used, This fact has lead to those who support large public works expenditure to argue for a low value of discount rate whereas those who wish to reduce public works expenditure often argue for a high value.

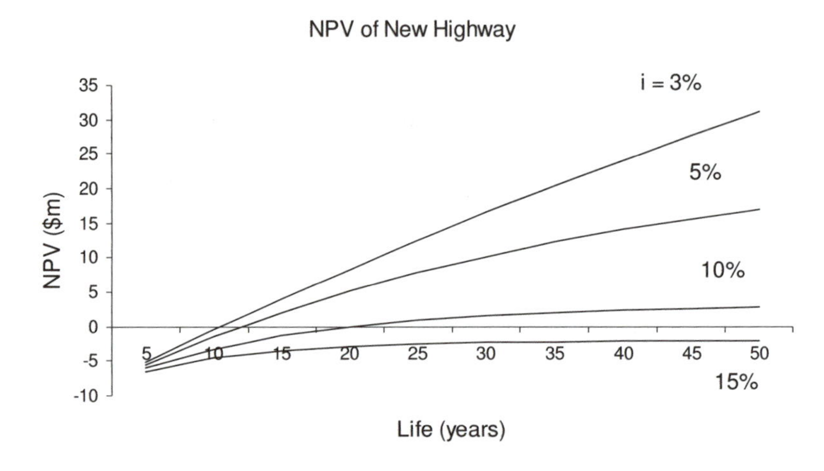

Figure 8.14 The effect of discount rate and life on the NPV of a road project.

The three most common theoretical bases for the choice of a discount rate are:

- the average yield on long-term government bonds;
- the social opportunity cost; and
- the social time preference rate.

As there is no agreed method we shall consider the arguments for and against each one in turn.

The average yield on long-term government bonds

It has been argued that the interest rate paid on long-term government bonds is the rate at which the government borrows money from the public. It is therefore appropriate that it should be used as the discount rate for evaluating public sector investments. In Australia this rate is typically 3–4% p.a. above the inflation rate. However, if we examine the federal budget as a whole it is clear that most government revenue is raised by taxes, not by the sale of government bonds. In fact, bonds are used by the government to regulate the money supply rather than to raise revenue. The argument to link the discount rate to the government bond rate is therefore questionable.

Another argument which supports the use of the long-term government bond rate is that this represents the risk-free interest rate in the private sector. Most investments carry some element of risk, due to uncertainties in the future economic climate. Even banks and building societies have been known to default on interest payments during major depressions. The interest on government bonds is guaranteed by the government and so carries no risk. Any private investment should return at least this interest rate, otherwise investors would withdraw their money and put it into government bonds. It is argued that it is therefore appropriate to use it for public sector investments.

The social opportunity cost

It is argued that funds invested by the government in major capital works have basically been displaced from the private sector and should, therefore, earn a rate of return at least as high as that prevailing in the private sector (based on the opportunity cost of capital concept). However, not all revenue raised by government taxation is invested in the private sector. In fact a high percentage of government revenue is consumption expenditure and produces no investment return. This fact must be taken into account when assessing the social opportunity cost. A further difficulty is in determining the appropriate marginal rate of return in the private sector. A typical rate for the social opportunity cost is 10–15% above inflation.

The social time preference rate

It is argued that, by undertaking investments in the public sector, society is effectively forgoing consumption in the current time period in order to achieve increased consumption in the future. The choice of a discount rate is equivalent to placing a weight on benefits and costs incurred in future years when compared with the present. Some economists consider that society is free to choose whatever stream of future consumption that it likes, and accordingly it may choose any set of relative weights on benefits and costs in future years.

The discount rate corresponding to this chosen set of relative weights is the social time preference (STP) rate, which will in general differ from the interest rate prevailing in the private sector. The STP rate may be any rate chosen by society (even 0%). Proponents of the STP rate usually argue for a low rate (in the order of 3–5%) on the grounds that government investments provide basic community infrastructure and therefore benefits over a long time scale should be considered.

Discount rate for public sector projects

Ultimately the choice of the discount rate for public sector projects tends to be a political decision. Often sensitivity analyses are carried out in which the NPV for all projects is calculated using a low, medium, and high value of the discount rate. For example, the NPV of all projects can be determined using discount rates of 5, 10 and 15% p.a. and the results compared.

Project life

Many engineers consider the life of an engineering project to be equal to the physical life of the constructed facilities. The physical life of a system ends when it can no longer perform its intended function. For example, a building is at the end of its physical life when it is no longer fit for habitation. A road pavement is at the end of its physical life when it is suffering total break-up and can no longer be repaired. However, two other concepts of project life are relevant in economic evaluation. These are the economic life and the relevant period of analysis.

The economic life of an engineering system ends when the incremental benefits of keeping the system operating for one more year are less than the incremental costs of maintenance and repair for one more year. As most engineering systems can be kept operating almost indefinitely with suitable replacement parts, the economic life is usually less than the physical life.

The relevant period of analysis is the maximum period beyond which the system performance has virtually no effect of its present value of benefits and costs. For high values of discount rate the NPV of a project does not significantly increase beyond a certain time. This time is the relevant period of analysis. For example, if a discount rate of 15% is being used it is pointless arguing whether the life of a building will be 30 years or 100 years as it will make no difference whatsoever in its economic evaluation. For lower values of discount rate more care must be taken in choosing the project life.

In general, the project life is the smallest of the physical life, the economic life and the relevant period of analysis. Some typical lives used for civil engineering projects are given in Table 8.8.

Table 8.8 Lives of some engineering systems.

Engineering system	Life
Vehicles	5–10 years
Roads	20–30 years
Bridges, buildings	30–50 years
Dams	50 years
Tunnels, cuttings, embankments	100 years

PROBLEMS

8.1 $2000 is invested at 5% p.a. for 20 years. What sum will be retrieved at the end of the investment period?

8.2 What is the present value of $2,000 that will be received in 5 years if the interest rate is 6% p.a.?

8.3 What is the present value of a series of annual payments of $10,000 over 10 years if the interest rate is 8% p.a.?

8.4 What are the annual repayments for a loan of $100,000 if the interest rate is 8% p.a. and the period of the loan is 45 years?

8.5 Following on from Question 8.4, how much is still owed after the first annual repayment on a $100,000 loan taken out at an interest rate of 8% p.a. over a 45 year period?

8.6 What are the monthly repayments for a loan of $100,000 if the annual interest rate is 6% p.a. (computed monthly) and the loan is over 50 years?

8.7 A project has initial costs of $100,000 and annual benefits of $10,000. What is the net present value of the project assuming an 8% p.a. discount rate over 25 years? Assume all costs and benefits occur at the end of the year.

8.8 If the benefit cost ratio is defined as B/(O+K) where B are the benefits, K the initial costs and O the ongoing costs, what is the benefit cost ratio in the case where initial costs are $20,000, ongoing costs $5,000 p.a. and benefits $10,000 p.a. for a discount rate of 6.8% p.a. and over a period of 20 years ? Assume all costs and benefits occur at the end of the year.

8.9 If the benefit cost ratio is defined as (B-O)/K where B are the benefits, K the initial costs and O the ongoing costs, what is the benefit cost ratio in the case where initial costs are $20,000, ongoing costs $5,000 p.a. and benefits $10,000 p.a. for a discount rate of 6.8% p.a. and over a period of 20 years ? Assume all costs and benefits occur at the end of the year.

8.10 As engineer for a company developing high efficiency solar panels you have an annual research and development budget of $500,000. Each year you call for submissions from within the company for projects that could be done. As part of this the proponents must estimate the cost of the proposal and what benefits would come from a successful outcome. This year you have eight proposals to consider, and one criterion for funding is economic benefits. The details of the proposals are given in Table 8.9. Any calculations that you do should assume a project life of 5 years and a discount rate of 20% p.a.

(a) Use the B/C ratio to determine which projects you would select. In this case the best proposals are chosen from a ranked list.

(b) Use NPV to determine which projects you would select. In this case the problem can be considered as one that could be solved using linear programming, but you may attempt to find a solution by inspection. As an additional exercise it would be good to set the problem up in EXCEL and use Solver to verify the solution obtained by hand.

Table 8.9 Summary of economic factors associated with the eight research proposals to be assessed.

Project	Initial Cost($)	Annual Benefit($)
A	100,000	55,000
B	200,000	100,000
C	50,000	24,000
D	10,000	8,000
E	20,000	20,000
F	345,000	120,000
G	100,000	34,000
H	250,000	160,000

8.11 A manufacturing company has space for one additional automated machine in its factory. It can purchase either Machine A which produces gadgets or Machine B which produces widgets. Economic data for the two machines are presented in Table 8.10.

Table 8.10 Costs and benefits for two machines.

	Machine A	Machine B
Initial Cost ($)	310 000	370 000
Revenue ($ pa)	70 000	120 000
Maintenance ($ pa)	15 000	20 000
Life (years)	10	5
Scrap value ($)	20 000	0

(a) Compute the net present value, equivalent annual worth, B/C ratio for each machine using a discount rate of 7% p.a. and a planning horizon of 10 years.

(b) Which machine should be chosen and why?

8.12 A car which was purchased 10 years ago for $20,000 is now thought to be worth $5,000 as a trade-in on a new model. Due to its age the old car now costs $2,000 per year in repairs alone, which it is assumed a new car would not need. Based on economics, should a new car be purchased for $30,000 if it is intended to keep it and evaluate it over a 10 year planning horizon? Assume that at the end of the 10 years a car purchased now would be worth $5,000 and the existing car would be worthless.

REFERENCES

Australian Bureau of Statistics (2006) www.abs.gov.au/ausstats (downloaded 19/7/06)

Adeoti, O., Ilori, M.O., Oyebisi, T.O. and Adekoya, L.O., 2000, Engineering Design and Economic Evaluation of a Family-Sized Biogas Project in Nigeria. *Technovation*, 20, 103–108.

Best, R., 2000, Low Energy Building Design: Good for the Environment and Good for Business. The Environmental Engineer. *Journal of the Environmental Engineering Society*, I.E.Aust., 1(1), 12–15.

Dreizin, Y., 2006, Ashkelon Seawater Desalination Project – Off-Taker's Self Costs, Supplied Water Costs, Total Costs and Benefits. *Desalination*, 190, 104–116.

Forkenbrock, D.J. and Foster, N.S.J., 1990, Economic Benefits of a Corridor Highway Investment. *Transport Research Part A: General*, 24A(4), 303–312.

Jorge, J.-D. and de Rus, G., 2004, Cost-Benefit Analysis of Investments in Airport Infrastructure: A Practical Approach. *Journal of Air Transport Management*, 10, 311–326.

Morgan, P. and Elliott, S.L., 2002, Mechanical Destratification for Reservoir Management: An Australian Innovation. Water, *Journal of the Australian Water Association*, 29(5), 30–35.

Rawlinsons, 2004, *Rawlinsons Australian Construction Handbook 2004*. (Rawlhouse Publishing Pty. Ltd.), 912pp.

Talvitie, A. (2000) Evaluation of Road Projects and Programs in Developing Countries. *Transport Policy*, 7, 61–72.

CHAPTER NINE

Sustainability, Environmental and Social Considerations

This chapter explains how environmental and social considerations play an important role in the planning and design work for engineering projects. Today, many governments use the concept of sustainability as a basis for the assessment of development projects, while community values place a strong emphasis on the environmental and social implications of engineering projects. Methods for taking account of social, environmental and sustainability issues are briefly discussed here.

9.1 INTRODUCTION

Prior to the 1970s, engineers focussed primarily on the technical and economic aspects of projects, without giving due concern to the social and environmental impacts of their work. To a large extent this was a reflection of the priorities of society at the time. Today it is recognised that engineers have a special role to play in the wise use, conservation and management of the earth's resources and to ensure that the needs of future generations are taken into account in the evaluation of all engineering work.

In 1828 Thomas Tredgold of the Institute of Civil Engineers described engineering as the art of directing the great forces of nature for the use and convenience of man. As we have seen already in Chapter 1, Nature at that time tended to be regarded as a powerful adversary to be tamed. Today such views are tempered by the realisation that the world has finite resources and the emphasis now is on the wise use of resources, not only to protect and conserve the environment, but also to achieve a sustainable way of life into the foreseeable future.

The concept of sustainable development has been discussed extensively in past decades, with the best known definition being that of the Brundtland Commission of 1987 (WCED, 1987):

> *Development that meets the needs of the present without compromising the ability of future generations to meet their own needs.*

This concept has developed further, and engineering systems, policies, designs and plans can be considered and evaluated today within the framework of sustainability.

In Section 9.2 of this chapter we look at the concept of sustainability and in Section 9.3 at the techniques which are used by engineers in planning and design to deal effectively with sustainability as an over-riding goal of engineering work.

While environmental and social issues are clearly inter-related with considerations of sustainability, they are looked at separately in this chapter. Environmental considerations and assessment programs are introduced in Sections 9.4 and 9.5, while social issues and assessment tools are dealt with in Sections 9.6 and 9.7.

9.2 SUSTAINABILITY

The systems approach provides a simple and consistent basis for investigating sustainability at all levels of society, from the global scale down to the individual. Gilman (1992) used systems concepts to provide the following definition of sustainability:

> *The ability of a society, ecosystem, or any such ongoing system to continue functioning into the indefinite future without being forced into decline through exhaustion or overloading of key resources on which the system depends.*

Based on this definition, Foley et al. (2003) have shown that to achieve sustainability it is necessary to manage appropriately all the resources that a system relies on. These include the natural, the financial, the social and the man-made infrastructure resources that are important to the functioning of the system.

Roberts (1990) has shown that self sustaining systems in nature are generally closed loop systems that have evolved gradually over time. In the past when humans have developed production systems, they have relied on an open loop, once-through use of resources, which results in much waste. To be sustainable it is necessary to use closed loop systems. Figure 9.1 shows a model of the initial extraction or use of resources, followed by the processing, transportation and consumption of the modified resources as a closed loop system that can evolve over time.

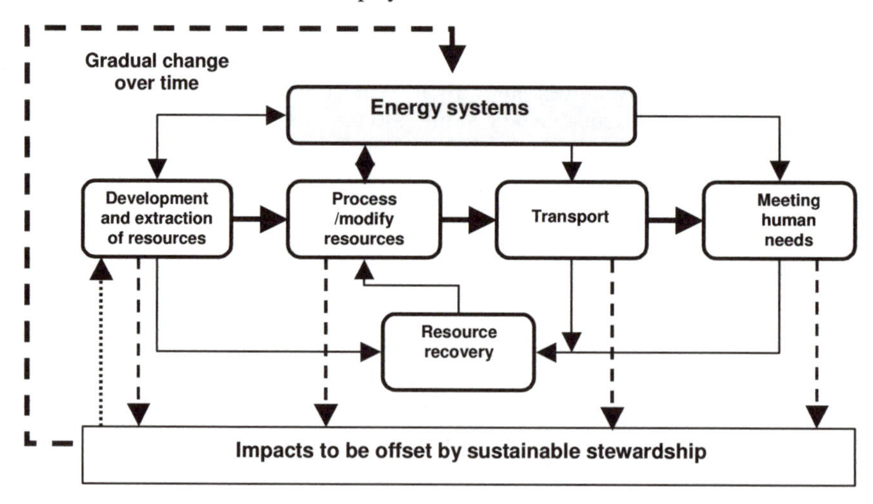

Figure 9.1 Closed Loop System representation (adapted from Roberts, 1990)

A key element of the model is the stewardship of the natural environment that is needed in order to provide a balance to the processes of extraction, modification and transportation. The main element of this closed-loop system is the recycling of waste from manufacture and consumption within a continuous cycle. Energy inputs are a crucial element in the closed loop system. For many natural systems the sun is the natural source of energy, but this is not so for man made systems at the present time.

A model that can be used to assess the sustainability of a system or of sustainable system development (Foley et al., 2003) is shown in Figure 9.2. It outlines the flow of resources within the system. The model identifies infrastructure and other human-made resources (I) as a key element of sustainability. Infrastructure for urban development includes buildings, and the water supply system, as well as systems for waste, transport and energy. Such a systems approach provides a good platform for assessing development and sustainability, where infrastructure and resource flow are principal considerations.

Each subsystem within the larger development system can be modelled (e.g. a single house within a city development). The flow of resources such as water, energy and finance to and from the system can also be included in a more holistic way than in many other currently available tools.

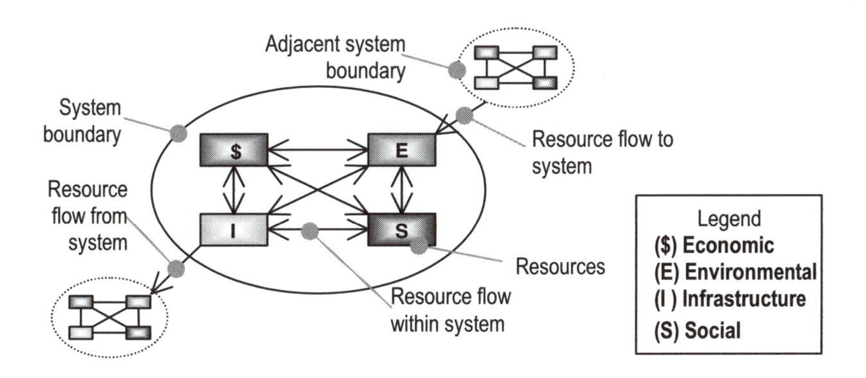

Figure 9.2 System representation (adapted from Foley et al., 2003)

Of course, a systems approach to the assessment of sustainability is not new. It was applied in early computer modelling studies by Forrester and Meadows in the 1970s, and in more quantitative studies which have continued up to the present. A major conclusion from such systems modelling, suggested by Suter (1999), is "that humankind needs to re-evaluate its exploitative attitude towards humans and the earth itself".

The term, sustainability, has been adopted in the business world to connote the principles of social and environmental responsibility. Many business organisations have recognized that profits alone do not guarantee continuity of existence for their companies and that sustainability adds to the long-term viability. Investors also increasingly see sustainability issues, such as controlling greenhouse gases, as priority concerns to be addressed by management. Stephan Schmidheiny (1992) of the World Business Council on Sustainable Development has suggested that population growth and economic development will eventually be constrained by environmental and social pressures. The question of what level of population is sustainable for the world is the focus of much debate amongst many scientific researchers and has been since the time of Thomas Malthus in the late 1700s. The key to the answer is in the level of resource use which is required for a standard of living that can be perpetuated for future generations. Sustainability has thus been explained as being the state for which all social well-being and development are within the Earth's biological capacity (Wackernagel et al., 1997).

Club of Rome

The Club of Rome was founded in 1968 and has been described as "a think tank and a centre of research and action, of innovation and initiative". It is composed of a group of scientists, economists, business people, senior public servants, Heads of State and former Heads of State from around the globe and addresses questions confronting the global society. The term "problematique" was developed by Club members for the inter-related issues that were to be examined. The Club of Rome commissioned the book "Limits to Growth" which was published in 1972. Twelve million copies of the book were sold world wide, in 37 languages. The book challenged a basic assumption of economic theory that the resources of planet Earth were infinite. The book was based on computer modelling of the consequences of the growth of global population. Various scenarios researched pointed to a major economic crisis occurring in the early 1990s. This did not happen. Subsequent updates on the book have been undertaken and in retrospect, one needs to ask, was "Limits to Growth" a fair or a false warning?

Adapted from Suter (1999)

9.3 PLANNING AND DESIGN FOR SUSTAINABILITY

The time dimension is a key component of sustainability. This has been highlighted by Fleming (1999) who has suggested that sustainability needs to be considered within cycles of continuous improvement. In evaluating the sustainability of engineering systems, projects, processes and operations, a key aspect to consider is the stream of resource usage. Various tools can assist in such evaluations. Some of these are listed in Table 9.1.

The continuous improvement methods employed by industry, such as Cleaner Production and Total Quality Management, evaluate the level of goal achievement over time. Many of these approaches use a systems approach to examine resources so that there is minimal waste with the conversion of linear type, once through systems to closed loop systems to reduce waste. Some of these processes and procedures are discussed later. These are mainly improvement processes used by industry to become more sustainable at a particular industry level, as well as processes used by Local Government such as Local Agenda 21, whereby the principles emphasizing sustainable development agreed to at the Earth Summit in Rio de Janeiro in 1992 can be undertaken at the local government level.

Table 9.1 Sustainable Improvement Processes

• Cleaner Production	• Total Quality Management
• Zero Emissions	• Life Cycle Analysis
• Environmental Impact Assessment	• Environmental Management Systems
• Factor 4	• Local Agenda 21
• Factor 10	• Eco-Labelling
• Green Procurement	• Triple Bottom Line Reporting
• Natural Capital	• Ecological Footprints
• Water Sensitive Urban Design	• Compass of Sustainability

In many governments and organizations, the assessment of projects, government processes and product development is against sustainability criteria. Included within this assessment is a greater interaction with the stakeholders and community groups through participatory and consultation programmes.

A major push for reduction of the use of resources comes in programmes such as Factor 10, which has the goal to lower the use of natural resources for generating material wealth using new technology, by an average factor of ten within 30 to 50 years for the purpose of approaching sustainability. This is equivalent to increasing resource productivity tenfold over the same time period. (Schmidt-Bleek, 1993). Sustainability assessment requires the development of a clear set of criteria against which the assessment can be conducted. This can be done for each of the elements of the environment, economics and social considerations. A number of the approaches listed have attempted to do that. The Triple Bottom Line (TBL) approach establishes individual indicators to show how well a company is doing against the economic, environmental and social goals which have been adopted by a corporate organization for improvement against sustainability criteria. Usually in this method indicators are chosen for a number of criteria for each element and then presented using a set of bars of different colours or stars to indicate how well a company is doing against particular criteria. The establishment of principles-based criteria for assessment of sustainability can improve the standard TBL approach.

A framework of assessing the natural capital within a region or within an organization can be developed with the use of systems such as the "Compass of Sustainability" to stress the *inter-connected* nature of the elements of environment, economics, and social well being (AtKisson and Hatcher, 2001, Hargroves and Smith, 2005). Over time the results of the planning and engineering of cities are measured by such indicators. Predictions of changes to the development of the city infrastructure and the resource input and output for a city's existence can be made against these criteria. The design of individual buildings will involve choices of materials to be used, lighting, air conditioning and water use. Each of these choices for achieving sustainability goals will involve assessment of embodied energy within materials used, energy use over the life of products, the ability of the product to be recycled and satisfaction of performance criteria.

Sustainability assessment does not replace Environmental Impact Assessments or functional decision making based on Life Cycle Analysis or Social Impact Analyses or combinations thereof. A sustainable assessment tool needs to be incorporated within the decision making framework to ensure that decisions in planning are, in fact, sustainable (Pope et al., 2004). Sustainability assessment can and should be applied to evaluating proposed and existing processes and projects at all levels of governmental and project decision making. Incorporating sustainability through the design and implementation phases of a development could involve the integration of social and environmental effects of a longer time horizon into analyses by the proponents of projects. Sustainability assessment requires defining clear societal goals which can be translated into criteria, against which assessment is conducted. It is essential that the assessment method is able to discern sustainable outcomes from unsustainable ones.

Ecological Footprint

Ecological Footprint (EF) is a measure of humanity's dependence on natural resources. For a certain population or activity, the EF measures the amount of productive land and water required to sustain the current level of resource usage for the production of goods and services and the assimilation of waste required by that population or activity.

In 2001, the world average EF was 2.55 global hectares per person, >30% above the current global capacity (Venetoulis and Talberth, 2005). The lifestyle of the average global citizen can therefore be considered as unsustainable. Natural resources are being used faster than they can be regenerated.

The size of an EF can change over time, depending on population, consumption levels, technology and resource use. EFs are measured in global area (hectares or acres). Wackernagel et al (1997) defined biologically productive areas as (a) arable land; (b) pasture; (c) forest; (d) sea space (used by marine life); (e) built up land; and (f) fossil energy land (land reserved for carbon dioxide absorption). The current global biologically productive area is 10.8 billion hectares, of which 21% is productive ocean and 79% is productive land. This represents less than one-quarter of the Earth's surface.

An individual's resource consumption is not restricted to local resources. The resources used by an individual are from around the world, clothes from China, cars from Korea and food from many different places. From a comparison of the Ecological Footprint to the existing biologically productive area, the sustainability of an activity, lifestyle or population can be determined.

An ecological footprint of Greater London, prepared by Best Foot Forward Ltd (http://www.bestfootforward.com/reports.html) found that London's EF was 42 times the current capacity, or 293 times the size of London. This equated to 49 million global hectares, twice the size of the UK, and roughly the same size as Spain. Is this sustainable?

Australia has a usage of 7.0 global hectares per person, making it the 5th highest in the world. There are many websites that enable calculation of your individual footprint on the earth (http://ecofoot.org/). The Global Footprint Network is trying to get countries to adopt the footprint as an international metric and over the last two years the assessment behind the footprint has changed (Venetoulis and Talberth, 2005).

The Matrix Evaluation of Sustainability Achievement is one such method (Fleming and Daniell, 1995; UNEP, 2002) which uses a weighting technique against a set of sustainability criteria and combines them with fuzzy set analysis. The important step is the definition of specific criteria which can be obtained to achieve the overall project goal. In the case of a building this might be to achieve a reduction of 50% of projected CO_2 emissions over an existing conventional design.

Therefore the inclusion of sustainability as a project goal could totally change the final solution from that resulting from a short term economic goal.

Global Reporting Initiative

The Global Reporting Initiative (GRI) is a multi-stakeholder process and independent institution cooperating with UNEP to develop and disseminate globally applicable Sustainability Reporting Guidelines. The GRI reporting framework now provides a framework for reporting on an organization's economic, environmental, and social performance. The Framework consists of a set of Sustainability Reporting Guidelines, Technical Protocols and Sector Supplements (Global Reporting Initiative, 2006).

Environmental Performance: The dimension of the environment when considering sustainability involves an organization's impacts on ecosystems of land, air, and water. The environmental indicators cover the related performance of inputs (material, energy, water) and outputs (emissions, effluents, waste).

Economic Performance: The economic dimension of sustainability concerns the organization's impacts on the economic conditions of its stakeholders and on economic systems at the local, national and global levels. The economic indicators illustrate the flow of capital amongst stakeholders and impacts of the company throughout society.

Social Performance: The social dimension of sustainability concerns an organisation's impacts on the social systems within which it operates. The GRI social performance indicators identify key aspects of labour practices, human rights, and broader issues affecting consumers, the community, and other stakeholders in society.

9.4 ENVIRONMENTAL CONSIDERATIONS

History of Environmental Concerns

While it is true that the environment and sustainability have become much more of a focus for concern among people since the 1970s it would be quite wrong to think that prior to that time there were not problems and concerns. In ancient Greece some 2500 years ago, Plato lamented the consequences of excessive logging and grazing in the mountainous region of Attica, near Athens. Similar concerns were raised 1000 years ago in Japan where a large sediment delta of eroded granite built up in Lake Biwa, due to logging for the construction of a temple (Parker, 1999).

Environmental legislation also dates from ancient times. Julius Caesar prohibited all wheeled vehicles in Rome between sunrise and two hours before sunset to improve the situation for pedestrians. This law fell into disuse, and in the 3rd century AD, writers complained about the level of noise pollution generated by traffic in the streets. The city of Florence had laws governing the polluting of the rivers Arno, Sieve, and Serchio as long ago as 1477 (Higgins and Venning, 2001) and in 1810 Napoleon was issuing decrees with regard to the types of industries that could be undertaken in the cities, thus eliminating polluting industries from the centres of cities.

In Australia some of the earliest laws dealt with environmental issues. There were laws to protect the quality of the stream which supplied water for the original settlement and it was prohibited to fell trees within 50 feet (15 metres) of the stream. People were not allowed to throw rubbish into these streams or have pigsties within a prescribed distance of them. In 1839, just three years after the city of Adelaide was settled, the state of the Torrens River was so bad that laws were passed to prevent, among other things, people driving cattle through it. This was the city's only water supply and yet it was treated with scant regard by many. There are many countries today where this still happens but as the world becomes more urbanised there is a need to be aware of the impact of human activity on the state of the planet earth.

The Tragedy of the Commons

The tragedy of the commons has played itself out worldwide at various times and to varying degrees. It can be explained as follows: when a group of herdsmen have access to a common pasture it is each individual's best interests to increase the size of their herd without reference to the overall carrying capacity of the land. However, this leads to the destruction of the common pasture, so that everyone loses. This type of situation occurs in many ways, such as when an increasing population results in increased waste disposal into the commons: rivers, lakes, oceans, and atmosphere. Sustainability can be explained in terms of ensuring the commons are maintained and are available for many generations to come.

A major issue in city development involves the concept of the commons in regard to the decreasing quality of air and water due to emissions from transport systems and runoff from road systems. Another issue linked to the commons on a global scale is the use of energy by the developed nations which has resulted in producing CO_2 emissions which have contributed to global warming. Those elements of society which are considered to be necessary for a high quality of life: food; energy and transport systems; communication systems; the supply of infrastructure; supply and maintenance of water and waste water systems; health systems; and the management of waste all need energy and hence generate CO_2 emissions. There is much discussion on the linkages between CO_2 production, climate change and quality of life. The linkage between use of energy and quality of life can be examined by inspection of emissions of CO_2 and energy use in industrialised countries in the world. The UN Human Development Index (HDI) is a measure of poverty, literacy, education, life expectancy, childbirth, and other factors for countries worldwide. It is a means of measuring well-being and has been used since 1993 by the United Nations Development Programme (UNDP, 2005). To compare the impacts of development across nations with respect to emissions, the Human Development Index (HDI) has been used in conjunction with both CO_2 emissions and electricity consumption per capita in Table 9.2.

Energy Use

Energy use in undeveloped nations is based on traditional systems of wood and coal, and not electricity. As these nations convert to electricity and develop industries that use more energy, there will then be a major increase in CO_2 production across the globe. The interesting aspect of Table 9.2 is that electricity production per capita depends on climate and not specifically on HDI ranking for the developed nations.

The developing nations are still using less than 5% of the per capita consumption of electricity of the developed nations. The growth in energy use of the developing nations is only just beginning and can be expected to increase 20 fold in the next 20–30 years. The need to develop better technologies for the sustainability of cities and lower CO_2 emissions are two goals which engineers will need to address very soon if there is to be equity among nations.

Of course the future can not be predicted with certainty, but engineers will help build it. The goals that are now chosen for all development will dictate the future. In the past, individual goals were pursued independently of each other, leading to the present crises for social well-being, the environment and the economies of some countries. There is a need to develop a balance between the use of resources, environment and social well-being to achieve sustainability into the future.

Table 9.2 World CO_2 Emissions and Electricity consumption (per capita)

HDI rank 2005	Country	Carbon dioxide emissions per capita (tonnes) 1980	Carbon dioxide emissions per capita (tonnes) 2002	% of world total CO_2 emissions 2000	Electricity consumption per capita (kW hrs) 1980	Electricity consumption per capita (kW hrs) 2002
1	Norway	10.6	12.2	0.2	22,400	26,640
2	Iceland	8.2	7.7	(.)	13,838	29,247
3	Australia	13.9	18.3	1.5	6,599	11,299
4	Luxembourg	29.1	21.1	(.)	10,879	10,547
5	Canada	17.2	16.5	1.9	14,243	18,541
6	Sweden	8.6	5.8	0.2	11,700	16,996
7	Switzerland	6.5	5.7	0.2	5,878	8,483
8	Ireland	7.7	11.0	0.2	3,106	6,560
9	Belgium	13.3	6.8	0.4	5,177	8,749
10	United States	20.0	20.1	24.4	10,336	13,456
85	China	1.5	2.7	12.1	307	1,484
127	India	0.5	1.2	4.7	173	569
158	Nigeria	1.0	0.4	0.2	108	148

Source: United Nations Development Program, 2005

9.5 ENVIRONMENTAL ASSESSMENT PROGRAMS AND TECHNIQUES

Environmental Impact Assessment is now an integral part of the planning of an engineering project, just as economic, financial and social issues and technical analyses are. A definition of Environmental Impact Assessment (EIA) that looks at the consequences is as follows: "Environmental Impact Assessment is a tool designed to identify and predict the impact of a project on the bio-geophysical environment and on society's health and well-being, to interpret and communicate information about the impact, to analyse site and process alternatives and provide solutions to sift out, or abate/mitigate the negative consequences on man and the environment" (UNEP, 2003).

Numerous countries have implemented EIA regulations from the late 1960s onwards. The origin of EIA stems from the National Environmental Policy Act 1969 (US), from which the practice has spread around the world. The Environment Protection (Impact of Proposals) Act 1974 (Commonwealth of Australia) was the first dedicated EIA legislation in the world and has now been replaced by Environment Protection and Biodiversity Conservation Act 1999 (Commonwealth of Australia). Within Australia there is state legislation as well. For example, in Victoria the Environment Protection Act 1970 provides the legislative framework for environment protection in Victoria, including the principles of environment protection, but is supplemented by other acts such as the Environmental Effects Act of 1978 for performing Environmental Effects Assessment.

Environmental impact assessment (EIA) is thus the means of including environmental factors with both economic and technical considerations at the planning stage of a project. The process of EIA:

- identifies the potential environmental effects of undertaking a project;
- presents these environmental effects with the advantages and disadvantages of the proposed project to the decision makers; and
- forces stakeholders to consider the environmental effects and informs the public of the project while giving them the opportunity to comment on the proposed project.

EIAs are necessary for large projects and are required under relevant planning legislation. They are necessary for mining projects, new transport corridor projects, major infrastructure development, large industrial factories or expansion of existing facilities. The EIA is a means of highlighting potential environmental and ecological disturbances which would be more difficult and more expensive to correct after their occurrence than before. One major benefit of EIAs is the assessment of habitats of many rare species of flora and fauna which might never have been investigated had it not been for the requirement of the EIA.

In project planning, environmental issues are now addressed at all stages of the planning process, from conception to closure and rehabilitation. There are alternatives that can be considered at an early stage to counter adverse assessments, such as:

- abandoning a project or a process;
- proposing alternatives to projects which have extreme detrimental impacts on the environment; and
- proposing alternatives for projects not economically or financially viable.

Very few projects have been deemed not viable merely because of increased costs of environmental controls. For example, at a new paper pulp mill the additional cost has been assessed to be less than 3% of the initial investment (UNEP 2003). The environmental controls that need to be put in place will vary from industry to industry, with significant higher costs for some industries. Major research projects are being undertaken on the sequestering of carbon dioxide from energy production. A detailed assessment of the practices and methods involved in

EIAs is contained in Canter and Sadler (1997) but the key elements of an EIA include:

- Scoping: identify key issues and concerns of interested parties;
- Screening: decide whether an EIA is required based on information collected;
- Identifying and evaluating alternatives: list alternative sites and techniques and the impacts of each;
- Forwarding measures that deal with risks and uncertainties of the proposed project and to minimise the potential adverse effects of the project; and
- Issuing a report called an environmental impact statement, which covers the findings of the EIA.

Environmental Impact Statements (EISs)

The EIA process produces a document, called the EIS, which provides information on the existing environment and predictions about the environmental effects which could flow from the proposal. Whether an EIS is to be done in its entirety or whether some other report is warranted by the proposal is decided by government legislation. In South Australia there is an independent statutory authority, the Major Developments Panel which controls the level of reporting required: an Environmental Impact Statement which is required for the most complex proposals; or a Public Environmental Report which is required for a medium level of assessment, sometimes referred to as a "targeted EIS"; or a Development Report required for the least complex level of assessment.

The EIS is the document that reports the findings of the EIA and depending on a country's legislation, is now often required by law before a new project can proceed. A typical EIS has three parts with different levels of detail: an Executive Summary in a style that can be understood by the public; the main document containing relevant information regarding the project; and a volume containing the detailed assessment of significant environmental effects. If there are no significant effects either before or after mitigation, this volume will not be required. The EIS should:

- describe the proposed action as well as alternatives;
- predict the nature and magnitude of the environmental effects;
- identify the relevant human concerns;
- list the impact indicators and determine the total environmental impact; and
- make recommendations for inspection procedures and alternatives to the plan.

As part of the EIA, as shown in Figure 9.3, there is usually a review of the EIS by the public and government to consider its accuracy and to recommend whether the proposal should go ahead or not. Different authorities and governments at the local, state or federal level have their own processes and procedures indicating time frames for each review.

The EIS is normally carried out by the developer or proponent of the proposed development. The developer generally uses a consulting firm that would assemble a multidisciplinary team to undertake the EIS. Although it might be argued that the developer might be biased, there are advantages in that the developer must pay for the EIS to be carried out, and since the developer has the relevant information it is more efficient for the developer to undertake the EIS. There is also the advantage that the proponent can modify the design to reduce adverse impacts during the preparation of the EIS.

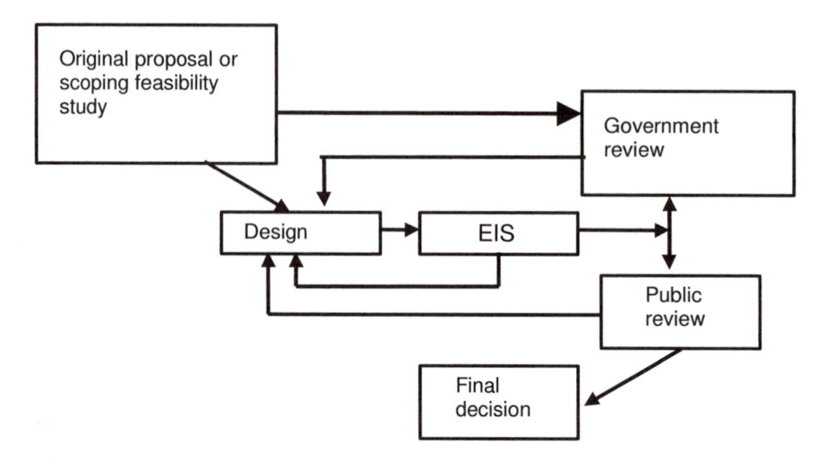

Figure 9.3 Flow chart for Environmental Impact Assessment

Most governments have regulations regarding EIAs and the reporting thereof, and give a formal procedure for the requirement of an EIA. For any significant project, an EIA would be required as part of the process of getting development approval. Generally a preliminary investigation, feasibility or scoping stage is performed prior to an EIA. It is very difficult to compare projects in terms of their environmental and social effects because of the many diverse factors involved. This contrasts with an economic comparison in which a single criterion such as NPV (see Chapter 8) gives an unambiguous ranking of projects. A number of techniques have been developed to assist in evaluating environmental and social factors. Some of these are listed in Table 9.3. Two of these that will be discussed here are the Battelle environmental index method and the Leopold matrix display technique. The Battelle method can be considered to be a checklist method with scoring and weights.

The Battelle method

The Battelle method or Environmental index method is an example of a checklist method. It was first designed for water resource development, (Dee et al., 1972; Dee et al., 1973; Martel and Lackey, 1977) but can easily be adapted for evaluation of a variety of types of projects. The principle lies in splitting the environmental impacts into four major categories: ecology; physical/chemical pollution; aesthetics; and human interest. These categories are divided into thematic data, as shown in Table 9.4. The full list is included in the Appendix to this chapter.

These thematic data are divided into environmental indicators. For example, in a project discharging wastewater, the water pollution could be represented by: BOD; dissolved oxygen; faecal coliforms; inorganic carbon; pH; temperature; total dissolved solids; and turbidity of the receiving water and/or of the waste stream.

Table 9.3 Main Advantages and Disadvantages of Impact Identification Methods (adapted from UNEP EIA Training Manual Edition 2, 2002)

Impact Methods	Advantages	Disadvantages
Checklists -simple ranking and weighting	Simple to use and understand. Good for site selection. Good for priority setting.	Do not distinguish between direct and indirect impacts. Do not link action and impact. The process of incorporating values can be subjective, hence controversial.
Matrices	Links action and impacts. Good method for visual display of EIA results.	Difficult to distinguish the direct and indirect impacts. Potential for double-counting of impacts.
Networks	Links action to impacts. Useful in simplified form for checking for second order impacts. Handles direct and indirect impacts.	Can become very complex if used beyond simplified version.
Overlays	Easy to understand. Good display method. Good siting tool.	Address only direct impacts. Do not address impact duration or probability.
GIS and computer expert systems	Excellent for impact identification and analysis. Good for examining different scenarios-'experimenting'	Heavy reliance on knowledge and data. Often complex and expensive.

Table 9.4 Thematic types for the Batelle method

Ecology (240)	Physical/ Chemical (402)	Aesthetics (153)	Human interest (205)
Species and populations	Water pollution	Land	Educational/scientific packages
Habitats and communities	Air pollution	Air	Historical packages
Ecosystems.	Land pollution	Water	Cultures
	Noise pollution	Biota	Mood/atmosphere
		Man made objects	Life patterns

Note: the numbers in brackets are the weightings for each category.

Once the environmental indicators are chosen using the table in the appendix, the method follows three steps:

First step: transform **environmental** indicators into an environmental quality (EQ) rank. This is usually done by using expert advice to convert the environmental measurement to a scale of 0 to 1 (0 for poor quality and 1 for good quality). Thus it is possible to quantify environmental deterioration or improvement for the given project. A sample transformation is shown in Figure 9.4 for Turbidity measured in Nephelometric Turbidity Units, transformed to the environmental quality index between 0 and 1.

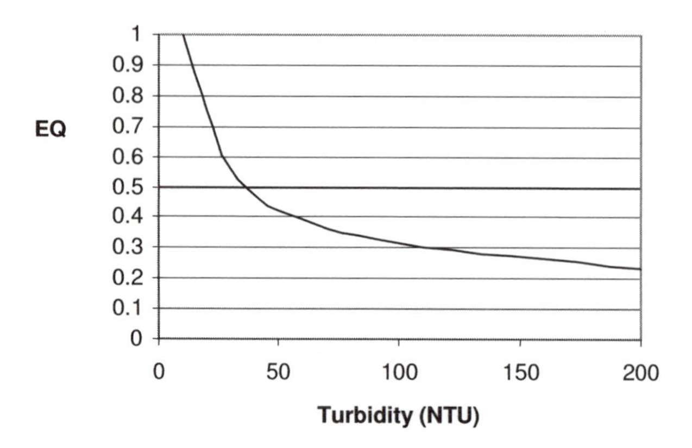

Figure 9.4 Environmental Quality transformation

Second step : use the distributed Parameter Importance Units (PIU) which were developed by Dee et al. (1972). The relative importance of each parameter is reflected by distributing a total of 1,000 points among the indicators (e.g. a total of 240 for Ecology)

Third step : an analysis is done for both the situation with and without the project in Environmental Impact Units (EIU). It can even reflect benefits or losses in terms of environmental conditions. The environmental impact is calculated using:

$$EI = \sum_{i=1}^{m} w_i \left[(V_i)_1 - (V_i)_0 \right] \qquad (9.1)$$

where: V_i is Environmental Quality (EQ) for each indicator i; the 1 subscript refers to the EQ with the project, and the 0 to the EQ without the project; w_i is the relative weight of indicator i; and m is the total number of indicators.

The major advantage of this method is that it gives a comparative analysis of alternatives. Thus it is effective when a choice is to be made between projects. The problem of dealing with qualitative data is encountered in all assessments, and synthesis of information depends on the experience and professional judgment of the team undertaking the assessment and the stakeholders who have been consulted.

The Leopold matrix display technique

The matrix display technique involves the construction of a large matrix which summarizes all of the environmental and social impacts of a proposed project. Often this information is given in a quantitative form. One example of this technique is the Leopold matrix developed by Leopold et al. (1971). One difficulty with the matrix display technique is the difficulty in absorbing all of the information contained in a matrix method.

The matrix lists a set of project actions across the top and a set of environmental characteristics down the left side. Table Part A in the attached Appendix lists the project actions, while Table Part B lists the environmental characteristics. Other actions or environmental characteristics can be added as appropriate to give over 8800 possible interactions. For any particular project only some of the possible interactions will be relevant, but the actions and environmental characteristics do provide a useful checklist to ensure that all possible effects have been considered.

Each cell of the matrix for which an impact is likely to occur is identified and divided by a diagonal line. An assessment is then made of the magnitude of the likely impact and a number between 1 and 10 placed in the upper left-hand corner. 1 represents the smallest magnitude and 10 the greatest (Leopold et al., 1971). This should be a factual assessment which should not include value judgements. A number between 1 and 10 is placed in the lower right-hand corner of each cell to indicate the importance of the impact. This is based on the value judgement of the evaluator. Leopold used the following example of a proposal that recommends construction of highways and bridges. One consideration was that bridges may cause a large amount of bank erosion because geologic materials in an area are poorly consolidated. This may lead to the investigator to mark the magnitude of impact of highways and bridges on erosion at 6 or more. If, however, the streams involved have high sediment loads and appear to be capable of carrying such loads without appreciable secondary effects, the effective importance of bridges through increased erosion and sedimentation might be considered to be relatively small and marked 1 or 2 in the lower right hand of the box. This would mean that while the magnitude of impact is relatively high, the importance of impact is not great.

It should be emphasized that the matrix is only one way of summarizing the environmental impact of a proposed project. It must be supported by an environmental impact statement which discusses all impacts identified in the matrix. For example, Figure 9.5 shows a reduced impact matrix for a proposed phosphate mining lease in Los Padres National Forest, California (Leopold et al., 1971). The mining will be an open-cut operation with ore processing on site, including crushing, leaching, and neutralization. The most important impact of the proposal is its likely effect on the California condor, a rare and endangered species which exists in the region. The primary actions of concern are blasting and the increase in truck traffic, both of which are likely to disturb the nesting of the condor. In addition, sulphur fumes from mineral processing could prevent the birds from landing to catch prey and hence present a danger to them.

These effects are shown in the row labelled "rare and unique species" as having a magnitude of 5 and an importance of 10. Other effects considered to be of moderate importance include the impact of industrial sites and buildings, highways and bridges, surface excavation, trucking, and the placement of tailings on the "wilderness qualities" of the area. The values in the boxes of a Leopold matrix are on an ordinal scale and cannot therefore be added or averaged. However, two or more projects can be compared in terms of the entries in an individual cell in the matrix to see which project will have the greatest impact on that particular characteristic. Another approach, instead of numbers, is to use colour coding of the rankings and highlight those of importance with red flags.

Project actions

Environmental characteristics	Industrial sites and buildings	Highways and bridges	Transmission lines	Blasting and drilling	Surface excavation	Mineral processing	Trucking	Emplacement of taillings	Spills and leaks
Water quality					2/2	1/1		2/2	1/4
Atmospheric quality						2/3			
Erosion		2/2			1/1			2/2	
Deposition, sedimentation		2/2			2/2			2/2	
Shrubs					1/1				
Grasses					1/1				
Aquatic plants					2/2			2/3	1/4
Fish					2/2			2/2	1/4
Camping and hiking					2/4				
Scenic views and vistas	2/3	2/1	2/3		3/3		2/1	3/3	
Wilderness qualities	4/4	4/4	2/2	1/1	3/3	2/5	3/5	3/5	
Rare and unique species		2/5		5/10	2/4	5/10	5/10		
Health and safety							3/3		

Figure 9.5 Abbreviated Leopold matrix for a phosphate mining lease (Leopold et al., 1971)

Multi Objective and Multi Criteria Assessment Approaches

Multi-objective approaches incorporate economic, environmental and social aspects and use project evaluation methods that are capable of handling multiple objectives. These methods can make use of qualitative data in an integrated way and have the capability of dealing with trade-offs between objectives.

Sustainability principles have yet to permeate the full range of methods and techniques used for project evaluation and assessment but they are becoming increasingly important. In planning and design situations where more than one objective is relevant, it is not usually possible to identify a single best solution. A design which is better in terms of its economic performance might not be as good in terms of its environmental and social impact. A design which has low environmental impact may involve low economic and social benefits. Compromise is the essence of good planning and design and is reflected in the push towards sustainability.

The role of the analyst in multiple objective planning is to identify the most efficient designs (in terms of all objectives) and elucidate the tradeoffs between them. A central concept is that of "inferior" and "non-inferior" designs or plans. Suppose two objectives are relevant in a particular planning problem. These may be national economic development and environmental quality. All feasible designs could be evaluated in terms of these two objectives (e.g. Environmental Quality and National Economic Development) and plotted in two-dimensional space, as shown in Figure 9.6.

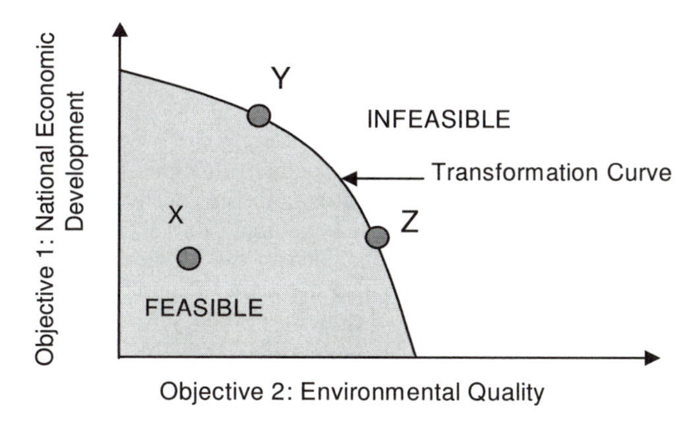

Figure 9.6 Multiobjective tradeoff between alternatives

By definition, an alternative A is inferior to alternative B if B ranks higher or equal to A for all objectives and higher for at least one objective. If A ranks higher than B for one objective and B higher than A for another they are both non-inferior (for this pairwise comparison).

In Figure 9.6 design X is clearly inferior to design Y and can be discarded from further analysis. Designs Y or Z are non-inferior to each other and should therefore both be retained in the evaluation. Designs which are not inferior to any other are called "non-inferior" designs. A line joining all non-inferior designs is called the transformation curve (or transformation surface if there are more than two objectives). The transformation curve represents the boundary between feasible and infeasible designs in objective space.

Clearly the preferred design should lie on the transformation curve, provided all relevant objectives have been included in the analysis. The choice of a final design involves a value judgment as to the relative importance of the objectives and, in many cases, will involve a political decision.

The role of the analyst in multiple objective planning is to identify the most efficient designs in terms of all the objectives, and recognize the tradeoffs among them. There are many techniques and software available to do this for different kinds of developments and an indication of the development of a Multi Objective Decision Support System (MODSS) for transport situations is described in some detail in Thoresen et al. (2001).

9.6 SOCIAL IMPACT

The environmental impact statement usually includes social impacts as well as effects on the physical and biological components of the environment. On the other hand, the Principles and Standards (USWRC, 1973) considered social well-being as a separate category of objectives. The social objective was defined as follows:

> *To enhance social well-being by the equitable distribution of real income, employment and population, with special concern for the incidence of the consequences of a plan on affected persons or groups; by contributing to the security of life, health and property; by providing educational, cultural and recreational opportunities; and by contributing to national security.* (USWRC, 1973)

If a full economic, environmental and social evaluation of a project is carried out, it is most important to avoid the **double-counting** of benefits or costs under more than one account. If, for example, an economic value is placed on lives saved through accident reduction and this is included in the *economic* evaluation of the project, it would not be appropriate to also highlight the number of lives saved in the *social* evaluation. Other examples of double-counting are discussed later in this section. The following types of social effects may need to be considered for a particular project:

- distribution of income;
- population distribution;
- employment;
- life, health, and safety;
- educational, cultural, and recreational opportunities;
- national security and emergency preparedness;
- displacement and relocation; and
- neighbourhood disruption and intrusion.

As with environmental effects, each social parameter can have its own ordinal or cardinal scale of measurement. The above effects are considered in turn.

Distribution of Income

In theory it is possible for a government to redistribute the benefits and costs of any public sector project among the various groups in society. This being the case, it can be argued that the role of the engineer or planner is to devise projects which maximise the total net benefits to the community and let the government decide how these benefits (and costs) will be distributed. However, this redistribution of benefits and costs is often not practicable because of the difficulties in identifying all affected groups and because of the administrative cost of carrying out the transfers. Redistribution of the benefits of engineering projects rarely occurs in practice.

It is therefore important that the distributional consequences of any major project be explicitly considered. Of particular relevance is the effect of the project on the income levels of certain target groups. These target groups may be distinguished on the basis of income (e.g. those living below the poverty level), race, sex, or geographic region.

It has been suggested by Weisbrod (1972) that the distributional question can be approached by placing explicit weights on the benefits and costs received by each group in the community. Traditional benefit-cost analysis gives the same weight to a dollar of benefits received by a pauper and by a millionaire. It is conceptually possible to give a higher weight to benefits or costs incurred by low-income earners and thus develop a single index of social welfare, using the following formula:

$$SW = \sum_{i=1}^{m} w_i [B_i - C_i] \tag{9.2}$$

Where: SW is the index of social welfare; B_i is the present value of benefits received by group *I;* C_i is the present value of costs incurred by group I; w_i is the "social" weight of benefits or costs to group *i;* and *m* is total number of community groups. The difficulties of deciding on a set of weights and implementing this approach are apparent.

Population Distribution

It has been argued that the redistribution of population throughout a region or the nation may be a major objective of government policy. This is particularly true in sparsely populated nations such as Australia. Major engineering projects in remote areas may contribute positively to the decentralization of population. The benefits of decentralization are related to enhanced national security and the benefits of rural versus urban living. High concentrations of population and industry are particularly vulnerable to attack by conventional or nuclear weapons. On the other hand, the settlement of remote areas leads to additional infrastructure costs for electricity supply, roads, railways, water supply, ports, and airfields.

Projects such as major irrigation developments which contribute to maintaining the size of the rural community may be supported on the grounds of the "healthier" lifestyle of country living. Continued growth of large cities at the expense of the surrounding country areas may also involve considerable increases in infrastructure costs in the city, as well as increases in the crime rate and other social problems. The negative aspects of population redistribution on individual families also need to be considered. These include the weakening of friendships and family ties and the stress associated with moving, setting up a new home, and adjusting to a changed lifestyle. The benefits of urbanization versus decentralization can vary from family to family depending on the independence of the family members.

Employment

One of the benefits cited by governments when new public works are announced is the effect that it will have on unemployment. A large engineering project may involve considerable employment in the construction phase and also in the operations phase. Some people consider this an achievement in itself. In an economy with little or no unemployment, workers employed on a specific project will be displaced from productive activities elsewhere in the economy. Therefore, the opportunity cost of their labour will be assessed as a *cost* to the project in the economic evaluation. In such a case, the number of people employed on the project is not a social benefit.

In an economy with high unemployment, some of the workers employed on the project may have been previously unemployed. Previously unemployed labour has an opportunity cost to the project considerably less than the wage rate. If this effect is correctly taken into account in the economic evaluation of all projects, there is no need to further describe the employment implications of the projects, as this would be a double-counting of the same benefit.

The foregoing applies to public sector projects only. In the economic evaluation of private sector developments all labour is costed at the actual wage rate paid. Thus, at times when there is less than full employment in the economy, the true net benefits of the project to society will be understated. In such cases it *is* valid to include the additional employment opportunities created by the project as part of the social evaluation.

The distributional consequences of employment opportunities may also be important; for example, in the creation of employment opportunities in depressed regions of the country or among disadvantaged groups such as ethnic, or low-income groups.

Life, Health, and Safety

Many engineering works contribute to saving lives, improving health, or increasing safety in the community. Typical examples are flood mitigation works, water treatment plants, sewage treatment, improved road alignments, grade separation, and coastal protection works. Attempts have been made to include the benefits of lives saved into the economic evaluation of public sector projects by placing an economic value on human life. Various methods have been suggested for estimating the economic benefits of saving human life. These include:

- The present value of the person's expected future earnings;
- A value imputed from political decisions which involve public expenditure aimed at reducing the number of deaths. For example, the government spends $50 m in equivalent annual worth to upgrade railway crossings which will save an estimated 100 lives per year from fatal accidents. The implied value of saving one life is greater than or equal to $50 m/100 (i.e. $0.5 m); and
- A dollar amount which people are prepared to accept as compensation in order to put up with the additional risk of death involved in the project. For example, a proposed new chemical plant will increase the probability of death per year to each person in an adjoining town by 1 in 10,000. The economic cost of this increased risk is the sum over all people in the town of the minimum level of compensation which each person is prepared to accept in order to tolerate the increased risk. The same basis can be used for projects which *reduce* the probability of death for part of the community.

Practical and theoretical difficulties exist with all of these economic approaches. An alternative approach is to estimate the number of lives saved per year and treat this as a social benefit. Similar arguments apply to projects which increase or reduce the risk to public health or the chance of accidents occurring. Either a dollar value may be used in the economic evaluation or an estimated reduction in the risk may be included in the social evaluation.

Educational, Cultural, and Recreational Opportunities

It has already been stated that the distribution of income, and in particular its spread to certain disadvantaged groups, is an important social effect. Similarly the distribution of educational, cultural, and recreational opportunities among regions and socio-economic groups may be an important consideration in evaluating a major project. For example, a major dam may provide a new recreational resource for fishing, boating, picnicking, and swimming (depending on the final end use of the water). The social benefits of these recreational uses may be as important as the economic benefits of the water provided by the dam. In some cases it may be advantageous to choose a dam site close to a population centre in order to increase the recreational opportunities for a particular group of people.

Educational and cultural values may be enhanced by engineering projects providing improved access to sites of historical, archaeological, or scientific interest. For example, a new road along the coast may provide improved access to some unique geological formations which are of considerable interest to the public, but controls need to be enforced; otherwise the project could lead to the ruin of the formation.

National Security and Emergency Preparedness

A new road or railway could have important consequences for national security. Such benefits are difficult to quantify in economic terms and, in fact, only become apparent in times of armed conflict. However, they may be an important social consideration in the evaluation of a project.

Emergency preparedness requires, for example, the provision of a flexible water supply, electricity grid, and road and rail network. The provision of a single dam to supply water to a city or a single power station to supply electricity may be economically efficient under normal circumstances but could be disastrous in the event of a catastrophe such as major technical failure, earthquake, hurricane, nuclear explosion, or act of sabotage. The provision of some redundancy in all engineering systems is a good rule to follow.

Displacement and Relocation

Engineering projects in urban areas may involve the displacement and relocation of some houses, and commercial or industrial buildings. A typical example arises from the land acquisition associated with an improved transport link, such as a freeway, arterial road, tramway, or railway. It is common practice for the households or firms displaced to be suitably compensated for the loss of their land, home, shop, or factory.

It is often argued that the loss of physical assets is suitably compensated for by the payment of fair market value. However, such an argument ignores the fundamental distinction between value in exchange and value in use. Consider the market for houses of a particular age and quality. If the current market price for such houses is P_o, this may be called the value in exchange of the property. Clearly all homeowners who are prepared to sell at this price or less will put their houses on the market. However, there are many homeowners who are not willing to sell at the price P_o. To these people, the value in use of the house exceeds its value in exchange. If a road authority were to compulsorily acquire such a house and pay compensation equal to P_o, the homeowner would clearly suffer a loss of utility. Full compensation should cover the value in use of the asset and would usually exceed P_o by a significant margin. A similar argument applies to commercial and industrial properties.

The difference between value in use and value in exchange of a house includes an allowance for the stress of moving, disruption of social ties, the cost of finding a new house, and changes in accessibility and travel costs at the new location. In practice such costs are difficult to assess and considerable negotiation may be required with the property owners.

The compensation costs for displaced households and firms are a true opportunity cost to a project and should be included in the economic evaluation, rather than the social evaluation.

Neighbourhood Disruption and Intrusion

An engineering project may involve significant disruption and/or intrusion into a community. This is particularly true of transportation corridors which can sever a cohesive community or produce intrusive impacts in the form of noise, vibration, air pollution, or aesthetic degradation. A sense of community is often very strong and may be centred on facilities such as schools, shopping centres, the town hall, or parks. Proposals which isolate one part of the community from another may receive considerable opposition from local community groups. Transport corridors should be planned to provide the minimum disruption to communities and ensure that adequate safe crossings such as overpasses or underpasses are provided for pedestrians and cyclists.

The principle of compensation described in the previous section can be applied to neighbourhood disruption and intrusion, but it is very difficult to define the loss of utility for an entire community. Attempts have been made to assess the economic effects of noise due to airports, highways, or railways by assessing the lower property values adjacent to these facilities. If such an approach is taken, these social effects can be included as a cost in the economic evaluation.

9.7 SOCIAL ASSESSMENT TOOLS AND METHODS

An extremely important facet of undertaking any project analysis now involves the participation of interested stakeholders. This is an area where many engineers will be called on to manage and participate with specialists who work in the social assessment area. A basic knowledge of the methods available for undertaking such project analyses are given in the UNEP EIA training manual (UNEP, 2002). A number of methods are described below which allow social concerns to be discovered and to determine local knowledge concerning a proposed project.

Analytical tools

Stakeholder Analysis is an entry point to Social Impact Assessment (SIA) and participatory work. It addresses strategic questions, e.g. Who are the key stakeholders? What are their interests in the project or policy? What are the power differentials between them? What relative influence do they have on the operation? This information helps to identify institutions and relationships which, if ignored, can have negative influence on proposals or, if considered, can be built upon to strengthen them.
Gender Analysis focuses on understanding and documenting the differences in gender roles, activities, needs and opportunities in a given context. It highlights the different roles and behaviour of men and women. These attributes vary across cultures, class, ethnicity, income, education, and time; and so gender analysis does not treat men or women as a homogeneous group.
Data Review of information from previous work is an inexpensive, easy way to narrow the focus of a social assessment, to identify experts and institutions that are familiar with the development and context, and to establish a relevant framework and key social variables in advance for the project.

Community-based methods

The Participatory Approach aims to ascertain local knowledge and actions. It uses group exercises to enable stakeholders to share information and to develop plans. These techniques have been employed successfully in a variety of settings to enable local people to work together to plan community-appropriate developments. Attributes of self-esteem, associative strength, resourcefulness, action planning and responsibility for follow-through are important to achieve a participatory approach to development. Generally there is a philosophy of empowerment of the stakeholders to enable people to adopt responsibility for outcomes. It can best be described as development of teambuilding skills and learning from local experience rather than from external experts. Other participatory consultation methods include selecting a sample of stakeholders to ensure that their concerns are incorporated into the assessment. This selection is for the purposes of giving voice to the poor and other disadvantaged stakeholders.

Other Participatory Methods

Role Playing helps people to be creative, open their perspectives, understand the choices that another person might face, and make choices free from their usual responsibilities. This exercise can stimulate discussion, improve communication, and promote collaboration at both community and agency levels.
Wealth Ranking (also known as well-being ranking or vulnerability analysis) is a visual technique to engage local people in the rapid data collection and analysis of social stratification in a community (regardless of language and literacy barriers). It focuses on the factors which constitute wealth, such as ownership of or right to use productive assets/resources, their relationship to locally powerful people, labour and indebtedness.

Mapping is an inexpensive tool for gathering both descriptive and diagnostic information. Mapping exercises are useful for collecting baseline data on a number of indicators as part of a beneficiary assessment or rapid appraisal, and can lay the foundation for community ownership of development planning by including different groups and making them aware of the implications of the project.
Needs Assessment draws out information about people's needs and requirements in their daily lives. It raises participants' awareness of development issues and provides a framework for prioritising actions and interventions. All sectors can benefit from participating in a needs assessment, as can trainers, project staff and field workers.
Tree Diagrams are multi-purpose, visual tools for narrowing and prioritising problems, objectives or decisions. Information is organized into a tree-like diagram. The main issue is represented by the trunk, and the relevant factors, influences and outcomes are shown as roots and branches of the tree. Other techniques such as mind mapping, as discussed in Chapter 6, can also be used for this purpose.

Observation and interview tools

Focus Group Meetings are a rapid way to collect comparative data from a variety of stakeholders. They are brief meetings - usually one to two hours - with many potential uses, e.g. to address a particular concern; to build community consensus about implementation plans; to cross-check information with a large number of people; or to obtain reactions to hypothetical or intended actions.
Workshop-based methods encourage participatory planning and analysis throughout the project life cycle. A series of stakeholder workshops tend to be held to set priorities, and integrate them into planning, implementation and monitoring. Building commitment and capacity is an integral part of this process (UNEP, 2002). These stakeholder workshops are in common use for many infrastructure and resource projects throughout the world.

9.8 SUMMARY

The engineer has a responsibility to both the client and society when developing infrastructure and this entails evaluating environmental and social effects of the project. Many consulting engineering firms are required to perform environmental impact assessments for projects that they plan and design. In many cases the engineer will be actively involved in stakeholder consultations, giving presentations to community groups as well as professional groups. To do this, communication skills and knowledge of the environmental assessment process are essential. Future work for both private development and for government infrastructure will be undertaken using sustainability principles to guide the development of the project. The role of the engineer in planning is to identify the most efficient designs, in terms of all the objectives, and recognize the tradeoffs among them.

Many professional engineering institutions world wide have embedded sustainability within their charters, but progress to achieve these ideals has in the past been thwarted by little political and legislative assistance, but this is changing. The production of CO_2 and the linkage to climate change are the main driving forces for the way industry and government are viewing sustainability. Energy is a major input into how societies maintain their level of well being. The integration of sustainable energy systems with a process of closed loop systems for resource use is seen as a key element in approaching sustainability.

PROBLEMS

9.1 A study is being carried out on alternative energy sources for the future generation of electricity in South Australia. The energy sources include coal, natural gas, nuclear, solar and wind energy (and combinations thereof). List and describe the environmental and social effects which would need to be considered when comparing these alternatives. Illustrate use of the Leopold matrix display technique by deriving the matrix relevant to the use of coal as the primary energy source for Adelaide, mined from Leigh Creek in the north of South Australia and transported by rail to Adelaide.

9.2 A winery is proposing to expand its operations with the development of a new processing facility. The winery is confident with its proposal since it includes some additional treatment of the waste water before it is discharged into a nearby waterway, although the machinery used will create increased noise in the quiet rural setting and some venting of NO and NO_2 (reported as nitrous oxides, NO_x). Details of the air and water discharges, and the noise produced are given in Table 9.5. The results of an expert panel assessing the environmental quality of the relevant parameters are included in Figure 9.7.

Use the Environmental Index System to suggest whether, on the basis of the environmental information provided, the winery should be allowed to expand its operations. List, and discuss in 20 words or less (each), the six most important social issues that should be considered in the environmental assessment of the proposal.

Table 9.5 Discharge estimates for winery and its proposed development.

Water Quality Parameter	Existing Operations	Option 1
Turbidity (NTU)	30	20
Temperature (°C)	20	22
DO (mg/L)	8	6
pH	8	7
Noise (dB)	5	20
Nitrous Oxides (NOx) (μg/L)	5	15

9.3 In 1974, Ehrlich and Ehrlich published a book, *The End of Affluence*, in which they proposed a "nutritional disaster that seems likely to overtake humanity in the 1970s (or, at the latest, the 1980s). Before 1985 mankind will enter a genuine age of scarcity" in which "the accessible supplies of many key minerals will be nearing depletion." Comment on why this did not happen and whether it will occur by 2020 as many environmental activists are predicting?

9.4 A wave energy farm is to be constructed 1 km off shore to supply both electrical energy and desalinated water to a coastal community of 2,000 people. You are to prepare a scoping study which would include the activities of the development with the potential environmental impacts and suggestions of mitigating solutions to these impacts. Hint : There are a number of wave energy systems being trialled around the world. Choose two of these systems and compare with each other.

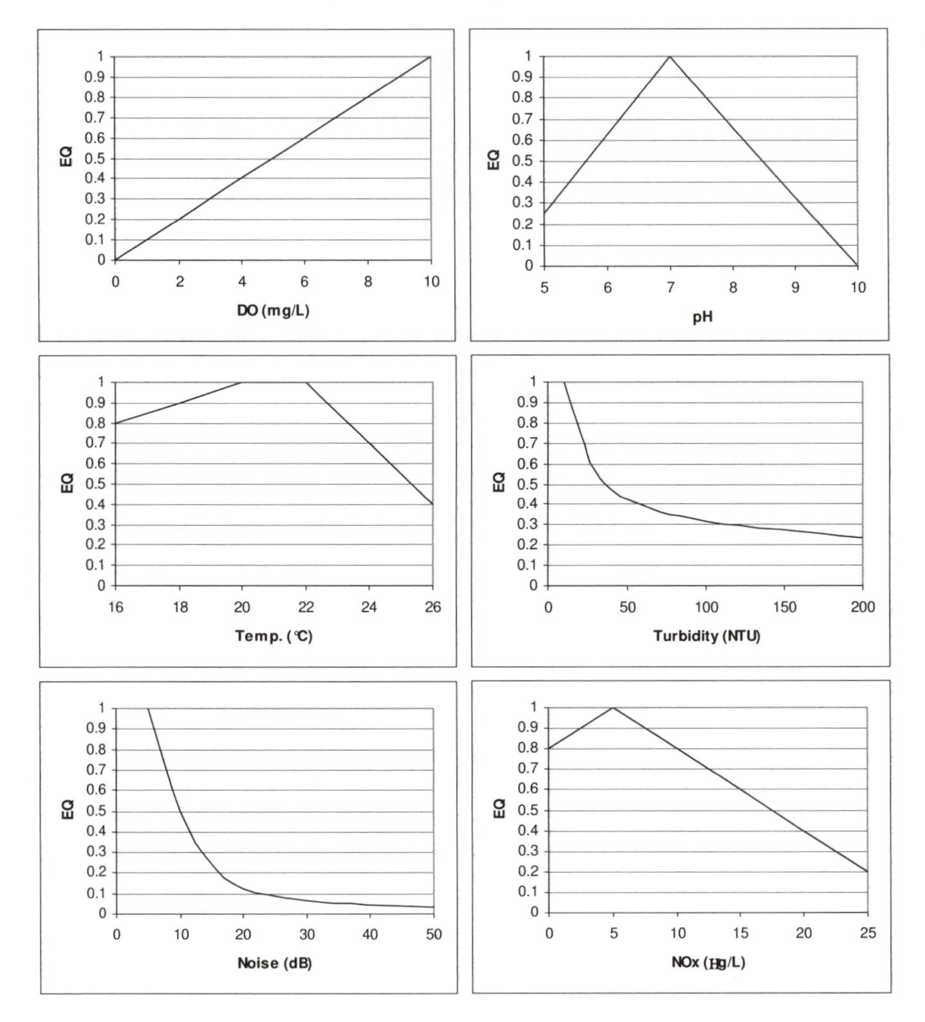

Figure 9.7 Environmental quality of various environmental parameters.

REFERENCES

AtKisson, A. and Hatcher, R.L., 2001, *The compass index of sustainability: Prototype for a comprehensive sustainability information system*, Journal of Environmental Assessment, Policy and Management, 3(4), 509–532.

Best Foot Forward Ltd, 2001, The use of Ecological Footprint and Biocapacity Analyses as Sustainability Indicators for Sub-national Geographical Areas: A Recommended Way Forward, http://www.bestfootforward .com/reports.html, accessed June 2006.

Canter L. C. and Sadler, B.B. 1997, *A Tool Kit for Effective EIA Practice — Review of Methods and Perspectives on their Application*, International Association for Impact Assessment, Fargo, North Dakota, U.S.A.

Dee, N., Baker, J.K., Drobny, N. L., Duke, K.M. and Fahringer, D.C., 1972, Environmental Evaluation System for Water Resource Planning, Final Rep., Batelle Columbus Labs, Columbus, Ohio, USA. 188pp.

Dee, N., Baker, J., Drobny, N. and Duke, K.M., 1973, An Environmental Evaluation System for Water Resource Planning, Water Resources Research 9, 523–535.

Fleming, N.S., 1999, *Sustainability and Water Resources Management for the Northern Adelaide Plains, South Australia*, PhD Dissertation at the University of Adelaide.

Fleming, N.S. and Daniell T.M., 1995, Matrix for Evaluation of Sustainability Achievement (MESA): Determining the Sustainability of Development, National Environmental Engineering Conference 1995, Towards a Sustainable Future: Challenges and Responses, The Institution of Engineers Aust. Preprints of Papers, 99–106.

Foley, B.A., Daniell, T.M., Warner, R.F., 2003, *What is Sustainability and can it be measured?* Australian Journal of Multidisciplinary Engineering, 1(1), 1–8.

Gilman, R., 1992, *Sustainability,* from the 1992 UIA/AIA Call for Sustainable Community Solutions. Retrieved 4 February 2003. http://www.context.org/ ICLIB/DEFS/AIADef.htm

Global Reporting Initiative, 2005, Reporting Framework, http://www. globalreporting. org/ ReportingFramework/, accessed June 2006.

Hargroves K., and Smith, M. H., 2005, *The Natural Advantage of Nations (Vol. I): Business Opportunities, Innovation and Governance in the 21st Century,* The Natural Edge Project, Earthscan/James & James.

Higgins J. and Venning, J. (eds), 2001, *Towards Sustainability: Emerging Systems for Informing Sustainable Development,* Sydney, Australia: University of New South Wales Press, 240 pp.

Leopold, L. B., Clarke, F.E., Hanshaw, B.B., and Balsley, J.R., 1971, *A Procedure for Evaluating Environmental Impact.* US Geological Survey Circular 645.

Martel, Gary and Lackey, Robert, 1977, A computerized Method for Abstracting and Evaluating Environmental Impact Statements, Bulletin 105 Virginia Water Resources Research Center, Virginia Tech, Dec.

Meadows, D.H., Meadows, D.L., Randers, J. and Behrens III, W.W., 1972, *The Limits to Growth*, A Report to The Club of Rome, Universe Books, New York. Available at: http://www.clubofrome.org/docs/limits.rtf, 20/05/04.

Parker, G., 1999, Progress in the Modeling of Alluvial Fans. *Journal of Hydraulic Research,* 37(6), 805–825.

Pope J, Annandale D, Morrison-Saunders A., 2004, Conceptualising sustainability assessment, *Environmental Impact Assessment Review*, 24, 595–616.

Roberts, D.V., 1990, *Sustainable Development: A Challenge for the Engineering Profession*. Paper presented at the International Federation of Consulting Engineers Annual Conference, Oslo, Norway, September.

Schmidheiny, S., 1992, *Changing Course: A Global Business Perspective on Development and the Environment*; MIT Press, Cambridge, MA.

Schmidt-Bleek, F., 1993, Wieviel Umwelt braucht der Mensch – mips, das ökologische Mass zum Wirtschaften , Birkhäuser, Basel, Boston, Berlin, English version in www.factor10-institute.org under the title "The Fossil Makers", accessed 23 Sept 2006.

Sharma, S., 1995, *Landscape and memory.* Knopf Publishers, New York.

Suter, K., 1999, About the Club of Rome, Australian Broadcasting Commission, Retrieved Sept 11, 2005. http://www.abc.net.au/science/slab/rome/rome.htm.

Thoresen T., Tsolakis, D., Houghton, N., 2001, *Austroads Project Evaluation Compendium,* ARRB Transport Research, Austroads Project No. RSM.F.N.021, Austroads Publication No. AP–R191/01

United Nations Development Program (UNDP), 2005, Human Development Report 2005, Available at http://hdr.undp.org/reports/global/2005/.

UNEP, 2002, EIA Training Resource Manual, 2nd Edition 600pp.

UNEP, 2003, Environmental Management Tools, Life Cycle Analysis, Retrieved 21 Sept 2005, http://www.uneptie.org/pc/pc/tools/lca.htm.

USWRC, 1973, *Principles and standards for planning water and related land resources*. United States Water Resources Council, Federal register, 38, Sept. 10, Washington D.C.

Venetoulis, J., Talberth, J., 2005, Ecological Footprint of Nations, 2005 Update, Redefining Progress, www.RedefiningProgress.org , 16p.

Weisbrod, B.A., 1972, Deriving an implicit set of governmental weights for income classes. In Cost Benefit Analysis, R. Layard (ed.), 395–428, London, Penguin.

Wackernagel, M. and Rees, W., 1997, *Our Ecological Footprint: Reducing Human Impact on the Earth*, New Society Publishers.

World Commission on Environment and Development (WCED), 1987, *Our Common Future*. New York: Oxford University Press.

APPENDIX 9A THE BATTELLE CLASSIFICATION AND LEOPOLD MATRIX

Table 9.6 The Battelle Environmental Classification for Water-Resource Development Projects (Dee *et al.*, 1973). The Bracketed Numbers are the Distributed Parameter Importance Units i.e. Relative Weights.

ECOLOGY (240)
Terrestrial Species & Populations
-Browsers and grazers (14)
-Crops (14)
-Natural vegetation (14)
-Pest species (14)
-Upland game birds (14)
Aquatic Species & Populations
-Commercial fisheries (14)
-Natural vegetation (14)
-Pest species (14)
-Sport fish (14)
-Water fowl (14)
Terrestrial Habitats & Communities
-Food web index (12)
-Land use (12)
-Rare & endangered species (12)
-Species diversity (14)
Aquatic Habitats & Communities
-Food web index (12)
-Rare & endangered species (12)
-River characteristics (12)
-Species diversity (14)
Ecosystems
-Descriptive Only

PHYSICAL/CHEMICAL (402)
Water Quality
-Basin hydrologic loss (20)
-Biochemical oxygen demand (25)
-Dissolved oxygen (31)
-Faecal coliforms (18)
-Inorganic carbon (22)
-Inorganic nitrogen (25)
-Inorganic phosphate (28)
-Pesticides (16)
-pH (18)
-Streamflow variation (28)
-Temperature (28)
-Total dissolved solids (25)
-Toxic substances (14)
-Turbidity (20)
Air Quality
-Carbon monoxide (5)
-Hydrocarbons (5)
-Nitrogen oxides (10)
-Particulate matter (12)
-Photochemical oxidants (5)
-Sulphur oxides (10)
-Other (5)
Land Pollution
-Land use (14)
-Soil erosion (14)
Noise Pollution
-Noise (4)

AESTHETICS (153)
Land
-Geologic surface material (6)
-Relief & topographic character (16)
-Width and alignment (10)
Air
-Odour and visual (3)
-Sounds (2)
Water
-Appearance of water (10)
-Land & water interface (16)
-Odour and floating material (6)
-Water surface area (10)
-Wooded /geologic shoreline (10)
Biota
-Animals -domestic (5)
-Animals -wild (5)
-Diversity of vegetation types (9)
-Variety within vegetation types (5)
Man-Made Objects
-Man made objects (10)
Composition
-Composite effect (15)
-Unique composition (15)

HUMAN INTEREST /SOCIAL (205)
Education/Scientific
-Archeological (13)
-Ecological (13)
-Geological (11)
-Hydrological (11)
Historical
-Architecture and styles (11)
-Events (11)
-Persons (11)
-Religions and cultures (11)
-'Western Frontier' (11)
Cultures
-Indians (14)
-Other ethnic groups (7)
-Religious groups (7)
Mood/ Atmosphere
-Awe/inspiration (11)
-Isolation/solitude (11)
- Mystery (4)
-'Oneness' with nature (11)
Life Patterns
-Employment opportunities (13)
-Housing (13)
-Social interactions (11)

Table 9.7 The Leopold Matrix (Leopold *et al.*, 1971). Part I Lists the Project Actions; Part 2 Lists the Environmental Characteristics and Conditions

PART 1: Project Actions (Horizontal Axis of Matrix)

A. MODIFICATION OF REGIME
a) Exotic flora or fauna introduction
b) Biological Controls
c) Modification of habitat
d) Alteration of ground cover
e) Alteration of ground-water hydrology
f) Alteration of drainage
g) River control and flow codification
h) Canalization
i) Irrigation
j) Weather modification
k) Burning
l) Surface or paving
m) Noise and vibration
B. LAND TRANSFORMATION AND CONSTRUCTION
a) Urbanization
b) Industrial sites and buildings
c) Airports
d) Highways and bridges
e) Roads and trails
f) Railroads
g) Cables and lifts
h) Transmission lines, pipelines and corridors
i) Barriers, including fencing
j) Channel dredging and straightening
k) Channel revetments
l) Canals
m) Dams and impoundments
n) Piers, seawalls, marinas, & sea terminals
o) Offshore structures
p) Recreational structures
q) Blasting and drilling
r) Cut and fill
s) Tunnels and underground structures
C. RESOURCE EXRACTION
b) Surface excavation
c) Sub-surface excavation
d) Well drilling and fluid removal
e) Dredging
f) Clear cutting and other forestry
g) Commercial fishing and hunting
D. PROCESSING
a) Farming
b) Ranching and grazing
c) Feed lots
d) Dairying
e) Energy generation
f) Mineral processing
g) Metallurgical industry
h) Chemical industry
i) Textile industry
j) Automobile and aircraft
k) Oil refining
l) Food
m) Lumbering
n) Pulp and paper
o) Product storage
E. LAND ALTERATION
a) Erosion control and terracing
b) Mine sealing and waste control
c) Strip mining rehabilitation
d) Landscaping
e) Harbour dredging
f) Marsh fill and drainage
F. RESOURCE RENEWAL
a) Reforestation
b) Wildlife stocking and management
c) Ground-water recharge
d) Fertilization application
e) Waste recycling
G. CHANGES IN TRAFFIC
a) Railway
b) Automobile
c) Trucking
d) Shipping
f) River and Canal traffic
g) Pleasure boating
h) Trails
i) Cables and lifts
j) Communication
k) Pipeline
H. WASTE DISPOSAL AND TREATMENT
a) Ocean dumping
b) Landfill
c) Emplacement of tailings, spoil and overburden
d) Underground storage
e) Junk disposal
f) Oil-well flooding
g) Deep-well emplacement
h) Cooling-water discharge
i) Municipal waste discharge including spray irrigation
j) Liquid effluent discharge
k) Stabilization and oxidation ponds
l) Septic tanks
m) Stack and exhaust emission
n) Spent lubricants
I. CHEMICAL TREATMENT
a) Fertilization
b) Chemical de-icing of highways, etc.
c) Chemical stabilization of soil
d) Weed control
e) Insect control (pesticides)
J. ACCIDENTS
a) Explosions
b) Spills and leaks
c) Operational failure
OTHERS
a)
b)

PART 2: Environmental Characteristics and Conditions (Vertical Axis of Matrix)

A. PHYSICAL AND CHEMICAL CHARACTERISTICS

1. *Earth*

a) Mineral resources
b) Construction material
c) Soils
d) Landform
e) Force fields & background radiation
f) Unique physical features

2. *Water*

a) Surface
b) Ocean
c) Underground
d) Quality
e) Temperature
g) Snow, Ice, & permafrost

3. *Atmosphere*

a) Quality (gases, particulates)
b) Climate (micro, macro)
c) Temperature

4. *Processes*

a) Floods
b) Erosion
c) Deposition (sedimentation, precipitation)
d) Solution
e) Sorption (ion exchange, complexing)
f) Compaction and settling
g) Stability (slides, slumps)
h) Stress-strain (earthquake)
f) Recharge
i) Air movements

B. BIOLOGICAL CONDITIONS

1. *Flora*

a) Trees
b) Shrubs
c) Grass
d) Crops
e) Microflora
f) Aquatic plants
g) Endangered species
h) Barriers
i) Corridors

2. *Fauna*

a) Birds
b) Land animals including reptiles
c) Fish & shellfish
d) Benthic organisms
e) Insects
f) Microfauna
g) Endangered species
h) Barriers
i) Corridors

C. CULTURAL FACTORS

1. *Land use*

a) Wilderness & open spaces
b) Wetlands
c) Forestry
d) Grazing
e) Agriculture
f) Residential
g) Commercial
h) Industrial
i) Mining & quarrying

2. *Recreation*

a) Hunting
b) Fishing
c) Boating
d) Swimming
e) Camping & hiking
f) Picnicking
g) Resorts

3. *Aesthetics & Human Interest*

a) Scenic views and vistas
b) Wilderness qualities
c) Open space qualities
d) Landscape design
e) Unique physical features
f) Parks & reserves
g) Monuments
h) Rare & unique species or ecosystems
i) Historical or archaeological sites and objects
j) Presence of misfits

4. *Cultural Status*

a) Cultural patterns (life style)
b) Health and safety
c) Employment
d) Population density

5. *Man-Made Facilities and Activities*

a) Structures
b) Transportation network (movement, access)
c) Utility networks
d) Waste disposal
e) Barriers
f) Corridors

D. ECOLOGICAL RELATIONSHIPS

a) Salinization of water resources
b) Eutrophication
c) Disease-insect vectors
d) Food chains
e) Salinization of surficial material
f) Brush encroachment
g) Other

E. OTHERS

a)
b)

CHAPTER TEN

Ethics and Law

In this chapter we discuss how ethics and legal obligations influence decisions that engineers make. Situations during an engineer's career do evolve which are complex and difficult to comprehend and depend on one's integrity, skills and values to enable an ethical course of action. An individual as a professional engineer must consider obligations to society, company and fellow employees in making responsible decisions. There are a myriad of laws controlling the work of an engineer and generally litigation will only be a last resort solution to a problem. The difference between the civil laws of tort and contract law will be discussed.

10.1 AWARENESS OF ETHICS

The interpretation of moral values, laws and principles which sway the decisions and behaviour of an individual or a group can be defined as ethics. Ethics is quite commonly termed moral philosophy, which attempts to distinguish between that which is right and that which is wrong, depending on an individual's set of values. It is used to analyse, to evaluate and to debate questions of morality which include (Newall, 2005):

- What is *right* and what is *wrong*?
- What are rights? Who or what has them?
- Is there an ethical system that applies to everyone?
- What are the differences between community interests and self interest? and
- What do we mean by honour, integrity and dignity?

The list can relate to the principles of any code of ethics. Many professions, such as the medical profession, the engineering profession and the legal profession, have their own codes of ethics. Some codes of ethics are entrenched in law by the licensing organisation of a profession with violations of the code subject to a civil or penal penalty, or an organisational penalty such as loss of licence or membership. Other codes are merely advisory or subject to enforcement by the promulgating institution. The engineering professions have established codes of ethics to guide the behaviour of engineers but these are still dependent on a choice by the individual.

However, philosophers also debate the issues of moral reasoning, normative ethics, and analyse the terms used in moral discourse incorporated in analytic ethics or metaethics which then leads to being able to rationalise the differences between normative and analytic ethics. Analytic ethics question whether moral judgments are even possible or do moral values exist objectively or only subjectively or are moral values linked to culture or individuals? "The purpose of analytic and normative ethics is to enable us to arrive at a critical, reflective morality of our own choosing" (Taylor, 1972).

Code of Ethics of The Institution of Engineers, Australia (2005)

All members of the Institution of Engineers, Australia, in the practice of the discipline of engineering, are committed and obliged to apply and uphold the Cardinal Principles of the Code of Ethics, which are:

- to respect the inherent dignity of the individual;
- to act on the basis of a well informed conscience; and
- to act in the interest of the community.

These principles are encapsulated within and established by the Tenets of the Code of Ethics.

The Tenets of the Code of Ethics:
1. Members shall place their responsibility for the welfare, health and safety of the community before their responsibility to sectional or private interests, or to other members;
2. Members shall act with honour, integrity and dignity in order to merit the trust of the community and the profession;
3. Members shall act only in areas of their competence and in a careful and diligent manner;
4. Members shall act with honesty, good faith and equity and without discrimination towards all in the community;
5. Members shall apply their skill and knowledge in the interest of their employer or client for whom they shall act with integrity, without compromising any other obligation to these Tenets;
6. Members shall, where relevant, take reasonable steps to inform themselves, their clients and employers of the social, environmental, economic and other possible consequences which may arise from their actions;
7. Members shall express opinions, make statements or give evidence with fairness and honesty and only on the basis of adequate knowledge;
8. Members shall continue to develop relevant knowledge, skill and expertise throughout their careers and shall actively assist and encourage those with whom they are associated to do likewise;
9. Members shall not assist in or induce a breach of these Tenets and shall support those who seek to uphold them if called upon or in a position to do so.

An awareness of alternatives is essential when deliberating any decision as this is what enables us to express a freedom of choice. The principle of freedom is essential to a theory of ethics. People can not be responsible for their actions if there is no freedom of choice. Any action will determine a consequence which partly results from the set of laws and partly from the principles governing the decision.

There is a relativity in ethics which explains that conduct called good or bad and acceptable or unacceptable now, has varied across time and among societies. Therefore some people call things "right" that many others have a problem with.

The Dalai Lama stated in a radio interview, "Of course, there are different truths on different levels. Things are true relative to other things; "long" and "short" relate to each other, "high" and "low," and so on. But is there any absolute truth? Something self-sufficient, independently true in itself? I don't think so." (Thurman, 1997).

Much of the moral reasoning of ethics in Western countries in the past was influenced by the Greek philosophers and, in particular, Socrates, Plato and Aristotle. Other major ethicists and philosophers through the ages have included the instigators of the various religions: Buddha (B.C. 483-563) the Indian philosopher and spiritual leader spreading Buddhism right through Asia; Lao-zi, or Lao Tze, the founder of Taoism and the supposed author of Tao Te Ching,; and Confucius (Kong Fu-zi), whose philosophy and religion came to dominate China for more than two millennia; Mohammed with the Koran and Islam; the Hebrew Bible with Judaism; and Christianity spread from Jesus. There are a myriad of writings to direct us into establishing a base from which ethics can be discussed.

A person who can think and decide on the basis of his or her own thinking could be considered an individualist, even when the reasoning leads to society's norms. This is because the person has reasoned the decision and not just acted on society's value system.

Socrates (470–399 BC)

Socrates was a philosopher who took his own life by drinking hemlock after being found guilty by an Athenian people's court of interfering with the religion of the city. Socrates "was accused of corrupting the youth" because he taught his students to question thinking. In his method, questions were posed to assist a person or group to examine their beliefs and their knowledge. The Socratic method uses hypothesis elimination, by eliminating those hypotheses that lead to contradictions and hence leaving the better hypotheses. This method forced people to examine their own beliefs as well as those of authorities. Socrates seemed to believe that wrongdoing resulted from ignorance, in that those who did wrong knew no better.

He did not recant and was sentenced to death. He could have escaped through help of his friends but refused. In fact, Socrates said, "The life untested by criticism is not worth living." Ullmann(1986)

The case of "normative ethics" where the answer describes how individuals ought to respond when moral decisions are to be made is muddied by the technological facts. Ullmann (1983) condemns people, including some engineers, who act as the implementers of others' goals without questioning them, when these goals could be to design weapon systems of mass destruction, or gas chambers and crematoria of concentration camps. The supporting argument from the condemned parties could be "that I was only obeying orders or doing my job". It is necessary as an engineer to collect information for any decision and weigh up the consequences of the outcomes resulting from that decision.

The dilemma of climate change is an interesting case study. The demand for energy seems to be increasing in all walks of life. This demand is increasingly being met by the manufacture of more energy by fossil fuel industries, which in turn produce more CO_2. The fossil fuel industries are deemed to be contributing to climate change and, as such, are contributing to the changing face of the earth.

Should individuals and industries cut their energy use? Would this reduce the production of CO_2 in the immediate future? Is this a major ethical decision that needs to be made by all people and engineers now? Can individuals and organisations generate their own energy requirements using renewable energy sources? Does an engineer have a responsibility to ensure that what is developed now does not impinge on future generations by sacrificing their quality of lifestyle? Under a sustainability ethos do we have a responsibility to future generations? Individuals in different industries are responding in different ways. Because of the widespread consequences of some technological innovations, there is a need for engineers to recognise ethical problems, and then confront the issues in the best way possible. Under the **precautionary principle,** if the results of an action will be severe or irreversible, and if there is no scientific certainty of the results, then the burden of proof falls on those who direct and take the action. There has been a debate within the climate change fraternity for many years over this principle with regards to greenhouse gases.

Throughout the world there are many different codes of ethics for professional engineers that set standards that should be adhered to by them. In Australia, The Institution of Engineers Australia (2000) has an approved Code of Ethics, which has also been adopted by The Association of Consulting Engineers, Australia, and The Association of Professional Engineers, Scientists and Managers, Australia. Many professional organisations have codes of ethics, and access to more than 850 professional ethics codes can be made through the Centre for the Study of Ethics in the Professions (Illinois Institute of Technology, 2005). It is essential to know what ethical behaviour is within engineering profession so that engineers can adhere to it.

From the Code of Ethics it can be seen that an engineer has an obligation to society, to employers and clients, and to other engineers. The Code of Ethics establishes the standards which engineers are expected to adopt to regulate their work habits and relationships. The guidance notes appended to the Code elaborate on:

- relating to clients and employers;
- interacting with the community;
- interacting with colleagues;
- areas of competence and descriptions of qualifications;
- acting as an expert witness;
- making public comment or statements; and
- unauthorised release of information.

Within any choice there is a legal or moral standard that an engineer has to adhere to. As a professional engineer it is essential to know what the right thing to do is. There are many examples of bad choices littering the profession, but for every bad choice there are many thousands of good choices that do not get any publicity.

In the UK, Chartered Engineers and Incorporated Engineers are expected to observe the Code of Conduct of the engineering institution they join. Each UK Institution follows the Engineering Council guidelines for Code of Conduct detailed in this section. The code of conduct stresses that engineers should act with integrity in dealing with all issues associated with the profession, from dealing with the public, their fellow workers, the environment and all business operations.

UK Council of Engineering (2005) - Guidelines for Codes of Conduct

Each licensed UK engineering institution places a personal obligation on members to act with integrity, in the public interest, and to exercise all reasonable professional skill and care to:

1. Prevent avoidable danger to health or safety.
2. Prevent avoidable adverse impact on the environment.
3. (a) Maintain their competence.
(b) Undertake only professional tasks for which they are competent.
(c) Disclose relevant limitations of competence.
4. (a) Accept appropriate responsibility for work carried out under their supervision.
(b) Treat all persons fairly, without bias, and with respect.
(c) Encourage others to advance their learning and competence.
5. (a) Avoid where possible real or perceived conflict of interest.
(b) Advise affected parties when such conflicts arise.
6. Observe the proper duties of confidentiality owed to appropriate parties.
7. Reject bribery.
8. Assess relevant risks and liability, and if appropriate hold professional indemnity insurance.
9. Notify the institution if convicted of a criminal offence or upon becoming bankrupt or disqualified as a Company Director.
10. Notify the institution of any significant violation of the institution's Code of Conduct by another member.

It has been stated by Johnston et al. (1995) that, historically, codes of ethics can be seen to be stifling dissent within a profession. This is because codes of ethical behaviour, in general, emphasize refraining from criticizing fellow professionals and maintaining the profession's reputation. Situations where major disasters have occurred have led to the perception that there have been a lot more "whistle blowers" in recent years in engineering. Perhaps as technology has become more complex and environmental concerns have heightened, there is a greater concern about ethics in decision making. To examine a problem from many angles is a useful way to develop an ethical approach as the choice to take a certain action will only exist if we are aware of alternatives.

10.2 ETHICS AND WHISTLE BLOWING

When does the dilemma of obligation to society override the obligation to the company or to peers? This concern is usually highlighted in the professional code of ethics. How does one address the obligations that a professional engineer has? Is there a natural hierarchy of obligation? Generally it is acknowledged that an obligation to society takes precedence over the employer. What are the responsibilities, of both a personal and professional nature, which complicate any ethical decision? When it comes to "blowing-the-whistle", there are numerous concerns, not the least being career prospects and family life. The story of the

Russian mining engineer, Peter Palchinsky, may serve as an inspiration for engineers today to follow moral principles, perhaps with less drastic consequences. Graham (1996) in his book, *The Ghost of the Executed Engineer – Technology and the Fall of the Russian Empire*, relates how Peter Palchinsky (1875–1929) spoke up against the misuse of resources. The misuse of technology and squandering of human resources by the Russian government incensed the professional side of Palchinsky and he attempted to make facts of the misuse known. He was imprisoned, but was released shortly thereafter by the efforts of his wife and friends. Palchinsky continued his professional engineering career and went on to criticize Stalin's projects: the White Sea Canal; Dnieper River Dam; and the Steel City of Magnitogorsk. His days were numbered, and he was executed in May 1929. It was shown by Graham that there are different consequences of voicing morally correct ideas or demonstrating an obligation to society, from those of politically correct ideas depending on the political environment at the time. The book also shows the dangers of engineering major projects which ignore human values.

The whistle blowing exploits of engineer Roger Boisjoly, concerning the Challenger disaster of January, 1986, are still important for study on ethical decision making. The details about how the launch went forward in unusually cold temperatures against the recommendation of the engineers are well known. Examining the events in ethical terms leads us to the question of what defines "whistle blowing". The term portrays the sharp sound of a whistle, warning of harm or trying to halt actions which have gone out of bounds. Roger Boisjoly's action did not fit that scenario. What branded him as a whistleblower was his decision to relate to a Presidential Commission *after* the disaster the story of events that led to the disaster. The tragedy he had warned about had already occurred. He told the commission the history of the O-ring problems, and the decision to launch, even though he was being pressured by managers of his own company not to do so. Boisjoly's account was forward looking in view of the design of newer shuttles. It was really the high visibility of Boisjoly's disclosure about the failure that labelled his report as "whistle blowing".

When a discloser deliberately steps outside of approved organizational channels to reveal a significant problem, then this could be termed "whistle blowing". In this case the warning against launching in the meeting before the disastrous launch was acting within approved channels. Can it be said that all those decision makers above Boisjoly acted ethically? His testimony to the Presidential Commission did not contravene the approved channels, but rather, his testimony was honest and demonstrated the obligation he had to society.

Non-ethical behaviour seems to be driven by the advance of industry and the desire for success, power and money. Over time the need to choose between right and wrong, survival of a family or company, use of the environment now or preserving it for future generations is becoming more important for the individual in daily life. Ethical decisions reflect the conscious decision to do what is the right thing.

Generally people use ethics to impose self constraints on behaviour whereas law could be described as a system of externally imposed or legislated constraints on the behaviour of the individual and society. The question of "what to do?" arises often when there is conflict between ethics and law.

Pulp and Paper Mills Canada

The international agency responsible for monitoring Canada's environmental performance, the Commission for Environmental Cooperation (CEC), announced that it will investigate lax enforcement of federal anti-pollution laws against Canada's pulp and paper industry. The investigation comes as a result of a complaint submitted to the CEC on behalf of a coalition of environmental groups in 2002.

"Certain pulp and paper mills have been allowed to violate Canada's anti-pollution laws with impunity for more than a decade," said Sierra Legal Defence Fund lawyer, Robert Wright. "The CEC's investigation will help illuminate why the federal government turned a blind eye to these polluters and spur the federal government to use the full range of enforcement tools at its disposal to safeguard our environment."

The complaint asserted that that the federal government failed to enforce the *Fisheries Act* and the *Pulp and Paper Effluent Regulations* for mills in Ontario, Quebec and the Atlantic provinces - despite documentation of more than 2,400 violations of federal laws self-reported by the mills over a five-year period.

Sierra Legal Defence Fund (2003)

10.3 THE LEGAL SYSTEM

The timing of the first written laws seems to be around 2400BC in Sumeria where rulers of the Sumerian states had promulgated laws and regulations. According to Kramer (1963), "The Ur-Nammu law code was originally inscribed no doubt on a stone stele, not unlike that on which the Akkadian law code of Hammurabi was inscribed some three centuries later".

Roman Law was the legal system operating in the City of Rome and in the Roman Empire as it expanded through many centuries through Europe and beyond. On the demise of Roman rule over Europe, Roman Law was largely forgotten. However, Roman Law continued to be an influence on legal thinking and legal practice and through the ages has been revived, transformed and reinvented many times. For the first two centuries of the Common Era, Roman legal science was very fertile. This age is considered to be the classical period of Roman Law, as the law which was taught and practised best exemplified the Roman legal tradition (Rufner, 2006). From the 10th century, a renewed interest in the law of the Romans was shown. Study and teaching of Roman Law at universities led to Roman Law being applied mainly in the area of civil law throughout most of Europe with England being an exception. However the Roman rules, applying in Europe from the 16th century, were very different from the Roman Law of antiquity and were called *Ius Commune* (common law). The modern legal European systems are derived from the *Ius Commune* as it was interpreted and rewritten by medieval jurists.

National codes, developed in the 18th and 19th centuries, superseded the *Ius Commune*. The most important of these was the Code of Napoleon in 1804 in France, which became the basis of establishing the rule of law in Belgium, the Netherlands, Portugal, Spain, and their former colonies. However, in the German

Reich, Roman Law remained in most of the provinces until 1900 when the German Civil Code was introduced. The Roman-Dutch Law based on Ius Commune still remains the basis of the legal system in the Republic of South Africa. The influence on the English Legal system from teaching of Roman Law at Oxford and Cambridge, amounted to ways of reasoning and concepts developed by continental legal scientists.

The English legal system has been followed in New Zealand, Australia and other former member countries of the British Empire, as well as, to a large part, in the United States. It is a "common law" system which generally results from legal cases over a long period of time, whereas the codified civil law system on the European Continent, as described above emanated from Roman law. The nature of "common law" is that it loosely describes the whole of the English legal system and has been used to distinguish between common law rules and equity, but the term in its broadest sense now embraces equitable principles.

There are two major sources for law, one being case law from judgements and the other from legislation through Parliament where legislators can delegate the law making to other bodies. Both new and amended legislation and case law result in the law continually changing. In the building industry, the engineer must be familiar with acts, ordinances and regulations which control the various facets of planning and construction of all urban infrastructure, to some degree, depending on the roles engineers take. There are some statutory modifications to common law, such as establishing the requirement of written contracts for sale of real estate.

Acts that will impinge upon an engineer's work are numerous and relate to all facets of the planning and development of infrastructure in a city and environs. For different countries and states these tend to be diverse and are continually changing under different governments.

In Australia there are both Federal laws (covering everyone in the country) and State laws (covering people in the State or Territory which passes the law). A State or Territory Parliament can pass a law in an area where there is no Federal law, or it can pass a law that adopts what is said in a Federal law. If there is a clash between a Federal law and a State or Territory law, the Federal law overrides the State law. The Constitution of Australia outlines the laws the Federal Government can pass. Hence engineers will find themselves working under both Federal and State laws.

Most engineering and building works are subject to one or more of a range of Government Acts and Regulations, at the State and Federal level. These may affect areas of planning development, design, construction, materials used, employment, environmental impacts, occupational health and safety, water supply and sewerage, water resources, natural resources, public health, energy and transport.

Different states have different Acts in Australia but some codes such as The Building Code of Australia (BCA), which is a set of technical provisions for the design and construction of buildings and other structures, are common to all states. The BCA is produced and maintained by the Australian Building Codes Board (ABCB) on behalf of the Australian Government and each State and Territory Government.

Criminal law is the body of law which deals with crime and the legal punishment of criminal offences. It is that part of law that deals with something of public interest. For example, the public has an interest in seeing that people are protected from being robbed or assaulted. Before the 18th and 19th centuries, legal systems did not clearly define criminal and civil law. Codification of criminal laws progressed and separation of civil and criminal law occurred. Criminal law distinguishes crimes from civil wrongs such as tort or breach of contract where there is a dispute between private individuals or organisations. The origin of tort comes from the French for "wrong" and is a civil wrong, other than a breach of contract. In engineering work the law of main interest is civil law, although criminal law can be relevant in the case of negligence.

10.4 LAW OF CONTRACT

Basic legal concepts

A **contract** is an "agreement" or a "promise" that is recognized by the law. "The common law assumes that a contract is an agreement between parties of equal bargaining power" (Cooke, 2001). A contract may be simply defined as a voluntary agreement where one party agrees to do work for another for a stated payment. For a contract to be legally enforceable it must include the following elements:

(a) An offer must be made by one party to the other;
(b) Acceptance of the offer must be made by the second party;
(c) There must be recompense, such as a promise of payment; and
(d) There must be a mutual intention to create a legal relationship.

It is interesting to note that a contract will only be valid if the work to be done is legal and if both parties are legally competent. Therefore, no contracts for terrorism or illegal drug deals will hold up in a court of law.

Written contracts

Any contract that uses words, spoken or written, is labelled a verbal contract. An informal exchange of spoken promises can be binding and regarded as legally valid as a written contract, but will generally lead to more disputation if large sums of money are involved. A spoken contract is often called an "oral contract". Therefore, all oral contracts and written contracts are labeled verbal contracts.

Courts in the United States have generally ruled that if the parties have the same intent and they act as though there were a formal, written and signed contract, then a contract exists. However, most jurisdictions require a signed written form for certain kinds of contracts e.g. real estate transactions and some building legislation, e.g. Home Building Act (NSW).

In Australia, there is no requirement for the entire contract to be in writing, although there must be evidence in form of a note or memorandum of the contract. The note or memorandum must be signed. In England and Wales there is a "Statute of Frauds" legislation, but only for guarantees which must be evidenced in writing, although the agreement may be made orally. Other kinds of contract (such as for the sale of land) must be in writing or they are void. If a binding agreement or an oral exchange is not honoured by one or more of the parties involved in the contract, by non-performance or interference with the other party's performance, then a "breach of contract" is deemed to have been committed.

All engineering contracts should be in writing to avoid ambiguities, and to simplify enforcement, if necessary. In the engineering and construction industry the word "tender" is used when describing an offer by one of the parties. The contract in an engineering sense includes drawings and specifications which include details of the full works required by the client/owner from the contractor.

Liability apart from contract

The Code of Hammurabi (1752 BC) imposed the death penalty on any builder whose structure failed and killed the occupant. Under the current legal system an engineer is liable to be guilty of a criminal offence if a structure fails and kills or maims someone or if it is designed with a gross breach of duty of care, gross carelessness and reckless disregard for the consequences of their actions. However, action is more likely to arise under civil law than under criminal law.

The Code of Hammurabi (1752 BC) Criminal Negligence

Hammurabi ruled Babylon, the world's first metropolis in Mesopotamia. A code of laws was established In Hammurabi's reign (1795–1750 BC). This is one of the earliest examples of a ruler proclaiming an entire set of laws, so that citizens might read and know the consequences of their actions. The code was carved into a stone, eight feet high.

The code clearly regulated the consequences if a builder built a house for someone, not constructing it properly, and the house collapsed and killed the owner; the builder would be put to death. If it killed the son of the owner, the son of that builder would be put to death as well. Compensation for all the goods inside the house had to be made as well.

These punishments took no note of explanations, but only the fact of failure, with one striking exception. An accused person was allowed to jump into the Euphrates River and let the gods appease the crime. Apparently the art of swimming was unknown; for if the accused survived he was declared innocent, but if he drowned he was guilty. Halsall (1998)

Contract Termination

A contract between parties is said to be discharged when the agreement is terminated. This can be done in a number of ways and usually for different reasons.

Agreement: The parties terminate the contract on mutual agreement by either party waiving their rights and/or releasing the other party from any obligations.

Completion: The contract is discharged completely when it is has been completely fulfilled in every respect by both parties.

Breach: Whenever a party fails an obligation which was included in the contract, then a breach of contract has been committed. Within this context, of course, an action results in trying to determine who is the cause of the breach and/or how the contract can recover from it. Breaches can arise where a contract is becoming seriously behind schedule. These are likely to follow many site meetings from the contract administrator to improve performance, or work on site slows down or even stops. The usual cause will be that the contractor is coming into financial difficulty. Clauses in the contract will allow both the owner/client to terminate the contract or the contractor may determine that the owner has not provided certain certificates or refused to supply sufficient instructions on the work. In either case, if the contract is terminated without good cause, then there is the possibility of either party being able to make a claim for damages for wrongful repudiation of the contract. In this kind of case there are bound to be lengthy delays and cost escalations on the contract, as well as large legal fees. Within the contract process a notice of intention to terminate by either party needs to be made under the procedures of the contract (Cooke, 2001).

Lapse of Time: Actions on a contract must be commenced within a certain time period decided by law.

Impossibility of Performance: This may arise when the contract works change so markedly from what was originally envisaged that a party can be exonerated from their obligations.

Most engineering contracts are terminated by completion or agreement. However, if termination occurs by breach of the contractor, then contract clauses usually permit the owner to retain all equipment and materials on site, completing the work at the contractor's expense. If the owner is in breach, then the contractor would take out an action against the owner. Failure to complete the contract within a specified time exposes the contractor to a claim of liquidated damages which could be included as a provision in the contract as a specific sum per day. Liquidated damages can be construed as a debt owed by the contactor to the owner and can be deducted from contract payments. Obviously the owner cannot delay the works and then recover damages. Therefore, all obligations on both parties should be strictly observed if their rights under the contact are to be preserved. A common occurrence for contracts running late is due to lack of supply of certain materials, long periods of rainfall not allowing the contractor to undertake work on site, and disputes between subcontractors and the main contractor. In some cases this can lead to liquidated damages.

The engineer will, in many cases, be in the position of an arbitrator to settle minor variations on the original contract. If there can not be an amicable solution between the contractor and the administrator (acting for the client), then all the parties could become involved in a dispute which could lead to litigation.

It is essential, therefore, that detailed and accurate record keeping of all contract variations and actions by the contractor need to be maintained by the resident engineer or administrator of the contract.

10.5 LAW OF TORT

William Prosser (1971) in his treatise, *Handbook of the Law of Torts*, defined "tort" as "a term applied to a miscellaneous and unconnected group of civil wrongs other than breach of contract for which a court of law will afford a remedy in the form of an action for damages." Besides damages, tort law will tolerate self-help in a limited range of cases for example, using reasonable force to expel a trespasser. Furthermore, in the case of a continuing tort, or even where harm is merely threatened, the courts will sometimes grant an injunction to minimise the continuance or threat of harm. Some engineering disputes, where negligence or damages are involved, are decided under the law of tort.

Negligence

Negligence is the broadest of the torts, forming the basis of many personal injury cases. The engineer has to be aware that there is a minimum *standard of care* that must be achieved in everyday activities. The engineer owes a *duty of care* to the client and *breaches* that duty by doing or not doing certain actions which are considered to be negligent. This breach must be the cause of client's (plaintiff's/claimant's) loss or damage and it is generally considered to be fair and reasonable to order the negligent party to pay compensation to the plaintiff/claimant.

Damages in Tort

When claiming for damages in tort, from deceit or negligence, the plaintiff has to justify to the court the amount that is being claimed (Cooke, 2001). "The object is to restore the plaintiff to the position in which he would have been placed if the wrongful act had not been committed" (South Australia v Johnson (1982) 42 ALR 161 at 169–170).

A case of damages can drive improved design procedures.

In a case of excessive cracking in a house caused by negligent footing design the court found that 50% was the responsibility of the builder, 25% the local authority and 25% the consulting engineer who advised the builder on the footing design. The damages awarded in this case were the cost of demolishing and rebuilding. The judgement was appealed and overturned at the Supreme Court but the High Court disagreed and reinstated the original decision. Cases such as this have resulted in a strict set of guidelines for footing design in South Australia. These judgements have also set a precedent for other cases involving damages for defective building work (Thomas, 2004).

10.6 THE PROCESS OF DISPUTES

Disputes in the engineering construction industry are of many different types. Historically, most disputes related to defective design and poor workmanship. As projects have become more complex, the nature of disputes has involved more parties from different disciplines. Traditionally, the mechanism for resolving construction disputes has followed a 'tiered' process involving:

- negotiation;
- facilitated negotiation or mediation;
- assisted negotiation or conciliation;
- expert determination;
- arbitration; and
- litigation.

As the dispute escalates through the stages, then so do the time and resources required for resolving the dispute.

Negotiation is a relatively inexpensive approach to dispute resolution. It generally is an informal process where the parties involved want to achieve outcomes as expeditiously as possible.

Mediation or conciliation can be seen as the next most expensive approach and can offer distinct advantages, in that the duration of the process can be pre-determined with the parties involved deciding the rules under which they and the mediator will operate, and the informality of proceedings may make the necessity for legal representation less likely. The benefits are a resolution achieved more expeditiously and economically.

Expert determination is also favoured as a resolution process, particularly for disputes relating to a single issue of a technical nature. Compared with litigation or arbitration, expert determination is generally faster and less expensive. In the case of complex technical disputes it is arguable that a Tribunal, which does not have sufficient (or any) technical literacy, can appropriately handle such disputes. The Superintendent, most likely an engineer, is the administrator of the Contract between the Principal (client/owner) and Contractor, and gives decisions on disputes. Increasingly the Superintendent has been viewed as being potentially compromised in this role with disputing parties, particularly those aided by legal advisors.

Arbitration previously offered a number of advantages over litigation. These entailed: privacy, less rigid procedures and formality; a shorter hearing time; lower costs; confidence in using a technical arbitrator; and a significant degree of finality, since a sound award is generally difficult to overturn. However, in the present climate in Australia, arbitration has become burdened with procedures similar to litigation, involving legal representation and procedural formality. The key benefits of time taken to bring a dispute to conclusion, and the cost, have escalated within these processes.

Litigation is both costly and time consuming. Disputes that arise during a project and go to arbitration or litigation are now unlikely to be resolved until after construction is complete, because of the lengthy procedures.

Because arbitration and litigation have become so costly, alternative methods of resolution are being called for in the construction industry. In the meantime, conciliation or mediation is being used as an intermediate step, before embarking upon the more costly and formal arbitration procedures.

The role of the Expert Witness

An Engineer can find himself or herself in an arbitration or Civil court action, or even a Criminal court action. The expert witness is usually asked by one or other of the parties contesting the case to act as independent engineer to investigate a failure or inadequate design or just advise on some dispute where an expert engineering opinion is required. However, an expert witness's general duty is to assist the court impartially on matters relevant to the expert's area of expertise. The expert has a duty only to the court and not to the party retaining the expert (i.e. the client). Accordingly, an expert witness is not an advocate for the client. The expert witness is required to follow the appropriate Court Guidelines for Expert Witnesses in fulfilling his or her duty. After collecting evidence, analysing the data and preparing a report, the expert will produce an expert report. It is appropriate for lawyers and experts to collaborate on the form of an expert report, but not acceptable for litigants or their lawyers to influence the content. There is a need to take care and record all communications which are undertaken as an expert witness with the litigant, and the instructing lawyers. Generally, such communications are protected from disclosure until the expert report is served on the other party.

In essence, the expert, because of his or her independence and expertise, is in the privileged position of expressing opinions in evidence, in contrast to normal witnesses, who deal in facts.

Quite often, when a client receives an impartial report on a matter, he or she is inclined to withdraw the claim. It is likely also that when the prepared report has been viewed by the judge this can lead to an order to negotiate settlement outside the court.

10.7 LEGAL RESPONSIBILITY IN MANAGING STAFF

There is a responsibility of employers in many countries in the world to ensure that staff within their organisations have a safe place in which to work and do not suffer from discrimination. Therefore there is a need to understand how the laws of Equal Opportunity, Discrimination, Human Rights and Occupational Health and Safety work.

Equal Employment Opportunity and Human Rights

In most countries such as the US and European countries there are national laws for equal employment opportunity and human rights. In the US this is administered by the Equal Employment Opportunity Commission. In the UK the national legislation requires organisations to have a written equal opportunities policy with procedures covering equal opportunities in recruitment, promotion, transfer, training, dismissal and redundancy. In the US the Federal Laws are usually supplemented by State Laws. The Australian Constitution does not give any power to the Federal Government to make laws in the area of discrimination. However, once the Commonwealth of Australia signs a treaty with an international organization (such as the United Nations or associated bodies like the International Labour Organization), the Federal Government has an international duty to draw up laws based on the principles in the treaty. Federal laws based on principles of human rights include:

- The Racial Discrimination Act 1975;
- The Sex Discrimination Act 1984;
- The Human Rights and Equal Opportunity Commission Act 1986;
- The Privacy Act 1988;
- The Disability Discrimination Act 1992;
- The Racial Vilification Act 1996; and
- The Age Discrimination Act 2004.

In addition to these laws, there is the Federal law of Equal Opportunity for Women in the Workplace Act 1999 (Formerly the Affirmative Action Act). Similar acts apply in many countries.

Each state government has an Act to promote equal opportunity in organisations and communities. For example, in South Australia it is the Equal Opportunity Act (SA), 1984 which is an act to : "promote equality of opportunity between the citizens of this State; to prevent certain kinds of discrimination based on sex, sexuality, marital status, pregnancy, race, physical or intellectual impairment or age; to facilitate the participation of citizens in the economic and social life of the community; and to deal with other related matters."

Equal Opportunity is about a person's behaviour, or how decisions are made in public areas of life. Put simply, it is about the respect to be shown to fellow workers.

Anti-discrimination or Equal Opportunity means that all people have a right to be treated fairly regardless of irrelevant personal characteristics. Equal Opportunity laws only apply in areas of public life, and private life is excluded. It is unlawful under the Equal Opportunity Act (SA) 1984 for anyone to be treated unfairly on the basis of: age; sex; marital status; pregnancy; sexuality; physical or intellectual impairment; and race.

The work areas covered by the Act include: accommodation; advertising; clubs and associations; conferral of qualifications; disposal of land; education; employment; and provision of Goods and Services. For the employment area, the law applies to paid full time, part time and casual workers, as well as unpaid or contract employees. It applies to all stages of employment: job advertisements and applications; interviews; promotions; training; and dismissal. In employment the following personal characteristics are also covered by Commonwealth laws: religion; political opinion; medical record; irrelevant criminal record; social origin; and trade union activity. This means that some of the current State legislation would have to be read in conjunction with the Federal legislation.

Application of legislation to employment

Maria sought a promotion to Team Leader of the roadwork's team at a council depot. Although she was an excellent worker and had previous supervisory experience, the depot manager decided not to consider her application, believing that she would not be respected by the all-male staff. Maria complained to the Commissioner. A representative held discussions with the manager, and Maria was given an interview for the team leader job.
(Equal Opportunity Commission of South Australia, 2006)

Occupational Safety, Health and Welfare

In the South Australian Occupational Health, Safety and Welfare Act 1986, and accompanying Regulations 1995, all employers must ensure that each employee employed or engaged by the employer, is safe from injury and risks to health while at work by providing and maintaining a safe working environment, safe systems of work, plant and substances in a safe condition, adequate facilities for the welfare of employees; and must provide such information, training and supervision that are necessary to ensure that each employee is safe from injury and risks to health.

It is necessary, therefore, that all employees are aware of the procedures and processes involved in Occupational Health and Safety. An engineer in many situations is likely to be responsible for staff, to ensure that Occupational Health and Safety breaches do not occur. Section 23A contains provisions for designers and owners of buildings that are particularly relevant for engineers. For example, a person who designs a building is expected to ensure that the building is safe for people who might work in, on or about it, from injury and risks to health.

In a case where a contractor's employee, working in an excavated area, was injured due to the wall of the excavation collapsing, the company was convicted of a breach of safety regulations and fined. The company was deemed to have "failed to:

- provide a safe working environment by ensuring the excavation was of a safe design;
- conduct an adequate hazard identification and risk assessment;
- shore the walls of the excavation; and
- adequately batter or bench the walls of the excavation." (SafeworkSA, 2006)

In another case, a manufacturing company was convicted and fined after an employee injured his right index finger while operating a machine. It was deemed that the company "failed to:

- carry out a hazard identification and risk assessment in relation to the machine;
- provide safe operating procedures for the machine;
- prevent access to the dangerous moving parts of the machine;
- provide an emergency stop that was immediately accessible to operators;
- provide an emergency stop device which would bring the machine to a full stop as quickly as possible; and
- provide such information, instruction, training and supervision as was reasonably necessary to ensure that the employee was safe from injury and risks to health." (SafeworkSA, 2006)

It must be realised that the engineer involved in construction, in the design office, in manufacturing and industrial processes must ensure that all operations are carried out with the safety of the employees paramount in all workplaces. The legislation will need to be satisfied but, more importantly, a sense of safety application needs to be ingrained in employees through training and awareness programmes.

10.8 SUMMARY

The issues raised in this chapter are important to any engineer who is making decisions that impinge upon the community, is involved in choosing new technology or persisting with old technology, and implementing safety procedures in the workplace. The engineer needs to be aware of the precautionary principle and to be prepared to defend their actions. Engineering has found itself in the middle of some of the largest disasters that have occurred (e.g. Chernonbyl), both in terms of social distress and provision of a solution. The questions of what is justified as a whistle blower or what is obligatory under the code of ethics need to be answered by taking all the ramifications of the decision into account—a moral dilemma for the decision maker. As an engineer, it is necessary to be aware of any discrimination that might occur in the workplace and be mindful of the rights of the individual. It is essential that all engineers are aware of their legal responsibilities in the areas of Occupational Health and Safety and Equal Employment Opportunities for staff who are working for them.

PROBLEMS

10.1 "Relativity applies to physics, not ethics" (Albert Einstein). Discuss how this statement is important for ethical decisions a professional engineer might make.

10.2 An engineer learns by accident that his employer, a major chemical fertilizer company, has been disposing of toxic wastes in an area where the frequency of cancer and birth defects is on the rise. He also learns that a TV documentary involving local citizens will be aired next week and these citizens are preparing a legal suit against the company. He immediately sells all his shares in the company and following advice that they have been sold, then tells his CEO what he has learned. The legal suit is filed two days later. Has he done anything unethical? Please discuss. (based on an example in Ullmann, 1986)

10.3. How much should you change your behaviour to fit with the beliefs, values and norms of those with whom you are interacting with? Whose responsibility is it to change: the visitor; sojourner; newcomer; or the host? Simple example: What about the language one speaks? Should a person who comes to Australia be required to speak English, only English?

Issues: At what point do you give up your cultural identity and moral integrity? At what point does adoption of another culture offend or insult? It is necessary to be aware of one's own moral stance and at the same time the need to display respect for others. Discuss with someone and write a paragraph on your outcomes.
(adapted from http://www2.andrews.edu/~tidwell/bsad560/Ethics.html)

10.4 Manager's dilemma (from EOCSA, 2006)
An employee comes to you to complain about behaviour she has been putting up with from another co-worker. She insists, however, that she does not want to make a formal complaint. What are some possible steps you might consider? Consider and debate the following ideas:

(a) Ask her why she insists on not proceeding further with her complaint;
(b) Respect her right not to proceed with a formal complaint and inform her of your duty of care to protect her and other workers;
(c) Monitor the situation and follow up with her to see if the problem gets any worse; and

(d) Consider recirculating relevant policies or codes of conduct to all employees.

10.5 There are many resources available on the web. Search your local Equal Opportunity website and one in another country and compare the differences in rulings on a number of different aspects, such as religion, age and racial discrimination.

REFERENCES

Cooke, J.R., 2001, *Architects, Engineers and the Law*, Federation Press, 3rd Edition.

Dym, C.L. and Little, P., 2000, *Engineering Design. A Project-Based Introduction*. John Wiley & Sons.

Engineering Council UK, 2005, Chartered Engineer and Incorporated Engineer Standard, UK Standard for Professional Engineering Competence, 17pp. http://www.engc.org.uk/documents/ CEng_IEng_Standard.pdf.

Equal Opportunity Commission of South Australia, 2006, Understanding Equal Opportunity, http://www.eoc.sa.gov.au/default_multitext.jsp?xcid=170.

Graham L.R., 1996, *The Ghost of the Executed Engineer – Technology and the Fall of the Russian Empire*, Harvard University Press.

Halsall, P.M., 1998, *Ancient History Source Book, Code of Hammurabi* (c.1750 BCE). http://www.fordham.edu/halsall/ancient/hamcode.html#horne

Johnston S., Gostelow P., Jones E., Fourikis R., 1995, *Engineering and Society. An Australian Perspective*, Harper.

Kramer, S.N., 1963, *The Sumerians. Their History, Culture, and Character.* University of Chicago Press.

Newall, P., 2005, *Ethics*, Galilean Library, www.galilean-library.org/int11.html.

Oberg, J., 2000, Houston, We Have A Problem. *New Scientist*, 165(2234), 26–29

Prosser, W., 1971, *Handbook of the Law of Torts*. 4th ed. St. Paul: West Publishing Co., 1971.

Rufner, T., 2006, *Roman Law*, http://www.jura.uni-sb.de/Rechtsgeschichte/ Ius.Romanum/english.html.

Safework, South Australia, 2006, http://www.safework.sa.gov.au/ contentPages/ LegislationAndPublications /default.htm, accessed Sept15, 2006.

Sierra Legal Defence Fund, 2003, International watchdog agency to investigate Canada's lax enforcement of pollution from pulp and paper mills, Release December 15, http://www.sierralegal.org/m_archive/2003/pr03_12_15.html.

South Australia v Johnson, 1982, 42 ALR 161 at 169–170.

South Australian Government, 2005, *Occupational Health, Safety and Welfare Act* 1986, amended 2005. http://www.parliament.sa.gov.au/Catalog/legislation /Acts/o/1986.125.un.htm.

Taylor P. W., 1972, *Problems of Moral Philosophy*, 2nd ed., Dickenson Publishing.

The Illinois Institute of Technology, 2005, The Center for the Study of Ethics in the Professions, Accessed Feb 2006, http://ethics.iit.edu/codes/index.html.

The Institution of Engineers, Australia., 2000, *Code of Ethics*, 8pp.

Thomas B., 2004, The assessment of damages for breach of contract for defective building work, *The Building and Construction Law Journal*, August, 20(4), 230–256.

Thurman R., 1997, The Dalai Lama interview with Robert Thurman, 1997, http://www.motherjones.com/news/qa/1997/11/thurman.html.

Ullmann, J. E.(ed), 1986, *Handbook of Engineering Management*, John Wiley & Sons.

Ullmann, J.E., 1983, The responsibility of Engineers to their Employers in J.H Schaub and Karl Pavlovic, *Engineering Professionalism and Ethics*, John Wiley & Sons.

CHAPTER ELEVEN

Risk and Reliability

In this chapter we consider the concepts of risk and reliability and how they can be considered in engineering planning and design. The concepts of risk and reliability are defined and illustrated in relation to engineering problems. Reliability-based design and the selection of safety coefficients are explained in detail. The concepts of resiliency and vulnerability are also introduced. Methods for evaluating the reliability of engineering components and engineering systems are explained.

11.1 INTRODUCTION

Whether or not we realise it, risk is something that we face every day of our lives. Every time that we travel in a car or other vehicle, cross the street or go for a jog we risk injury or death. Fortunately these risks are very small in modern society, unlike the risks faced daily by our ancestors in pre-historic times. All engineering works have some associated risks in their construction, manufacture and use. Part of the role of the engineer is to manage those risks so that they lie within ranges acceptable to society.

Why can't engineers design and build systems and products that have zero risk of failure? We will consider structures such as buildings, bridges and dams as an example. In the first place, such structures are subject to loads that are not fully controlled by the engineer including the effects of tornados, earthquakes, floods, tsunamis, landslides and impacts by falling objects. All structures are designed to safely withstand certain design loads, but there is always some probability that these loads will be exceeded. For example, a building can be designed to resist the largest earthquake on record, but there is always some probability that a larger earthquake will occur. Secondly, it is not economic to attempt to reduce the probability of failure to zero. The large amounts of money spent in attempting to build "failure proof" structures would be better spent in reducing other areas of risk to human life, for example, building better roads, or providing more hospitals. A similar argument applies to engineering systems other than structures such as chemical plants, electrical equipment or pieces of machinery.

11.2 LEVELS OF RISK

Dougherty and Fragola (1988) define **risk** as the combination of the probability of an abnormal event or failure and the consequences of that event or failure to a project's success or a system's performance. This combination of probability (or likelihood) and consequences is used in tables of risk ratings such as Table 11.1.

Table 11.1: Risk Ratings (Reproduced with the kind permission of Neill Buck and Associates)

Likelihood	Consequences				
	Insignificant	Minor	Moderate	Major	Catastrophic
Almost Certain	Moderate	High	Extreme	Extreme	Extreme
Likely	Low	Moderate	High	Extreme	Extreme
Possible	Low	Moderate	Moderate	High	Extreme
Unlikely	Insignificant	Low	Moderate	Moderate	High
Rare	Insignificant	Insignificant	Low	Low	Moderate

In Table 11.1 catastrophic consequences would involve loss of life, major consequences include serious injury or major economic loss, moderate consequences include minor injuries or moderate economic loss and minor consequences include minor economic loss. Efforts to minimise risk should be concentrated on the activities that have an extreme or high risk rating.

As discussed previously, all humans face many risks on a daily basis. The probabilities associated with these risks have been estimated in a number of studies (US Nuclear Regulatory Commission, 1975; Stone, 1988; Morgan, 1990; BC Hydro, 1993). Some of these probabilities are given in Table 11.2.

Risk of Gastroenteritis Among Young Children Who Consume Rainwater in South Australia

Eighty-two percent of rural households in South Australia have rainwater tanks as their main source of drinking water. There are some risks associated with the consumption of rainwater as it may be contaminated by pathogens from birds or animals. A study was carried out to assess whether there was a significant difference in the risk of gastroenteritis among 4 to 6 year old children who drank rainwater compared to those who drank treated mains water.

Over 1000 children were included in the study which was carried out over a six week period. Parents of the children were asked to keep a diary of gastrointestinal symptoms. The occurrence of gastroenteritis was based on a set of symptoms (defined as a highly credible gastrointestinal illness or HCGI). It should be noted that most incidences of gastroenteritis were mild.

Based on the raw data, the incidence of HCGI was 4.7 episodes per child–year for those who only drank rainwater compared to 7.3 episodes per child-year for those who only drank mains water. However, when adjustments were made for other risk factors, there was no significant difference in the incidence of HCGI between the two groups. The study concluded that consumption of rainwater did not increase the risk of gastroenteritis compared to mains water for 4 to 6 year old children. One possible reason for this conclusion is the acquired immunity by the children who drank rainwater, as they had all done so for at least a year prior to the study and could have developed immunity to a number of microorganisms during this period.

Source: Heyworth et al., 2006

Table 11.2: Annual Probability of Death to an Individual

Source of risk	Annual probability of death	Source of information
Car Travel	1 in 3,500	Morgan (1990)
	1 in 4,000	US Nuclear Regulatory Commission (1975)
	1 in 10,000	Stone (1988)
Air Travel	1 in 9,000	Morgan (1990)
	1 in 100,000	US Nuclear Regulatory Commission (1975)
Drowning	1 in 30,000	US Nuclear Regulatory Commission (1975)
Drowning (UK average)	1 in 100,000	Morgan (1990)
Fire (UK average)	1 in 50,000	Morgan (1990)
Household Electrocution (Canada)	1 in 65,000	Morgan (1990)
Electrocution	1 in 160,000	US Nuclear Regulatory Commission (1975)
Fire (U.K. average)	1 in 50,000	Morgan (1990)
Lightning	1 in 2,000,000	US Nuclear Regulatory Commission (1975)
	1 in 5,000,000	Morgan (1990)
	1 in 10,000,000	Stone (1988)
Nuclear Reactor Accidents	1 in 5,000,000	US Nuclear Regulatory Commission (1975)
Structural Failure	1 in 10,000,000	Morgan (1990)

11.3 RELIABILITY BASED DESIGN

The **reliability** of an engineering component or system is defined as the probability that the system will be in a non-failure state. Failure needs to be defined for each particular system under consideration. For example, for a building, failure may be defined as the complete collapse of the structure or may correspond to excessive deflections that cause cracking of the walls. For a water supply system, failure could correspond to running out of water, although it is more likely to correspond to the imposition of extreme restrictions on household consumers and industry. It could also correspond to the provision of water of an unacceptable quality to consumers. For a freeway system, failure could correspond to extreme congestion associated with low vehicle speeds and long delays.

In this section, reliability-based design will be explained for structural elements as an example of engineering systems. The same principles apply to many other types of engineering systems.

Probabilities of failure are rarely calculated for engineering structures where the usual practice is to make use of safety factors or safety coefficients. This approach will be demonstrated for the design of simple structural members in tension. Examples of tension members are shown in Figure 11.1.

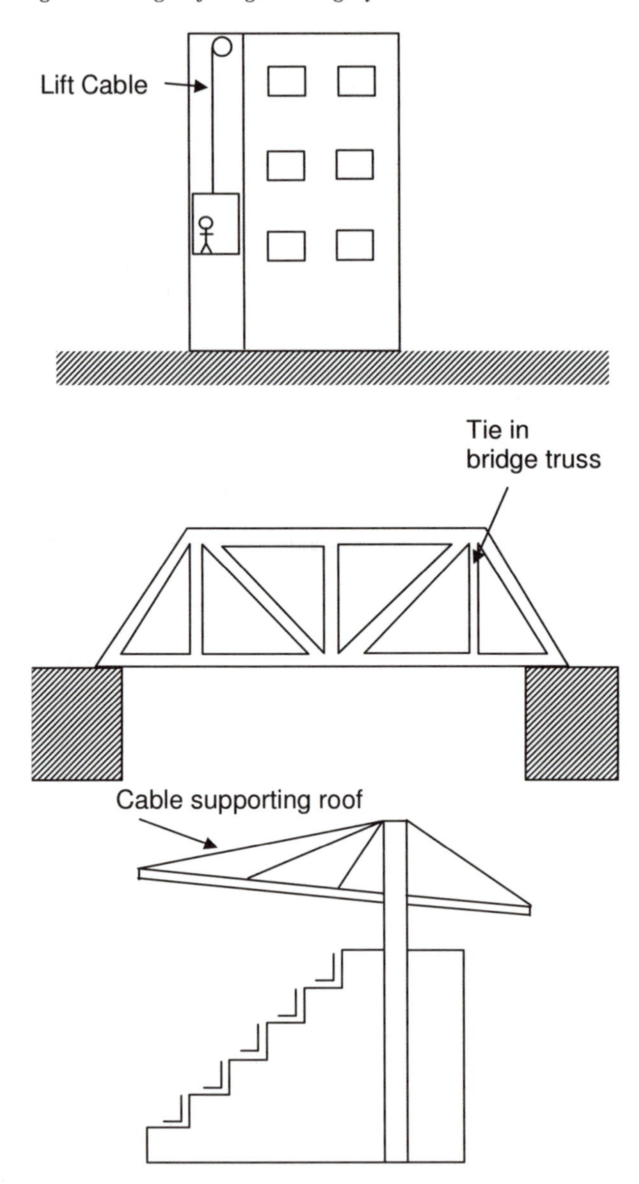

Figure 11.1 Examples of tension members (Lift cable, tie in a bridge truss, cable supporting a roof)

Consider the analysis of a steel lift cable of cross-sectional area, A. If the maximum force that will be applied to the cable during its lifetime, S, is known, the maximum stress in the cable, σ, can be calculated using the equation:

$$\sigma = \frac{S}{A} \tag{11.1}$$

where A = the cross-sectional area of the cable. If σ is less than the stress that will cause failure of the steel cable, σ_f, the cable will not fail. In the design situation, the failure stress of the steel and maximum applied force are assumed to be known, hence the required cross-sectional area of the cable, A, can be determined by rearranging Equation 11.1 to give:

$$A \geq \frac{S}{\sigma_f} \tag{11.2}$$

However, this approach is not conservative as it does not take into account uncertainty in S, σ_f, and A. Clearly the maximum force that will be applied to the cable during its lifetime is not known with certainty. There can also be variations in the failure stress of the steel that comprises the cable due to variations in the manufacturing process. Finally, the actual cross-sectional area of the cable may vary from the value specified in the design due to variations in production of the cable.

One approach that provides some margin of safety in the design is to use a **safety factor**. If we define T, the resistance of the cable, as:

$$T = \sigma_f A \tag{11.3}$$

we require T to exceed S by some margin to allow for uncertainties. Define the safety factor, γ as follows:

$$\gamma = \frac{T}{S} \tag{11.4}$$

where γ is typically in the range 1.5–3.0 depending on the loading conditions, material and likely mode of failure. The cross-sectional area of the cable is then determined by combining Equations (11.3) and (11.4) as follows:

$$A \geq \frac{S\gamma}{\sigma_f} \tag{11.5}$$

The safety factor approach is no longer commonly used in structural engineering design, but is still used in geotechnical engineering. Structural engineers now use a reliability-based approach to structural design. This approach explicitly recognises the uncertainties in the various factors involved in the design. In the reliability approach we define the difference between T and S as the **safety margin**, Z. i.e.:

$$Z = T - S \tag{11.6}$$

We expect variations in T and S due to some of the factors described previously. We will assume that T and S are normally distributed random variables that are independent of each other. Typical distributions of these variables are shown in Figure 11.2. This figure shows some small but finite probability that either T or S is negative. Clearly negative values of T or S are physically impossible, as the normal distribution is only an approximation to the true distributions of the resistance T and maximum applied force S.

Let the means of T and S be designated by $\mu(T)$ and $\mu(S)$ (respectively) and the standard deviations of T and S be designated by $\sigma(T)$ and $\sigma(S)$ (respectively). Then, if both T and S have normal distributions, so will Z. The mean and standard deviation of Z can be determined using Equations (11.7) and (11.8).

$$\mu(Z) = \mu(T) - \mu(S) \tag{11.7}$$

$$\sigma(Z) = \sqrt{\sigma(T)^2 + \sigma(S)^2} \tag{11.8}$$

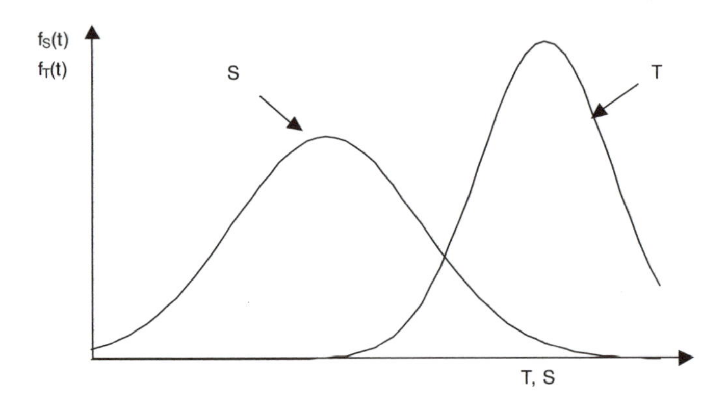

Figure 11.2: Probability distributions of resistance T and applied force S

The probability distribution of Z is shown in Figure 11.3.

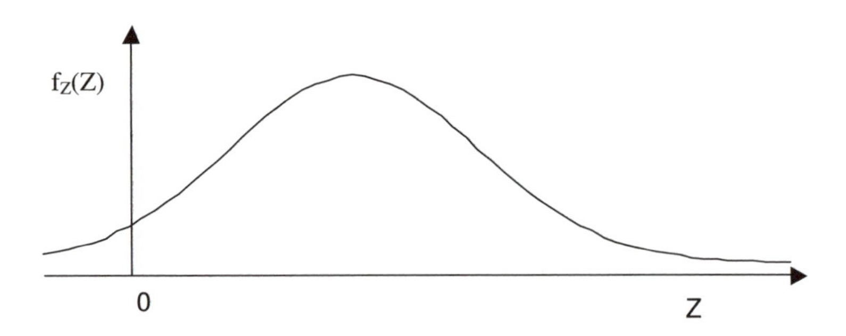

Figure 11.3: Probability distribution of the safety margin , Z

If Z is negative the cable will fail, if it is positive, the cable will not fail. The probability of failure is designated P_f which is given by Equation (11.9):

$$P_f = P[Z < 0] \tag{11.9}$$

As stated earlier, the reliability of a system, R is defined as the probability that the system will be in a non-failure state. Hence:

$$R = 1 - P_f = 1 - P[Z < 0] \tag{11.10}$$

A commonly used measure of safety is the safety index, β, which is defined as:

$$\beta = \frac{\mu(Z)}{\sigma(Z)} = \frac{1}{V(Z)} \tag{11.11}$$

where $V(Z)$ = the coefficient of variation of Z = $\sigma(Z)/\mu(Z)$. For the case where Z follows a normal distribution, β is related to the probability of failure as indicated in Table 11. 3.

Table 11.3: Relationship between safety index and the probability of failure (adapted from Warner et al, 1998)

Safety Index β	2.32	3.09	3.72	4.27	4.75	5.20
Probability of failure, P_f	10^{-2}	10^{-3}	10^{-4}	10^{-5}	10^{-6}	10^{-7}

As T and S are now considered to be random variables, the safety factor defined in Equation (11.4) may be expressed in terms of the mean values of T and S, i.e.

$$\gamma_0 = \frac{\mu(T)}{\mu(S)} \tag{11.12}$$

where, γ_0 is called the central safety factor. Warner et al (1998) show that β is related to γ_0 by the following expression:

$$\beta = \frac{\gamma_0 - 1}{\sqrt{\gamma_0^2 V(T)^2 + V(S)^2}} \tag{11.13}$$

where $V(T)$ and $V(S)$ are the coefficients of variation of T and S (respectively).

Hence in order to maintain the same safety index (and hence probability of failure) γ_0 must vary depending the values of $V(T)$ and $V(S)$. In a design situation it is desirable that all structural elements have the same probability of failure, so it is better to select a single value of β rather than various values of γ_0.

As noted previously, a major weakness of using γ_0 is that it does not take into account variations in T and S. This may be overcome to some extent by using conservative values of T and S. For example define S_k as the value of the applied load S that has a 5% chance of being exceeded. Furthermore, define T_k as the value of the resistance T that has a 95% chance of being exceeded. Then the "nominal" safety factor is given by the following equation:

$$\gamma = \frac{T_k}{S_k} \tag{11.14}$$

thus we use a high value of the applied load and low value of the resistance in assessing the nominal safety factor.

Figure 11.4 shows the relationship between the distributions of T and S and the values of T_k and S_k. Modern structural codes of practice use partial safety coefficients to allow for the variabilities in T and S separately. The partial safety coefficients are applied to T_k and S_k to obtain design values T_d and S_d as follows:

$$T_d = \frac{T_k}{\gamma_T} \tag{11.15}$$

$$S_d = \gamma_S S_k \tag{11.16}$$

Thus a safe design requires:

$$T_d > S_d \tag{11.17}$$

or:

$$T_k > \gamma_T \gamma_S S_d \tag{11.18}$$

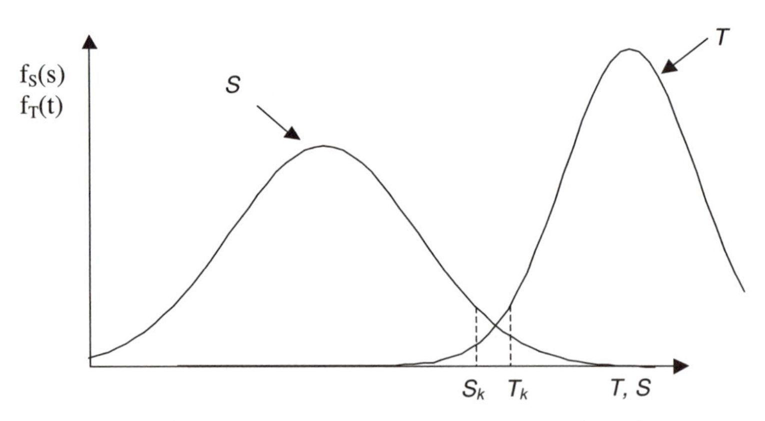

Figure 11.4: Relationship between T and S and the values of T_k and S_k.

Values of the partial safety coefficients, γ_T and γ_S are selected to ensure that that the probability of failure, P_f, given by Equation (11.9) lies within an acceptable range.

It should be noted that the approach outlined above is based on a number of simplifying assumptions such as that T and S are independent and have normal distributions. Furthermore, even in the simple case of a tensile member, the variability in T can be due to variations in materials properties as well as variations in the actual dimensions of the member compared to those assumed in the design. For this reason the probability of failure determined is usually referred to as a nominal probability of failure.

A similar approach to that outlined above can be used for other structural elements such as beams, columns and frames or other engineering components such as electrical transformers, items of machinery or distillation columns.

The above analysis assumes that structural failure will occur due to unusually high loads or low resistance of the structure itself (due to lower than expected material properties or dimensions). In reality it has been observed that most structural failures occur due to human error caused by factors such as:

1. gross errors in the design of the structure;
2. loads applied to the structure that were not anticipated in the design (e.g. an aircraft crashing into a building); or
3. inappropriate construction methods.

This underlines the need for proper quality control in engineering design and construction including appropriate work practices, supervision and checking.

11.4 SELECTION OF SAFETY COEFFICIENTS

As outlined in Chapter 3, most engineering design work is guided by codes of practice that specify minimum levels of safety to ensure that the frequency of failure is within limits that are acceptable to the community. The setting of these levels of safety involves a balancing act between overly conservative design on the one hand and unconservative design on the other. If the levels of safety chosen lead to overly conservative design, money will be spent on achieving these high standards that could be better employed elsewhere in the economy (for example building new hospitals or purchasing additional medical equipment). If the levels of safety are too low and design is unconservative, failures (and the consequent loss of life or economic loss) will occur more frequently than is acceptable to the community and there will be a public outcry.

Following Warner et al. (1998), it is appropriate to consider the choice of the optimum probability of failure for a particular class of structure (e.g. highway bridges). Clearly as the probability of failure, P_f increases the cost associated with the failure of all bridges also increases as shown in Figure 11.5. On the other hand, increasing the probability of failure is associated with reducing the margin of safety and hence the cost of constructing all bridges as shown in Figure 11.5. The total cost is the cost of failures plus the cost of construction of all highway bridges. The value of P_f that corresponds to the minimum total cost is the optimum probability of failure. In practice, this value may be quite difficult to identify and values of the partial safety coefficients, γ_T and γ_S are specified in codes of practice based on experience with previous successful design procedures and comparison with a number of standard design cases.

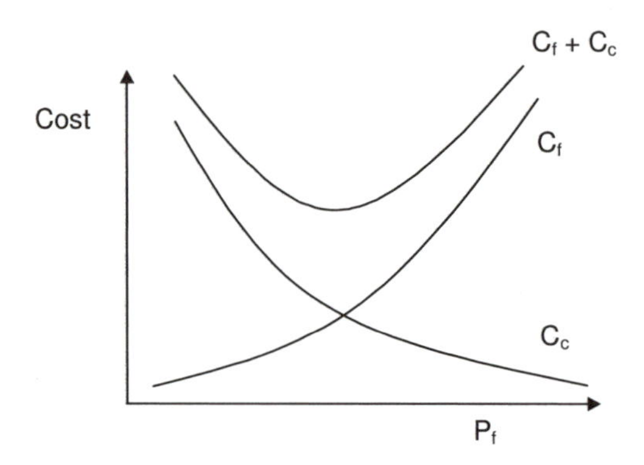

Figure 11.5: Tradeoff between cost and probability of failure, where C_c = cost of construction and C_f = cost of failure (adapted from Warner et al., 1998)

11.5 RELIABILITY, RESILIENCE AND VULNERABILITY

The **reliability** of an engineering system is defined in Section 11.3 as the probability that it is in a non-failure state at any point in time. Reliability is clearly an important performance indicator for most engineering systems. However, it is not the total picture. Hashimoto, Stedinger and Loucks (1982) identified the following three performance measures of a water supply system: reliability, resiliency and vulnerability.

Reliability is defined above. **Resiliency** is the probability of the system returning from a failure state to a non-failure state in a single time step. **Vulnerability** is the average magnitude of failure of a system, given that it does fail. Obviously it is desirable for a system to have high values of reliability and resiliency but a low value of vulnerability.

For example, Figure 11.6 shows the frequency distributions of water supply pressure at a critical node that will be delivered by two different designs of a water supply system. If P_c is the minimum acceptable pressure at the critical node, the reliability of each system is the probability that the pressure at the critical node is above this value. This is given by the area of the respective distribution above P_c. In this case, both systems have the same reliability and hence are equal according to this measure. Figure 11.6 also shows the vulnerabilities of the two systems i.e. the average pressure deficit for pressures below the critical value. Clearly System A is more vulnerable than System B and hence is inferior according to this measure. The final choice between the two systems will depend on these two measures as well as other factors such as their cost and environmental and social effects.

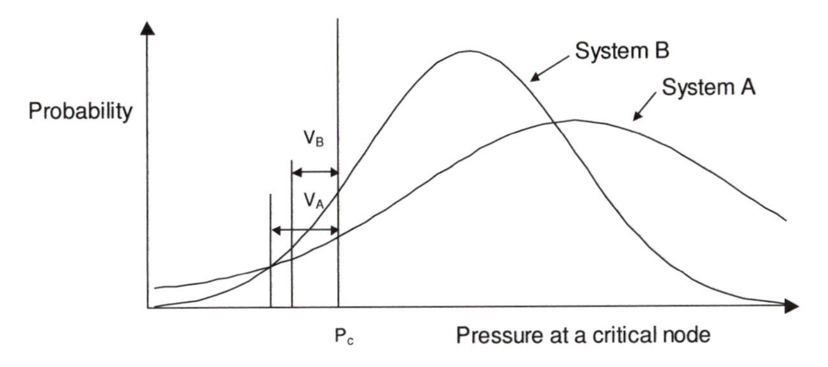

Figure 11.6: Distribution of pressure at a critical node for two designs of a water supply system

11.6 RELIABILITY OF ENGINEERING COMPONENTS

Many engineering systems contain a large number of components, any one of which can fail at a particular point in time with varying consequences to the overall system performance. For example a water supply system for a city may consist of thousands of pipes, hundreds of pumps and valves and tens of tanks. The impact of the failure of each one of these will vary depending on its function and location in the system, its capacity, the time of day and of the year, the number of customers affected and so on. We now want to consider how the reliability of system components may vary over time.

Figure 11.7 shows a commonly observed pattern for failure rates of engineering components (Agarwal, 1993; Hyman 1998). This is the so-called "bath tub" model of component failures.

The probability of failure per unit time is high initially during the "break in period" as some components fail due to poor materials or workmanship as they are placed under stress for the first time. During the mature phase of operation, a small number of failures occur due to unexpected causes such as occasional high loads or localised weak points. During this period the failure rate per unit time is approximately constant. As the components begin to reach the end of their useful life the failure rate starts to increase as the effect of corrosion and fatigue take effect. This is the so-called "wear out period".

A simple analysis can be carried out for the mature phase by assuming that the probability of a component failure per unit time is constant and equals λ. Under these conditions, Hyman (1998) derives the following equation for the reliability of a component:

$$R(t) = e^{-\lambda t} \tag{11.19}$$

where $R(t)$ is the probability that the component does not fail in the time period 0 to t.

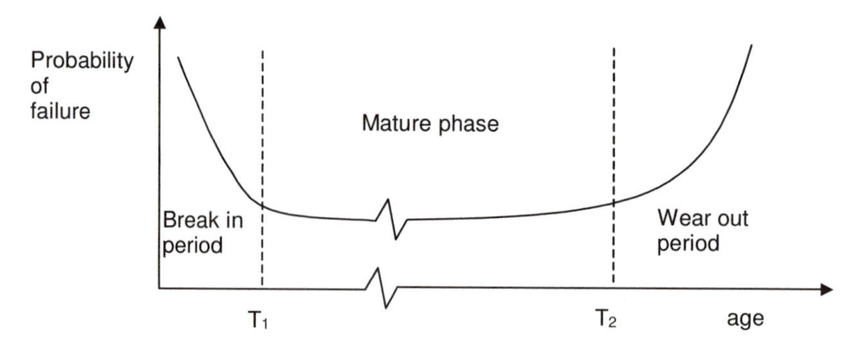

Figure 11.7: Typical failure rates for engineering components

Another important measure is the mean time to failure, *MTTF*, which is defined as the average time that a component will be in service before failure occurs. Hyman (1998) shows that the following relationship holds:

$$MTTF = \frac{1}{\lambda} \tag{11.20}$$

Example: Light bulbs manufactured by a particular company have a mean time to failure of 250 hours. Assuming that the bulbs have a constant probability of failure, what is their reliability for a period of 100 hours? 250 hours?
Answer: Applying Equation (11.20),

$$\lambda = \frac{1}{MTTF} = \frac{1}{250} = 0.004 \text{ hours}$$

and therefore:

$$R(100) = e^{-0.004x100} = 0.670$$
$$R(250) = e^{-0.004x250} = 0.368$$

Thus, only 36.8% percent of the light bulbs are expected to last to the mean time to failure and 67% will last 100 hours or more. The apparently high failure rate in the first 100 hours (33%) is a consequence of the assumption of a constant failure rate. In reality they are more likely to have a low initial failure rate that gradually builds up over time.

11.7 SYSTEM RELIABILITY

As noted earlier, an engineering system is composed of a large number of non-identical components that interact in some defined way. The failure of individual components can have different consequences on the performance of the system as a whole, depending on the nature of the components and how they are connected together. We will consider two extreme cases to demonstrate this effect.

Series System

Consider a system that consists of a set of components in series. For example, consider a chain comprised of n links. The failure of any one of the links will cause the chain to fail. From elementary probability theory, the probability that the system does not fail is the product of the individual probabilities of non-failure (assuming that the failure of each component is independent). i.e.

$$R_{series} = R_1 \text{x} R_2 \text{x} R_3 \text{x} \ldots R_i \text{x} \ldots R_n \qquad (11.21)$$

where R_{series} is the reliability of the series system and R_i is the reliability of component i. Therefore a series system has a reliability that is less than the lowest reliability of any of its components.

In the special case that each component has a failure rate that is constant with time, Equation (11.19) gives:

$$R_i = e^{-\lambda_i t} \qquad (11.22)$$

and hence Equation (11.21) becomes:

$$R_{series} = e^{-\lambda_1 t} \text{x} e^{-\lambda_2 t} \text{x} \ldots e^{-\lambda_i t} \text{x} \ldots e^{-\lambda_n t} \qquad (11.23)$$

hence:

$$R_{series} = e^{-\lambda_s t} \qquad (11.24)$$

where, the failure rate of the series system per unit time is given by:

$$\lambda_s = \lambda_1 + \lambda_2 + \ldots + \lambda_i \ldots\ldots + \lambda_n \qquad (11.25)$$

Thus, in a series system where each component has a constant failure rate, the system failure rate will also be constant and will equal the sum of the individual failure rates.

Parallel Systems

A parallel system will not fail unless every component of the system fails. An example is a set of parallel pipes supplying water to part of a city. Clearly if one pipe fails, water can still be supplied to the city (presumably at a reduced pressure or level of service). However, if failure is defined as no water being available, all pipes must fail before the system failure occurs.

For a parallel system with n components, the probability of system failure is the probability that all components fail. Therefore, (assuming independence of the individual components) the probability of failure for the system as a whole is given by:

$$P_{f.parallel} = P_{f1} \text{x} P_{f2} \text{x} \ldots\ldots P_{fi} \text{x} \ldots\ldots\ldots P_{fn} \qquad (11.26)$$

where, $P_{f.parallel}$ is the probability of failure of a parallel system and P_{fi} is the probability of failure of component i. As reliability is defined as one minus the probability of failure, Equation (11.26) becomes):

$$(1 - R_{parallel}) = (1 - R_1) \text{x} (1 - R_2) \text{x} \ldots\ldots (1 - R_i) \text{x} \ldots\ldots (1 - R_n) \qquad (11.27)$$

It can be shown that the reliability of a parallel system is always greater than the reliability of its most reliable component. Given this high level of reliability, why don't we design all engineering systems as parallel systems? The answer is that there is a high cost of providing redundancy. Parallel systems are used when high levels of reliability are required. For example, all hospitals have a emergency power supply consisting of a motor-generator set that will start if the regular supply of electricity fails. This ensures that patients on life support systems are not at risk in the event of a power failure. Most pumping stations are designed to have a number of pumps in parallel. This not only allows for the flow provided by the pumps to vary as the demand on the system varies, but also allows for redundancy should one of the pumps breakdown.

Unlike series systems, parallel systems do not have constant failure rates per unit time, even if all of their components do.

Mixed Systems

Many real engineering systems do not correspond to simple series or parallel systems. They are described as mixed systems. A mixed system can be analysed by breaking it into a set of series and parallel systems and calculating the reliability one step at a time. For example Figure 11.8(a) shows a mixed system. Figures 11.8(b) through 11.8(e) shows the steps involved in computing its reliability.

11.8 SUMMARY

Knowledge of the concepts of risk and reliability are fundamental to the planning and design of engineering systems. Risk involves two components: the likelihood of an abnormal event or failure occurring and the consequences of that event or failure. All humans face risks on a daily basis. While the probability of failure of engineering systems can't usually be reduced to zero, it is the role of the engineer to ensure that this probability is kept within bounds that are acceptable to the community.

The reliability of a system at a particular point in time is the probability that it will be in a non-failure state at that time. Engineering systems can be designed to achieve a specified level of reliability due to known hazards, although there are often unexpected hazards that are difficult to quantify. Identification of the appropriate level of reliability of an engineering system may be viewed as a trade-off between the cost of construction or manufacture and the expected cost due to failure.

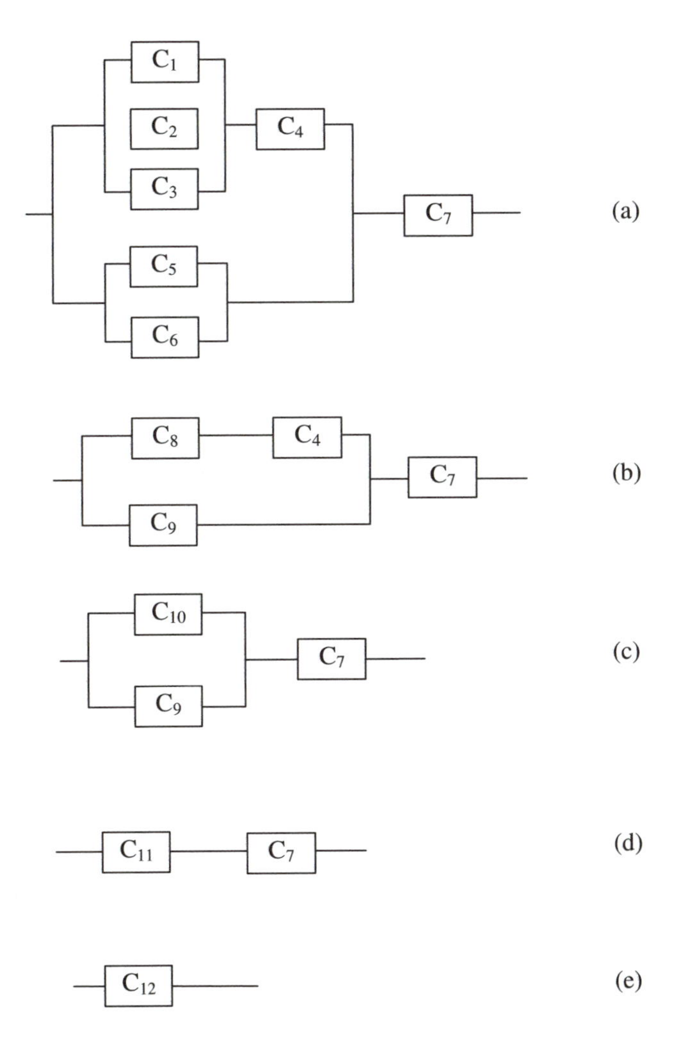

Figure 11.8: Steps involved in reducing a mixed system to a simple system

In addition to reliability, resiliency and vulnerability of a system may also be important in planning and design. Resiliency is the probability of the system returning from a failure state to a non-failure state in a single time step. Vulnerability is the average magnitude of failure of a system, given that it does fail. A system with a high reliability does not necessarily have a high resiliency or low vulnerability and vice versa.

Most engineering components wear out over time. Knowledge of this behaviour can be used to assess the reliability of a system or component over a specified period and hence to plan the repair or replacement of components in an optimal fashion.

An engineering system is usually composed of many different components, each with its own reliability. The reliability of the system can be determined based on the reliabilities of its individual components and the manner in which they are connected. Series systems and parallel systems represent special cases of engineering systems.

PROBLEMS

11.1 A cable is used to raise and lower a hopper in a mine shaft. The properties of the cable are given in Table 11.4.

Table 11.4: Properties of cable for mine hopper

Characteristic	Mean value	Standard deviation
Cross-sectional area (mm^2)	10	0.8
Failure stress (MPa)	300	30

The maximum load applied to the cable has a mean value of 2000 kN with a standard deviation of 400 kN. Assume all random variables are normally distributed and independent of each other.

(a) Estimate the mean and standard deviation of the resistance of the cable. Note if X and Y are two independent random variables with mean m_x and m_y and coefficients of variation V_x and V_y (respectively) and $W = XY$, then the mean of W, m_w, and the coefficient of variation of W, V_w, can be found using the following equations (Benjamin and Cornell, 1970):

$$m_w = m_x m_y$$

$$V_w^2 = V_x^2 + V_y^2 + V_x^2 V_y^2$$

(b) What is the mean and standard deviation of the safety margin, Z?
(c) What is the value of the safety index, β ?
(d) What is the probability of failure?
(e) If the coefficient of variation of the cross-sectional area remains constant at 8%, what is the mean value of cross-sectional area that corresponds to a probability of failure of 0.01? (Hint: Try trial-and-error using a spreadsheet).

11.2 A computer part has a per unit failure rate of 0.002 failures per day.

(a) What is the probability that this part will fail in it first year of operation?

(b) What value of the per-unit failure rate will ensure a probability of failure of less than 5% in the first year?

11.3 A power generating system has four components with probabilities of failure of 0.003, 0.0025, 0.004 and 0.001 (respectively). The system only fails if all of the components fail. What is the reliability of the system?

11.4 Four system components (A, B, C and D) have the following reliabilities: 0.99, 0.90, 0.95 and 0.92 (respectively). Determine the reliabilities of the following systems:

(a) A and B combined in series
(b) C and D combined in parallel
(c) A and B combined in parallel and the resulting system combined in series with C and D.

11.5 A water pump has a reliability of 0.85. How many pumps should be combined in parallel so that the total system has a reliability of 0.95?

REFERENCES

Agarwal, K.K., 1993, Reliability engineering. (Kluwer Academic Publishers, Dordrecht, The Netherlands).

Australian Institute of Company Directors, 2006, *Company Directors Course. Module 8: Risk- Issues for the Board,* (Sydney, AICD).

BC Hydro, 1993, Guidelines for Consequence-Based Dam Safety Evaluations and Improvements, Report number H2528, (BC Hydro: Hydroelectric Engineering Division).

Benjamin, J.R. and Cornell, C.A., 1970, Probability, statistics and decision for civil engineers. (McGraw–Hill, New York, NY.)

Bernstein, P.L., 1998, *Against the Gods. The Remarkable Story of Risk,* (New York, John Wiley and Sons).

Dougherty, E.M. and Fragola, J.R., 1988, *Human Reliability Analysis,* (New York, John Wiley and Sons).

Heyworth, J.S., Glonek, G., Maynard, E.J., Baghurst, P.A. and Finlay-Jones, J. , 2006, Consumption of untreated tank rainwater and gastroenteritis among young children in South Australia, *International Journal of Epidemiology*, 35, 1051–1058.

Hyman, B.,1998 *Fundamentals of Engineering Design,* (Upper Saddle River, New Jersey, Prentice-Hall).

Morgan, G.C., 1990, Quantification of risks from slope hazards, *Proceedings of GAC Symposium on Landslide Hazard in the Canadian Cordillera.*

Reid, S.G., 1989, Risk acceptance criteria for performance-oriented design codes, *Proceedings of ICOSAR'89, the 5th International Conference on Structural Safety and Reliability,* (San Francisco).

Stone, A. 1988. The tolerability of risk from nuclear power stations. *Atom 379,* May, 8–11.

U.S. Nuclear Regulatory Commission, 1975. *Reactor Safety Study: An Assessment of Accident Risks in U.S. Commercial Nuclear Power Plants*. WASH-1400 (NUREG 75/014).

Warner, R.F., Rangan, B.V., Hall, A.S. Faulkes, K.H. (1998), *Concrete Structures*, (Melbourne: Longman).

CHAPTER TWELVE

Engineering Decision-making

Decision-making is central to engineering work and an introduction to mathematical decision theory is presented in this chapter. Decision theory deals with simple, precisely defined (and therefore rather idealized) problems of choice. Real engineering decision situations are usually quite complex, and do not always lend themselves to the simplification needed to achieve one of the idealized situations for which a standard analysis is available. Nevertheless, the concepts of decision analysis provide a useful framework from which real problems may be approached. In particular, the identification of the alternative courses of action available, and the systematic evaluation of possible outcomes and consequences, are very important first steps in dealing with any real decision problem. Decisions in a competitive environment are particularly relevant to engineers and some of the basic concepts of game theory are also outlined to provide an introduction to this fascinating area of human behaviour.

12.1 INTRODUCTION

The space shuttle Columbia's 28th mission into space was originally scheduled for launch in January 2001, however, technical problems led to a number of postponements. In fact there were a total of 18 delays before the decision was taken to launch on 16 January 2003 (see Figure 12.1). As the launch day approached engineers were worried about the craft's external insulation sheets. In previous launches some had come loose and it was felt there was a potential for damage. Should the launch be abandoned until the issue was resolved? Some engineers thought so, but management was keen to go ahead. When insulation sheets were found to have come off during the launch and possibly damaged part of the wing, the decision was made to proceed with the mission and not to interrupt it so that an inspection could be carried out. During re-entry on 1 February 2003, with wing temperatures over 1,500 °C and at speeds in excess of Mach 24, Columbia disintegrated high over the United States killing all seven astronauts on board. Somewhere in the decision-making process there had been a miscalculation, an oversight or error. As a result, lives were lost and the space program was once again in the headlines for all the wrong reasons.

When engineers make decisions they are not always life and death decisions. Some may be quite trivial and some may appear to be trivial. One thing is certain: decision-making is a key part of engineering planning, design and management and it is important for engineers to be aware of the sorts of decisions that arise in their professional lives, and the way decisions are made. There is much more to decision-making than might first appear, and the work that is presented here aims to summarise results that workers in the field have developed over the last 50 years or so. In the main, decision theory covers simple and rather idealised situations: nevertheless, the methods of analysis have practical applications and even when the theoretical methods of analysis are not directly applicable to the complex situations which occur in engineering practice, they can provide a useful insight into the appropriate decision-making procedures.

Figure 12.1 The launch of the space shuttle Columbia, 16th January, 2003. Used with permission of NASA.

Before presenting the methods for dealing with decisions it is worth clarifying the elements in the decision-making process. A decision involves choosing between a number of alternate courses of action where the decision-maker is attempting to maximise some measure of satisfaction. The process involves identifying all the options, working out the consequences of each, and then evaluating and comparing the consequences. Some decisions are made based on certainty of outcome, where the difficulty is either in modelling the system accurately enough to produce meaningful results, or having a process that enables one to select an optimum solution from a large number of options. On the other hand decision theory is generally taken to deal with situations where some aspects of the consequences of the decision may be unknown. Decisions may be taken with risk of outcome where risk is defined as the situation where objective data exist which allow the estimation of the probability of each consequence (ReVelle et al., 2004). For best results the probabilities should be determined based on historical data or experiments. Another decision situation occurs under conditions of uncertainty of outcome where no objective data exist and therefore the decision is taken in ignorance of the chance of various outcomes occurring. An intermediate state is that of incomplete knowledge where some information exists about the probabilities of occurrence but not enough for exact specification (Kmietowicz and Pearman, 1981). Finally, decisions may be taken in the presence of competition and this can have a significant effect on the decision-making process. This is best dealt with using game theory and this will be introduced in Section 12.6.

12.2 DECISION-MAKING WITH CERTAINTY OF OUTCOME

Although not normally considered in decision-making theory it is worth mentioning that some decisions are taken based on complete knowledge (or assumed complete knowledge) of a deterministic system. For example, when NASA engineers plan their missions into space there can be a need to ensure the planets and moons are properly positioned to optimise travel time and the fuel loads that are necessary. In the case of missions to Mars, for example, there is a suitable month or two only every two years (the launch window) (Squyres, 2005). These sorts of decisions are completely deterministic and involve the following steps: gather the relevant information, make the necessary calculations and, assuming the criteria for the decision have been clearly defined, the decision appears. The decisions rely on an accurate model of the system, in this case using Newton's laws of motion, and a method to solve that model so that the required decision can be made.

A second type of decision under conditions of certainty of outcome involves the choice of an optimum solution from a range of mutually exclusive options where there is full knowledge of the consequences of each decision. For example, for NASA there is a need to choose between solid and liquid fuels in the rockets for each mission. Both options would satisfy each of the requirements to a different extent, and each requirement may be more or less important than the others. The aim is to use the information to find an optimum combination. For the Shuttle, solid rockets are used initially due to their compactness (which is important in reducing drag through the atmosphere), their simplicity, and the fact that they need not be shut down before the fuel is fully spent. Subsequent stages take advantage of the benefits of liquid fuel, with a higher power to weight ratio, and the fact that they can be shut down and restarted as necessary. Once again, the system is deterministic, but there are a number of objectives such as weight, volume, cost and operational factors that must be satisfied and the decision comes from modelling a number of different options and selecting one.

The thinking here goes back to the methodology outlined in Chapter 3 where it was shown that, in the solution to engineering problems, a key requirement is to develop as many promising concepts as possible and then to evaluate them based on the objectives and measures of effectiveness. Each potential solution will satisfy the objectives to a certain degree, and it will also be known that some of the specifications are more important than others. The question then is which to choose. Bazerman (1998), for example, suggests that a rational decision-making process involves a number of steps: define the problem, identify the criteria, weigh the criteria, generate alternatives, rate each alternative on each criterion, and compute the optimal decision.

Weighting and scoring techniques

In the weighting and scoring technique a set of objectives are derived that a solution must satisfy. Each objective is also given a weight that reflects its relative importance. Each potential solution is then given a score for each objective and an overall score can then be calculated as:

$$S = \sum_{i=1}^{n} w_i s_i \qquad (12.1)$$

where S is the total score for each option over the n objectives, w_i is the weight for the ith objective and s_i is the score for that objective.

The process is best illustrated with an example taken from Hubka et al. (1988) where students were required to design an automatic water pump suitable for small boats that were to be left for long periods of time moored in the water. In order to assess the designs the team developed a weighting scheme to set the relative importance of the various aspects of the solution. Factors such as price, pumping capacity, reliability and product life were all given a ranking on a scale of 1 to 5 based on how important each was. In the design described the most important factors were identified as low price, high reliability and a low maintenance effort. The least important factors were the weight or size of the device and its life.

The second part of the task was to determine how well each of the potential solutions (there were 8 that reached the final selection round) satisfied those criteria. To do this the designers developed a rating scheme for each property. For example, based on the expected requirements it was possible to give a score to the device based on its anticipated pumping capacity. A low rate of 1 litre per hour might score a 1 whereas a 3 litre per hour capacity might score a 4.

The next step was to estimate how well the particular solutions would perform and to use the rating curve to convert these performances to a score. For example, for an option involving osmosis as the primary method of removing water, it was estimated that the score for capacity would rank a 1, price a 3, servicing costs a 3, all-round use a 1, reliability a 2, weight a 3, installation a 3, life a 4, small disturbance to looks a 3 and low maintenance a 2. Multiplying the weight and score together and summing gave a final score of 87. This was lower than a number of the other options with the best scoring 127 overall. Full details of the process can be found in Hubka et al. (1988).

The weighting and scoring technique has a number of advantages. Firstly, it is useful to have a method that gives a quantitative answer following the comparison of the alternatives. Secondly, it is useful in the planning stage to be forced to consider the performance of the options and this method assists in that process. Thirdly, the method can be applied at a number of stages through the design and planning process. Unfortunately there are difficulties with the method in that the scores and weights may be hard to derive and can be quite subjective.

Hobbs (1980) sets out the characteristics that a proper set of weights must have if the procedure is to be theoretically valid. The weights must be proportional to the relative value of unit changes in their objective functions, meaning that weights cannot be simply given ordinal values of (for example) 1 for low importance, 2 for average importance and 3 for high importance. Nor can the most important factor be given (for example) a 10, the next most important a 9 and so on. A proper procedure for deriving and checking weights requires decision-makers be quizzed on their choices and trade-offs checked. For example, if $w_1 = 2$ and $w_2 = 4$ then an increase in the score of objective 1 by 1 unit should be viewed as identical in overall benefit to a change in value of objective 2 by 0.5 units. The weights are much more about relative trade-off values than importance. This process of determining weights may take some time but as Hobbs (1980) notes: "As a typical siting study [for power plants in the U.S.] costs tens of millions of dollars or more, it seems reasonable to require that an extra afternoon be taken to ensure that weights are theoretically valid."

Power supply for Baden-Wűrttemberg

Renn (2003) shows the application of a weighting and scoring technique in a survey that was carried out in Germany to provide power supply planning data for the medium to long term. Four scenarios were developed:

1. Utilization of technologies (nuclear power);
2. Protection of resources (sustainable, conservation);
3. New lifestyle (green, change in lifestyle); and
4. Business as usual.

Two groups of people (a group of engineers and a group of church representatives) were then surveyed to come up with a point value (weighting) for a total of 74 criteria and an expected utility (EU). The final value of each scenario was summed as the product:

$$EU = \sum p_i u_i$$

where p_i is the point value and u_i the utility of each of the n components.

The criteria covered, among other things, economic aspects (e.g. technical efficiency, security of supply, profitability, macro-economic effects), protection of environment and health (e.g. using environment as a waste sink, conservation of nature and ecosystems, health risks) and social and political aspects (e.g. protecting dignity of humans, political stability, social justice). Each criterion was allocated a score from 0 to 100 points based on the groups assessment of its likely outcome. Unfortunately, few details were available on the system used to allocate the weights, although it was noted that the final decision was sensitive to these. The final scores showed both the engineers and church representatives preferred Scenario 3 (new lifestyle) over Scenario 2 (protection of resources).

The results are interesting for a number of reasons. The engineers were actually surprised to find that their deliberations had led to them selecting Scenario 3. This was not what they had expected and the author suggests they had probably set some of the weights too low. Also the scoring of the various options indicated that people often employ a hugely non-linear scale. One of the options, for example, had energy savings of 40% yet led to a much smaller difference (20%) in the scores given to that scenario. It was suggested that this may have been partly due to people not believing the efficiency figures and discounting the benefits to a degree.

The scoring and weighting process allows decisions to be made based on chosen weightings and estimated satisfaction factors. Once those values have been selected there is no uncertainty to be dealt with and the final decision should be a dependable one. It may be argued that the procedure has a number of weaknesses: the weights are subjective and so too, therefore, are the estimated scores for each objective. However, these issues are largely due to the nature of engineering problems which are, as we have seen, often ill-defined, open-ended and complex. The benefit of the procedure is that it forces the decision-maker to think deeply about the objectives that are required for the solution and to consider also how well a range of possible solutions would satisfy those objectives.

There are a range of other decisions that are based around complete knowledge of the system where the solution involves finding an optimum solution. These are dealt with in some detail in Chapter 13 on optimisation.

Decision-making in the brain

Philosophers and researchers have long pondered the actual mechanisms of decision-making in the brain. A popular model has been to assume information is gathered and weighed up until there is sufficient evidence one way or another to tip the balance. Recent work by neurobiologists (Rorie and Newsome, 2005) has verified this as a model in an ingenious series of experiments involving monkeys. In the experiments a series of moving dots were displayed on a computer screen where there was an overall pattern of movement, either right to left or left to right. Additionally each test had different levels of visual noise, achieved by having a varying number of the dots moving at random or against the mean flow. The monkeys, viewing the screen, simply had to decide the mean direction, with this being determined by the researchers monitoring brain regions known to be associated with the preparation for eye movement. They also monitored neural activity using functional magnetic resonance imagery (fMRI) and found neurons that fired in response to left to right motion and others that fired in response to right to left motion. While all this was going on they searched for an area of the brain that was most active. This was found in the lateral intraparietal area.

The accuracy of the decision, and the time taken to make the decision both fitted a model where a single area of the brain integrates two signals, one signifying left to right motion and the other right to left motion, and makes a decision once a threshold has been passed. Animals really do gather information and make a decision once the balance has tipped.

12.3 DECISION-MAKING WITH RISK OF OUTCOME

We now move to decisions where there is a range of possible outcomes that follow a decision, and where it is possible to estimate the probability of each. As noted earlier, these are referred to decisions with risk of outcome. For example, when NASA decide on the actual launch date within the launch window they do so taking into account the various weather conditions that may occur at that time in the future, and the likely effect of attempting a launch under each of those conditions. The Challenger disaster in 1986, when all on board were lost in an explosion soon after take-off, was believed to be due largely to the cold conditions at the launch that had an adverse effect on rubber O-rings in the rocket assembly leading to their failure (Feynman, 1999). Engineers were aware of the issues associated with cold temperatures, and would have been able to estimate the probability of such conditions based on historical data. The final outcome highlights the importance of assessing decisions carefully.

Single-stage decisions with risk of outcome

In many decision-making situations the aim is to make a single decision using knowledge of the possible outcomes and their probability. A method to handle this single-stage type of problem is the decision tree and it is best illustrated with an example.

A building company is planning for a possible expansion by buying out a competitor. If they do not buy out the competitor, they know, based on past experience, that there is a 70% chance of making a $1m profit and a 30% chance of losing $400,000. If they borrow and buy out the competitor they estimate there is a 20% chance of a large loss ($10m) but an 80% chance of a good profit of $3m. What should they do?

The first stage in the solution is to realise that the basic decision in this case is to be made between two options: buy out or do not buy out the competitor. Following each decision option there are a number of consequences that may occur where the probability of occurrence is known. This can be represented in graphical form as a decision tree as shown in Figure 12.2, where each branch from the main trunk represents a decision option and subsequent branches show possible consequences of each decision option. In this example there are two options, but more could easily be shown and considered without any change in the basic procedure. The decision tree shows the options, the consequences of each option and the probability that they will occur. By convention decisions are shown emanating from a square whereas chance events emanate from a circle.

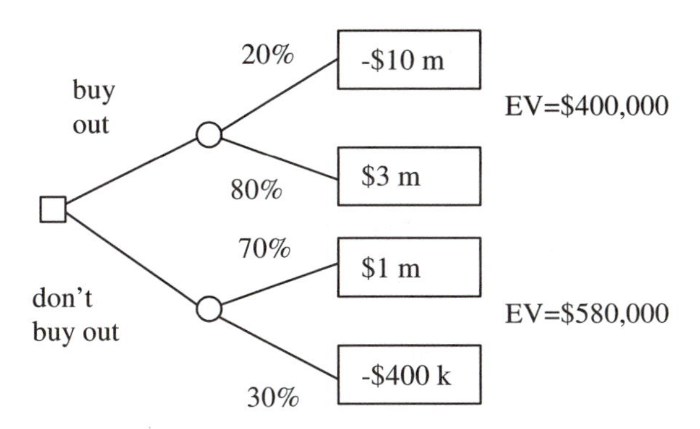

Figure 12.2 The decision tree for the building company decision.

In the next stage of the problem the expected value (EV) of each decision is calculated by multiplying probabilities and outcomes or consequences and summing them for each decision option:

$$EV(a_i) = \sum_{j=1}^{n} O_{ij} p_{ij} \tag{12.2}$$

where there are n possible courses of action a_i, and for each action there are a number of outcomes O_{ij}, where each has a probability p_{ij} of occurrence. This concept was first suggested by Bernoulli in 1738. In the case of buying out the opponent or doing nothing the expected values are calculated as:

$$EV = 0.20(-\$10m) + 0.80(\$3m) = \$0.4m = \$400{,}000$$
$$EV = 0.70(\$1m) + 0.30(-\$0.4m) = \$0.58m = \$580{,}000$$

From the decision tree it is evident that, using the expected value, it is better not to buy out the competitor and expect to make a superior profit each year. It should be noted that the expected values calculated ($400,000 and $580,000), are not to be considered as actual earnings, rather as statistical averages. If the decision was taken a large number of times (a many-off decision) and if the outcomes occurred with the probabilities quoted, these figures would be the average outcome for each decision.

Multi-stage decisions with risk of outcome

As a further complication to the decision-making process it is possible for situations to arise where a series of decisions must be made with each depending on the outcome of the following one. Once again, an example will be used to describe the method of analysis.

Suppose engineers planning to upgrade the water supply to a developing area are faced with two options: to install a small capacity pipeline (action a_{11}) or a large capacity pipeline (action a_{12}). In the first time period, the subsequent outcomes in either case are low demand or high demand. Having observed the states of nature, it is possible in the second time period to follow up action a_{11} with a second-stage decision, namely to do nothing (a_{21}) or to construct a second parallel pipeline to increase the capacity (a_{22}). This decision is again followed by two possible states of nature, i.e. low or high demand. Although the second stage of the process involving a_{21} could well be followed by subsequent stages, we consider in this simple example only a two-stage process. Note that there are no second-stage actions to be considered if the first-stage decision is to install a large-capacity pipeline (a_{12}).

The costs shown at the tips of the branches in Figure 12.3 take account of lost opportunity, as for example when the decision to install a large pipeline is followed by low usage.

In the first stage of the process the probabilities of the states of nature are 0.4 and 0.6 for low demand and high demand, respectively. However, in the second stage, the probabilities are 0.2 and 0.8 for low demand and high demand, in the case where high demand has previously been experienced in the first stage. In the case of low demand in the first stage, the second-stage probabilities are 0.7 for low demand and 0.3 for high demand. The calculations are carried out in terms of costs, so that the negative signs can be dropped. In order to analyse the problem, the expected cost is first calculated at node 3, as 0.7 (70) + 0.3 (325) = 146.5. Expected costs at nodes 3 to 8 are shown. The expected value at node 2 (which is also the expected cost of installing the large pipe originally) is the weighted sum of the values at nodes 7 and 8, i.e. 0.4 (218) + 0.6 (158) = 182. At node B the expected cost is the lower of the values at nodes 3 and 4, i.e. 146.5. Likewise at C the expected cost is equal to that for the preferred action a_{2l}, i.e. 316 at node 5. The preferred actions are indicated in Figure 12.3 by arrows emanating from the decision nodes. Finally, the expected cost at node 1 is obtained from those at B and C as 0.4 (146.5) + 0.6 (316) = 248.2. This is the expected value of action a_{11}, which is larger than for a_{l2}.

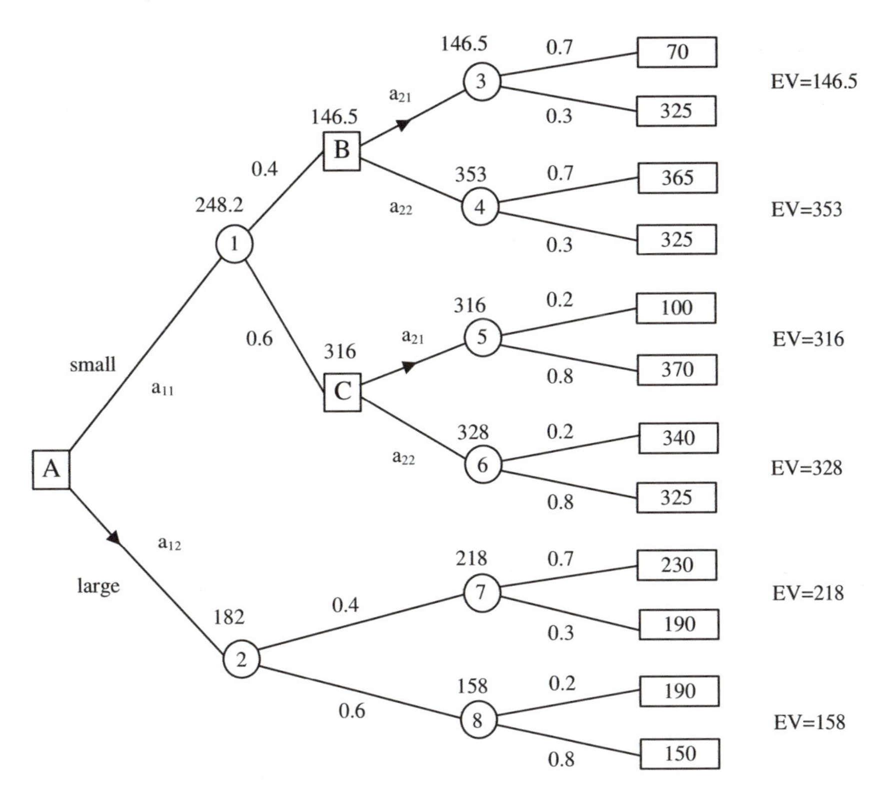

Figure 12.3 A multi-stage decision for a water supply expansion project.

The analysis in Figure 12.3 shows that, with the data given, action a_{l2} is the better at Stage 1. When the decision process has been analysed by the last-step-first procedure outlined above, the appropriate course of action is determined for each stage of the process, irrespective of the previous decisions and previous outcomes. Consider for example that action a_{11} is chosen in Figure 12.3, despite the outcome of the analysis. Although it will not be known in advance whether node B or node C will result, the appropriate course of action can be read directly from the decision tree for each possibility. These are indicated in the tree by the arrows leaving the (square) decision nodes.

Multi-stage decision processes can arise for which the decision tree becomes a very complex network. In cases where the multi-stage process has certainty of outcome, various network optimisation procedures such as linear programming and dynamic programming can be used to obtain a solution. These are dealt with in Chapter 13, although not specifically in relation to decision analysis.

Four dice paradox

Just in case it is thought that mathematics and expected values can lead to a reliable method of decision-making in all situations, consider the following where there are four dice with specially marked faces:

A – 4 on four faces, 0 on two faces	EV = (4 x 4 + 2 x 0)/6	= 2.667
B – 3 on six faces	EV = (6 x 3)/6	= 3.000
C – 2 on four faces and 6 on two faces	EV = (4 x 2 + 2 x 6)/6	= 3.333
D – 5 on three faces and 1 on three faces	EV = (3 x 5 + 3 x 1)/6	= 3.000

If the dice are rolled with the aim of one beating the other by showing a higher number A will beat B 2/3rd of the time, B will beat C 2/3rd of the time, C will beat D 2/3rd of the time and D will beat A 2/3rd of the time!

These are known as non-transitive dice and were designed by Bradley Efron of Stanford University to highlight a probability paradox (Peterson, 2002). In practice it means that a person could allow an opponent to select whichever die he of she liked, and still be able to pick one which, given a sufficient number of games, would lead to a reliable prospect of winning.

12.4 RISK AVERSION, RISK ACCEPTANCE AND UTILITY

The decision tree process outlined in the previous section allows rational decisions to be made, based on expected monetary outcomes. To show that this may be unrealistic for human decision-makers, consider for a moment the decision of whether to take out insurance. If one analyses the decision using a decision tree taking account of the chance of a loss, the size of the loss and the cost of the insurance it will be seen that the decision to insure will often have the higher expected cost. And yet, most people do use insurance for a range of purposes: to protect their house, car, life, bicycle, luggage, and health to name a few. The reason is that, as noted by Schwartz and Begley (2002) the decision, in these cases, is not purely one of maximising expected outcome. People generally allow their feelings about gains and losses to colour their behaviour. In some cases they show risk aversion: insurance is a good example where people will gladly pay to have the risk of a significant loss reduced. In other cases they will happily take a gamble, showing risk acceptance: the decision to buy a lottery ticket is a good example of that form of behaviour.

At this stage it is necessary to clarify one aspect of the decisions being discussed. In some cases the decision will be a one-off: this relates to a set of circumstances that are unique and that will not be repeated. The decision to take out life insurance is an example of a one-off decision. A many-off decision, on the other hand, is one that, although taken once, applies to multiple occurrences. For example, a firm considering insurance for its fleet of cars and trucks is making a many-off decision since the decision will apply to many individual vehicles. In the case of many-off decisions there is an argument for taking the decision that has the superior expected value since there are likely to be sufficient outcomes to give a final outcome close to the expected. For this reason large firms may choose to self-insure their vehicle fleet. A similar situation will arise with central governments. For example, the Israeli government planned to self-insure its proposed water desalination plants against the possibility of terrorism and war. Dreizin (2006) noted that one reason for this was that these sorts of exceptional events "may never materialize". The cost of the self-insurance was calculated and included in the cost benefit analysis that was carried out to determine the project feasibility.

Tendering to Repair a Dam Wall

Tabatabaei and Zoppou (2000) report on an engineering project where faulty construction work on a concrete dam wall was discovered during routine maintenance work. The concrete poured in 1912 near the base of the dam in the Australian Capital Territory was found to be of very poor quality – a situation with no easy or obvious solution method given the dam's location and role as a water storage facility for Canberra, the national capital. Tenders were called and this led to three submissions, each proposing a quite different solution, particularly in relation to handling possible high flow events that might occur while the repairs were being carried out.

The tenders were assessed using a series of objectives (the total cost, the flood risk, the demolition technique and relevant experience) which were weighted in the ratios of 100:50:30:10 respectively. Each of the tenders was given a score for each objective and the product of the score and weight summed. This led to the decision to award the tender to Tender T1.

The decision on whether to require a coffer dam during repair was based on a decision tree analysis where the cost of the coffer dam was compared to the expected cost of having no coffer dam based on the assumption of a 70% chance of a rain (potentially leading to a flood) occurring during the repairs, and where the probable number of rain days was determined using a statistical analysis that summed over all possible occurrences.

For one-off decisions, therefore, if we are to use decision trees to analyse real decisions, it will be necessary to factor in the attitudes towards risk. This brings us to the concept of utility, which is designed to be a measure of how a particular outcome is valued rather than the outcome itself. It may well be a function of the monetary outcome, but is unlikely to be a linear one and this then leads to the sorts of decisions that people make.

If utility is to be used in making decisions then the task is to define a relationship (see for example, Figure 12.4) between a sum of money (measured in dollars) and its utility (measured in arbitrary units of utility or satisfaction).

It is important to stress at this stage that this relationship is likely to vary from person to person and that even a single person will at times show risk acceptance and at other times risk aversion. It will therefore need to be defined on a case by case basis. According to Bazerman (1998) a common utility function shape is in the form of an S-curve as shown in Figure 12.4. The plot shows that as gains increase the person sees a reduced rate of rise in overall satisfaction and would therefore be averse to taking risks to increase any gain. By the same token, if there are to be losses, then in some circumstances an additional loss is not viewed as badly and the person might accept additional risk.

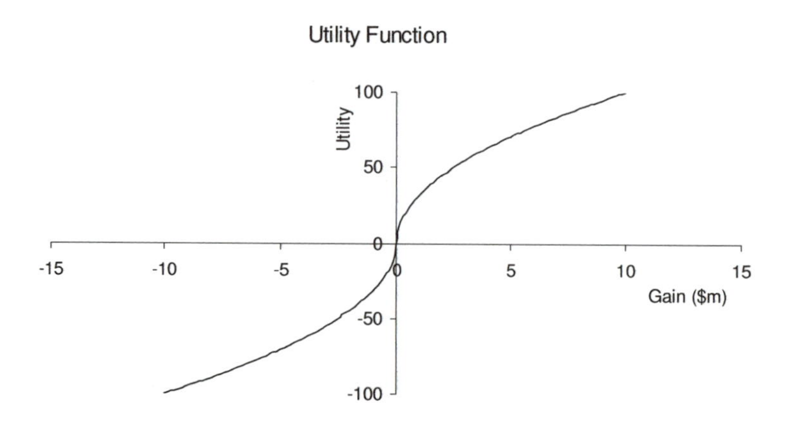

Figure 12.4 A utility function showing general aversion to risk associated with gains and acceptance of risk associated with losses.

The application of the utility concept is through the use of the decision tree. In this case, the outcomes and expected values are written in terms of utility rather than a monetary value. A re-analysis of the first problem faced by the managers deciding whether to buy out a competitor will be used to illustrate the method. We will use the utility function shown in Figure 12.4. Its derivation will be covered following the example. For the moment it is sufficient to note that a utility value can be determined for any dollar amount between a loss of $10m and a profit of $10m. The revised decision tree using the utility value in place of the previous gains and losses is shown in Figure 12.5.

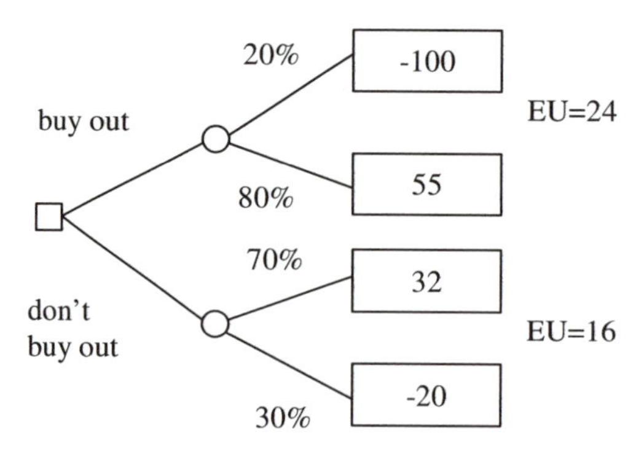

Figure 12.5 Final part of decision tree, the expected utility of each decision is calculated.

Note that the decision, based on maximising expected utility, is to buy out the competitor. This is the opposite of the previous one, and one that takes into account the attitude of the decision-maker to the various profits and losses possible following the decision.

To illustrate the development of the utility function for this problem we consider the positive side of the function with values ranging from a break-even point ($0) to a profit of $10 m. To start the process, a decision tree is set up, as shown in Figure 12.6, where a "game" is played. The person responsible for making the decision is quizzed to determine what probability, p, of an outcome they would accept to take a chance on making the profit by taking decision a_1 if the alternative was to do nothing, a_2, and make a guaranteed profit or loss of $\$x$, where the value of the outcome of a_2 is varied over the range of possible outcomes ($0 to $10 m in this case).

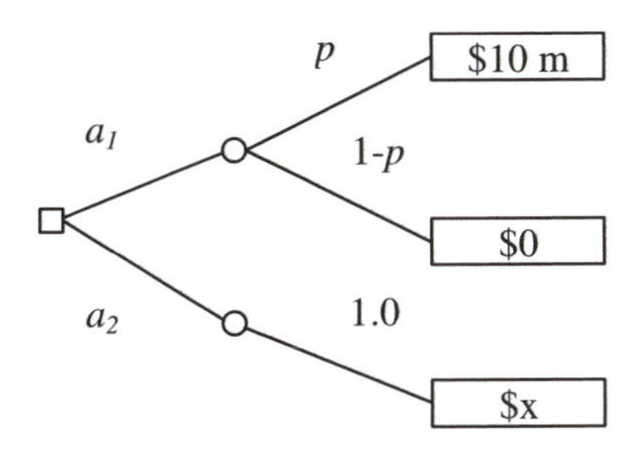

Figure 12.6 The decision tree used to derive a utility function.

For example, if x were $0 it is likely that the decision-maker would accept any probability, p, marginally greater than 0.0 to take action a_1 since any outcome cannot be worse than choosing a_2. The utility, on a scale of 0 to 100 is determined as $100p$. This then gives the first point on the utility plot ($0, 0.0). The value of x is then increased and the question put again. With $x = \$3$ m the decision-maker has to decide what chance of success they would require to take the gamble between a sure $3 m profit or a situation where they could make a $10 m profit but also might get $0. For this particular case the decision-maker decided that a 55% chance of the profit would make the gamble worthwhile. Note that this is considerably higher than the 30% chance that the dollar values would suggest. This leads to an assessment of the utility of $3 m as 0.55 (100) + 0.45 (0) = 55, i.e. the point ($3 m, 55). The value of x is then increased again and the decision re-taken. By the time the decision-maker is offered a sure $10 m profit it is clear they would require a 100% chance of winning the $10 m to take the gamble. This gives the final point ($10 m, 100) and the result is as shown in Figure 12.4. The negative limb of the plot can be derived using a similar procedure.

12.5 DECISION-MAKING WITH UNCERTAINTY OF OUTCOME

In the examples dealt with so far it has always been assumed that the probability of occurrence of a given event has been known. In some situations, however, it may not be possible to estimate the probability of occurrence and yet a decision may still be required. There are a number of ways to deal with problems of this type: each will be demonstrated with an example of a decision that is represented by a tree shown in Figure 12.7. In this decision, engineers must choose between keeping a dam deliberately low in the hope that good rains will come and fill it (action a_1) or choosing a more conservative course of action (a_2) where a higher level is maintained. In the situation where it was possible to estimate that there is a 20% chance of good rains, deliberately keeping the level low will lead to a good payoff (100 utility units) compared to the poor outcome (20 utility units) with poor rain. The conservative approach has a lower payoff in the event of good rains (50 utility units) but a reasonable payoff with poor rain (40 units).

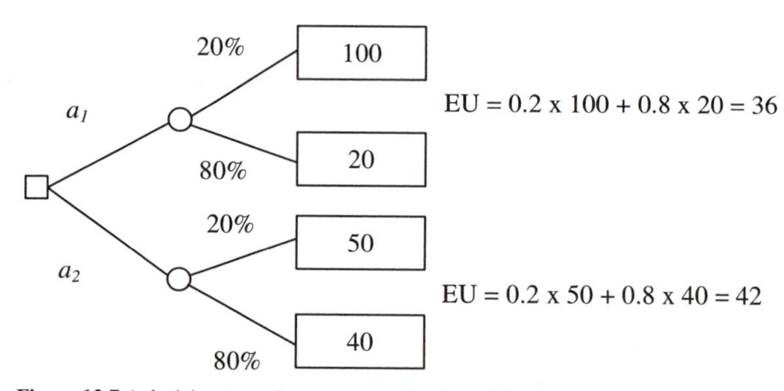

Figure 12.7 A decision tree where a course of action, either a_1 or a_2 is to be carried out.

Based on the probabilities and outcomes shown the decision taken would be a_2 since this has the higher expected value (42 compared to 36). Before we describe methods to handle the problem of missing probabilities it is worth noting that one way to handle uncertainty in regard to the probabilities is to carry out a sensitivity analysis based on the possible range of probabilities. In the work of Mankuta et al. (2003) probabilities were reported as a base value, but a low and a high option were also given. This meant that it was possible to investigate how sensitive the decisions were to the actual probabilities and the results of decisions could be viewed in this light. This idea of using a range of values is one of the features of fuzzy logic and fuzzy sets, where variables can have a range of values (often in the form of a triangular distribution) and the analysis is designed to handle this characteristic (e.g. Chan et al., 2000).

Laplace's principle

If the probabilities are unknown, Laplace's principle (actually first described by Bernoulli) involves assuming all probabilities are equal. The decision tree in that case is shown in Figure 12.8. In this case the decision taken would be a_1 since it now has the higher expected outcome (60 compared to 45).

Although application of Laplace's principle allows decisions to be calculated it is acknowledged that it generally leads to poor decisions since it gives equal weighting to all possible outcomes, even ones that are likely to be rare.

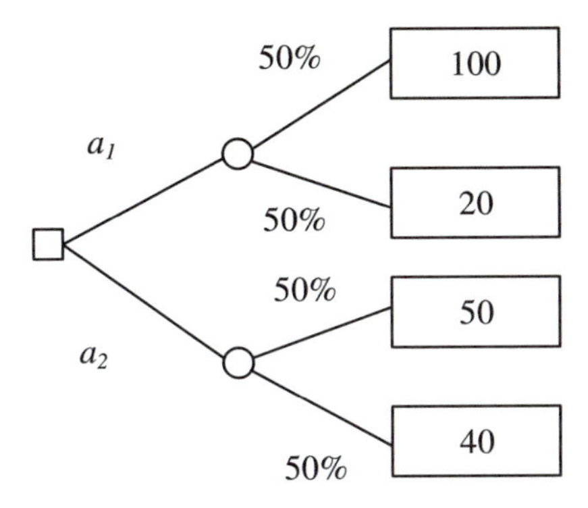

Figure 12.8 The decision tree for Laplace's method where all probabilities are assumed equal.

Pessimistic criterion

If the probabilities are unknown one way to choose is to assume the worst and minimise the potential losses. In this case the choice will be determined by the course of action that has the least worst outcome. Bennett (2001) has pointed out that the precautionary principle (in environmental theory) is equivalent to decision-making under uncertainty and in fact applies the pessimistic principle. Under a different name, the "minimax" strategy is a well known optimisation goal: minimise the maximum damage your opponent can do (Holland, 1998). Looking from the point of view of making profits rather than suffering losses ReVelle et al. (2004) refer to this as a 'maximin' strategy where the aim is to maximise the minimum payout. For the situation under discussion action a_2 would be selected since its worst outcome is better than that for action a_1 (40 compared to 20).

Optimistic criterion

The opposite of the pessimistic criterion is the optimistic approach where the best possible outcome is sought. ReVelle et al. (2004) call this the 'maximax' strategy, to maximise the maximum payout. According to Manhart (2005) humans in general have an optimistic bias in their decision-making, particularly when it comes to personal preferences but this may not assist them in a situation where an engineering type decision is being made. In the problem under discussion action a_1 would be preferred as it has the better possible outcome (100 compared to 50).

Minimising regret criterion

Another method for making decisions is based on minimising the regret that a decision will bring. This one is slightly more elaborate, but is a valid way to come to a decision. Once again, the method will be illustrated with the same problem with the decision tree redrawn in Figure 12.9. The possible outcomes are determined by which of two possible natural events n_1 or n_2 occur. It is assumed that the probability of these is unknown.

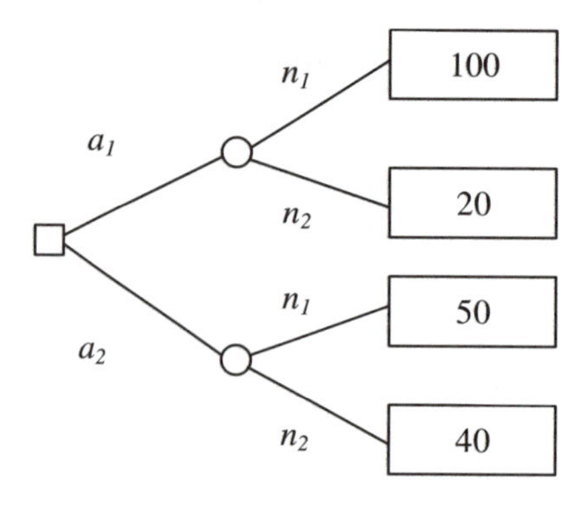

Figure 12.9 A decision tree to choose between actions a_1 and a_2, with unknown probabilities and natural events n_1 and n_2.

Firstly, consider the situation where the first of the two possible natural events, n_1, has occurred. Then, if action a_1 had been selected the decision-maker would have no regrets since they have chosen the action that gives them the best result for this case (100 compared to 50). If, however, they had chosen action a_2, the regret would be 50 since the best possible result with natural event n_1 is 100. Now, if the second outcome, n_2, had occurred they would have a regret of 20 if they had chosen action a_1 and no regret if they had chosen a_2. These have been compiled in Table 12.1. On the basis of attempting to minimise the maximum regret, action a_1 would be chosen since this has the smaller maximum regret.

Table 12.1 Regret outcomes for the problem under discussion.

Decision	Event n_1	Event n_2	Maximum
a_1	0	20	20
a_2	50	0	50

Using subjective probabilities

Even if probabilities are not known it is possible to estimate some based on the decision-maker's best estimate. Whereas this may appear to be a relatively simple procedure, recent research has shown that it is fraught with danger. As an example, Winman et al. (2004) suggest that if one wants to find out the probability of an event occurring quite different answers will occur if the person is asked "Estimate the lower and upper range of values for the probability of such and such occurring" or "What are the chances that the probability of such and such occurring will be between x and y?" Both questions seem to be asking much the same question. However, by asking it in different ways not only does one get different estimates, but the confidence associated with the answer is quite different. As a matter of interest, for this particular situation, the latter question leads to better answers than the former.

Buying information

The decision situations we have to date considered, listed in decreasing order of information available to the decision maker, are as follows:

- certainty of outcome; (full information on outcomes)
- risk of outcome;
- structured uncertainty; and
- unstructured uncertainty. (zero information on outcomes)

The acquisition of further information is always an important option to consider, and one that Savary and Parker (1997) cite as a rapidly emerging sector in engineering. The additional information may make it possible to move up one or two levels on the above scale. Even when the decision situation remains risky, additional information can improve decision making by improving the probability estimates or even eliminating some of the less likely outcomes from consideration.

In engineering decision-making, additional information may be acquired in various ways, for example through laboratory testing, pilot studies, theoretical analysis, field trials, or engaging a research agency. In a study of the tendering behaviour of construction companies Mochtar and Arditi (2001) found that around 50% of firms purchased information (sometimes, often or always) from research agencies in an effort to strengthen their bids, although the authors did note that this was not the preferred method given its cost.

12.6 DECISION-MAKING WITH COMPETITION

In August 1955 a decision was taken in Moscow to put a satellite into orbit around the earth before July 1, 1957. According to Schefter (2000) "[the] timing had only one purpose: to beat the United States of America into space". In the end, delays meant that it was not until October 1957 that the task was accomplished but the aim had been achieved; the Americans had been beaten.

Engineering, as a profession, is generally undertaken in a competitive environment. Jobs are often won or lost in a tendering system where a firm will describe what work it wants done in a call for tenders and firms will then bid for the opportunity to undertake the work. The bidding process is confidential and the bids are made with no certain knowledge of potential rival bids. The aim, of course, is to put in the most attractive bid which would often be selected on the basis of cost, as long as the firm is confident that the bidding firm is competent to undertake the work. Game theory was developed to cover situations such as these, and while it may not be possible to undertake sophisticated analysis of realistic complex situations, there is benefit in being familiar with some of the concepts that help to explain how humans react under these circumstances.

When engineers respond to a call for tender, they are doing so in an environment where other firms will be attempting to win the tender themselves, and therefore any chance of success must depend on how their competitors act. This complicates and changes fundamentally the decision-making process. For example, if two firms (Firm A and Firm B) are bidding for a tender they will each want to put in as high a bid as possible to maximise the profit. However, Firm A knows that too high a bid may mean that Firm B may put in a lower offer and win, meaning there is no profit at all. Therefore when Firm A bids, it does so with some knowledge of what it expects Firm B to bid. But there is another level: Firm A will know that Firm B's decision will be based on what it (Firm B) believes Firm A will believe about its intentions. This adds a level of complexity to these sorts of decisions and makes the problems all the more intriguing. One firm conclusion to come from studies of people making decisions is that they often do not act in a way that a decision tree analysis would predict.

John von Neumann, Oskar Morgenstern and the zero sum game

The foundations for game theory were set out in The Theory of Games and Economic Behavior by John von Neumann and Oskar Morgenstern in 1944. A fundamental breakthrough came when von Neumann was able to show that for a particular type of game, the two person zero-sum game, a strategy was possible to maximise rewards and that strategy was based on each player maximising his or her minimum payoff.

The zero-sum game is one where one person's loss is another person's gain and all payoffs sum to zero for all possible strategies. Poker, for example, when played by two players is a zero sum game. Add another player, or a banker who takes a cut of the pot, and it is no longer a two person zero sum game. The tendering process, even when undertaken between only two firms is unlikely to be zero sum since the loser will lose the cost of putting together the tender bid whereas the winner will make a profit from undertaking the job that it was bidding for. These two sums are likely to be quite different.

As a simple example of behaviour that is at odds with the concept of maximising expected value consider the following game. One dollar is to be divided between two people, A and B. A gets to choose how much of the dollar he or she wants, and therefore how much B will be left with. However, the dollar is only given to A to share if B agrees to the way the dollar is divided. What should A and B do? Knowing that B will be better off with anything, A should offer them 1c and pocket the other 99c each time. But this is not what happens! In practice most B players reject anything less than 25c. At this level they would prefer to get nothing themselves to get the satisfaction of seeing A get nothing too! Studies have shown that most societies have a code of fairness and most A players will offer around 50c. This of course is not the outcome that one would expect from a decision tree analysis, but it is behaviour that makes life so interesting.

Prisoner's dilemma

A classic problem in the early theory of games is a situation described as the prisoner's dilemma. The scenario is that two long-time thieves have been caught near the scene of a burglary. They have been taken to the local police station and are being interviewed separately. Each is told that if he confesses to the pair's guilt and his partner does not, then he will be given protection and escape jail, while his partner will get 20 years in jail. If both confess they have been told that they will be treated leniently and get 10 years jail. The thieves know that if neither confess then the worst that is likely to happen is that they will get 1 year in jail because there is little evidence on which to convict them. What should each man do?

There is certainly an argument for not confessing because that gives a good outcome, but only if the other doesn't too. In developing a solution the first task is the represent the range of outcomes as a payoff matrix, as shown in Table 12.2. Note first of all that it is not a zero sum game since the sum of each of the outcomes at each of the four possible strategies does not sum to zero. In fact none of them sum to zero. Table 12.2 presents all the options and the payoffs in terms of years in jail. In this case, therefore, the smaller the payoff the better.

Table 12.2 Payoff table for the prisoner's dilemma. Note: payoff is (Payoff to A, Payoff to B)

		Prisoner B	
		confess	silent
Prisoner A	confess	(10,10)	(0,20)
	silent	(20,0)	(1,1)

To determine the most likely outcome it is necessary to investigate the various options. Consider first Prisoner A. He considers his options in the light of the behaviour of Prisoner B. If Prisoner B confesses then A will either get 10 years if he also confesses or 20 years if he does not. Therefore, in that situation he should confess. On the other hand, if Prisoner B stays silent A will either get 0 years if he confesses or 1 year if he stays silent too. In that case he should confess. Therefore, no matter what B does A should confess. Prisoner B can go through the same process and arrive at the same conclusion. Therefore, rational prisoners would both confess and get 10 years in jail. Note that this is not the best outcome for them, but if they are acting rationally and are unable to confer, this is the one they would choose. There are of course a range of extensions to the sorts of games that can be contemplated. The concept of cooperative and non-cooperative games adds another level of complexity. For the treatment here, cooperative games will not be considered.

The argument used to come to a final rational decision with the prisoner's dilemma can be formalised by the definition of a number of strategies: that of dominant, strictly dominant, dominated and strictly dominated strategies. It will be seen that these allow a formal approach to be taken to the problem of making decisions with competition.

Dominant Strategy

In the prisoner's dilemma we saw that for Prisoner A the best option was to confess, no matter what Prisoner B did. This situation, where one particular course of action is better than all others, no matter what the competitor chooses to do, is called a strictly dominant strategy. If the course of action is equal to, or better than any other it is a dominant strategy. In game theory, a rational player will, by definition, always choose a strictly dominant or dominant strategy (if one exists).

By a similar argument, for Prisoner A staying silent is a strictly dominated strategy (worse in all situations) and a rational player will never choose a dominated or strictly dominated strategy.

Iterated dominant strategy

While the prisoner's dilemma shows a simple four option game, with a strictly dominant strategy for both players, many payoff matrices would be more complicated with no clearly dominant strategy. Table 12.3 shows an example of such a game. Note that it is a zero sum game since for each of the nine strategies the sum of the payoffs to the players is exactly zero. A general strategy to solve the problem will now be given.

Table 12.3 Payoff table for a two person zero sum game. Note: payoff is (Payoff to A, Payoff to B)

		Player B		
		b1	b2	b3
	a1	(-1,1)	(4,-4)	(-3,3)
Player A	a2	(1,-1)	(2,-2)	(2,-2)
	a3	(-2,2)	(13,-13)	(1,-1)

Looking first at B's options: it is clear that action b2 will not be chosen by a rational player since it is dominated; that is no matter what Player A does, Player B would be worse off if b2 was chosen since it results in payoffs of -4, -2 or -13 whereas Player B, by choosing a different strategy (b1) could improve the payoffs to 1, -1 or 2. Therefore option b2 is removed from the game. The payoff table is now as shown in Table 12.4.

Table 12.4 Payoff table for a two person zero sum game with b2 removed. Note: payoff is (Payoff to A, Payoff to B)

		Player B	
		b1	b3
	a1	(-1,1)	(-3,3)
Player A	a2	(1,-1)	(2,-2)
	a3	(-2,2)	(1,-1)

Turning now to Player A, it is evident that option a1 is strictly dominated by a2 and should also be removed. This leaves the payoff table shown in Table 12.5.

Table 12.5 Payoff table for a two person zero sum game with b2 and a1 removed. Note: payoff is (Payoff to A, Payoff to B)

		Player B	
		b1	b3
Player A	a2	(1,-1)	(2,-2)
	a3	(-2,2)	(1,-1)

Now, it is clear that Player B should choose b1 as this is strictly dominant (-1 > -2 and 2 > -1). Knowing this Player A will choose a2. The final solution is therefore a2 from Player A and b1 from Player B. The process whereby some alternative options were removed gives the procedure the name of an iterated dominant strategy.

Nash equilibrium

Rational players with dominant strategies will play them, and in some cases this will determine the outcome of a game. However, there are circumstances in non-zero sum games where there is no dominant strategy and in this case the question arises: what will rational players choose? Table 12.6 shows a payoff matrix that has no dominant or dominated strategies for either player. If Player A selects a2 then Player B should choose b1. If Player A selects a3 then Player B should choose b3. If Player B selects b1 then Player A should choose a2, and if Player B selects b3 then Player A should choose a3.

Table 12.6 Payoff table for a two person game. Note: payoff is (Payoff to A, Payoff to B)

		Player B	
		b1	b3
Player A	a2	(1,1)	(2,0)
	a3	(0,2)	(3,3)

This is an example of the type of situation studied by John Nash, and was part of the work for which he shared the Nobel prize for economics in 1994 (actually the Bank of Sweden Prize in Economic Sciences in Memory of Alfred Nobel). Nash's strategy can be explained quite simply: if Player B assumes Player A will select a2 then he or she should choose b1 (as this has the higher payoff); if Player A assumes Player B will play b1 then he or she should choose a2. Hence there is agreement and this joint strategy (a2, b1) is denoted as a Nash equilibrium point. A Nash equilibrium in a two person game can be defined as a set of strategies where neither would be better off switching strategies unless the other did also.

There is a procedure for determining Nash equilibrium points. For each player identify the best strategy for each possible play by the opponent. For B it is b1 if A plays a2 (a2, b1) and b3 if A plays a3 (a3, b3). Turning now to Player A, his or her best strategies for the two possible plays by B are (a2, b1) and (a3,b3). Since both players agree on both these positions, both are Nash equilibriums.

Now, the Nash equilibrium point is quite different from a dominant strategy and there is no guarantee that rational players will necessarily choose such a strategy but it is believed there are grounds for supporting the idea that they will. Some studies have attempted to verify the proposal but they have not always been able to do this: however, the concept of Nash equilibrium points is well established and worth considering.

It should be pointed out that in many games there will not be a Nash equilibrium point and in other there may be more than one. This then introduces the need to be able to decide between multiple Nash equilibrium points.

Focal point equilibria

Although not strictly part of rational behaviour, it has been found that in some situations with multiple Nash equilibrium points, one of them stands out as being better for some reason. It is believed that this explains a number of features of real life decision-making in a game environment and some examples quoted in Bierman and Fernandez (1998) highlight these. These are called focal point equilibria and appear to be important in separating various options in a decision.

Two people are asked to nominate heads or tails with the view that if they both give the same they will win money. There are two Nash equilibria for this (head-head or tail-tail). In tests, 86% chose 'heads' making it the focal point equilibrium. Two people are asked to write a positive number. If they agree they win money. With all the possible outcomes 40% chose the number 1, making it the focal point equilibrium.

12.7 THE HUMAN FACTOR IN DECISION-MAKING

In the previous sections we have seen a range of methods for making decisions and it would be comforting to think that human decision-makers would use them in some sort of rational process. Extensive research in human decision-making has shown that things are seldom so simple: for a start humans show preferences in the methods that they use in coming to a decision and secondly, they are not always rational in their behaviour.

According to Greening et al. (2004) when people attempt to make decisions they go through a range of alternative methods. First preference is for minimax type procedures where they consider possible gains and losses and attempt a decision on this basis. If a decision can be made using this approach then the process is halted. If a clear decision does not arise from this they move on to scoring and weighting techniques. If a decision is still not possible they then move on to mathematical modelling of the system to assist them in deciding what to do. Rational decision-makers also use the concept of satisficing. Satisficing sets an acceptable level of performance for each criterion and once that level has been surpassed they then move on to the next most important criterion. This can be linked with much quoted idiom that "near enough is good enough". These methods all rely on a rational decision-maker and there have been enough counter-examples reported in the literature to suspect that humans are not always so.

A fascinating investigation into one aspect of decision-making was reported by Shiv et al. (2005a, 2005b). An experiment was set up where patients in a hospital were given an initial $20 and then allowed to take part in a number of rounds of a game of chance. For each round of the game the person was offered an extra one dollar: they could either keep that dollar or gamble on the toss of a coin. If they elected to gamble and a head came up they lost that dollar, a tail meant they won $2.50. If one generates a decision tree for this situation (see Figure 12.10) it is clear that the best option is to take the gamble each time since it has the higher expected outcome. (This is especially true if the game is to be played many times, making it a many-off decision.) The researchers selected two groups of people for the study: the control group was selected randomly from patients at the hospital, while a second group was from patients who had suffered damage to a part of the brain that controls emotions. The researchers found that:

- people with damaged emotional centres in the brain did much better than people with no damage ($25.70 after 20 rounds compared to $22.80), and this was due to them electing to gamble in more rounds of the game (84% compared to 58% for those with normal brains); and
- normal patients were more likely not to gamble following a loss in the previous round (referred to as myopic loss aversion), whereas those with brain damage were not affected by a win or a loss.

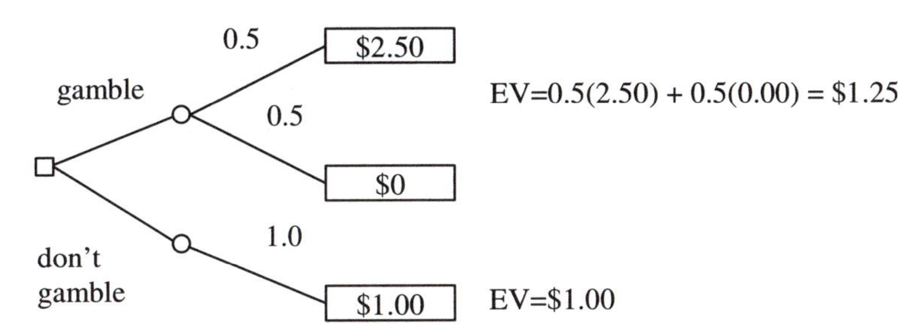

Figure 12.10 The decision tree used to decide whether to gamble the dollar.

The researchers stopped short of suggesting that everyone should look for investment advice from accountants with damaged emotional centres, and it must be remembered that this superior performance would only occur in a situation (or game) with a positive expected value. It does, however, raise some interesting questions about real decision-making.

Based on work by a number of researchers it is possible to draw the following conclusions about human decision-making:

1. People tend to avoid risk when seeking gains but choose risk to avoid losses. Paulos (1988) quotes an example from past research. "Choose between a sure $30,000 or an 80% chance of winning $40,000 and a 20% chance of winning nothing. Most people will take the $30,000 (even though the expected value is less than the other option). On the other hand if faced with a sure loss of $30,000 or an 80% chance of losing $40,000 and 20% chance of losing nothing most people will take the possibility of losing the $40,000 even though the expected loss is greater." This behaviour is illustrated with the S-shaped utility function shown earlier in Figure 12.4.
2. Although people use the concept of expected value in their decision-making it has been found (Knutson and Peterson, 2005) that quite different parts of the brain are involved in the calculation. One area is used in the determination of the probability, another determines the utility value in the case of a loss and a third area determines the utility value in the case of a gain. This may help to explain the quite different behaviour when considering gains or losses.
3. The availability heuristic distorts the estimation of probability by inflating values for situations that we can recall easily (are readily available in our

memory) and reducing estimates for situations we cannot. This effect may be significant, especially given that people would be unaware that they are being influenced in this way. At the time of the Challenger launch in 1986, for example, management at NASA believed the chances of failure were 1 in 100,000; engineers working on the mission believed the chances were 1 in 100 (Feynman, 1999). The engineers, by working very closely with the Shuttle hardware would have been well aware of all the minor faults and failings that came up on a regular basis and would have had little trouble in recalling potentially fatal flaws. Managers on the other hand may have been unaware of many of these and so had more trouble imagining or recalling such flaws.

4. When estimating values, people can be overly influenced by any starting value that is suggested. For example, Bazerman (1998) quotes a study where people were asked to estimate the percentage of African countries that were part of the United Nations and were given, as part of the question, a random starting value. They were then asked to estimate if the percentage was higher or lower and what they thought it might be. Groups that were given 10% as a start value estimated (on average) that 25% of countries were part of the U.N. Groups that were given a start value of 65% estimated (on average) that 45% of the countries were part of the United Nations. The random start value had a significant effect on the final value, and this effect is likely to be part of any real decision-making process, particularly where values must be estimated.
5. When a project consists of a number of complex tasks that must all be completed, humans underestimate the chance of failure to complete on time. Bazerman (1998) calls this the conjunctive events bias and suggests it results from people's inability to comprehend that any complex system will malfunction if only one of its components fails and even though the probability of failure might be small for each component, the overall probability can be much higher. He cites this as a warning, for example, when considering the development of nuclear reactors and in undertaking DNA research.
6. When part of a group, humans tend to become more confident in their estimation of unknown quantities. This behaviour is often referred to as "groupthink" and has been shown to be widespread. Although it is possible for groups to lead to a better estimation of an unknown quantity the disturbing feature is that, even if the estimate is not a good one, the confidence in it will be unrealistically high.
7. Dörner (1989) suggests that many problems in human decision-making can be explained by a failure to assess the impact of side effects on decisions, and an inability to appreciate the behaviour of non-linear systems. When pressed for time humans tend to over-react with any responses and to ignore safety rules, particularly if they had been violated before without apparent problems.
8. It is possible to determine the characteristics of good decision-makers based on the way they react to outcomes that come from their earlier decisions. In an extended simulation of decision-making Dörner (1989) found that good decision-makers tended to make more decisions and to make more complex ones. They also developed hypotheses about the system they were working with and tested those hypotheses as they went. They were more likely to ask more 'why' questions, attempting to understand the environment and tended to be less likely to continually change decisions once made. Good decision-

makers also anticipated side effects and tended to have more thinking and less action. They were more likely to react to changing conditions as they went rather than setting a course at the start and sticking to it. Poor decision-makers, on the other hand, often acted without prior analysis, failed to anticipate side effects, assumed that the absence of immediately obvious negative effects meant that correct measures had been taken, let over-involvement in projects blind then to emerging needs and changes, and were prone to cynical reactions if things went badly.

12.7 SUMMARY

Decision-making comes into many aspects of an engineer's work, and it is important to appreciate, firstly, that there are many different types of decisions that need to be made and, secondly, that there are methods suitable for each type. While the methods outlined have been applicable to somewhat idealised cases, the intention was to give some insight into the process, and to demonstrate procedures that can be applied to a range of decision types.

Some decisions are made with complete knowledge. These may be based on systems that are completely deterministic and some may involve the selection of a course of action from a number of competing ones where the choice is mutually exclusive. For these, a weighting and scoring technique is suitable. We have also foreshadowed the role of optimisation which will be dealt with in detail in Chapter 13.

When there is incomplete knowledge about the conditions that will prevail at the time the decision is implemented it is still possible to come to a decision and this has been illustrated both when the decision is based on outcomes measured in monetary units or when it is possible to take the decision-makers preferences and goals into account using the concept of utility.

Having introduced the concept of the expected value and assumed that decision-makers would act to maximise it we have found that in a competitive environment humans act in a quite different, but still rational manner.

Finally, an introduction to the human factor in decision-making has been given where it is shown that humans are to a large extent driven by a pre-programmed brain and this can lead to less than optimum decisions.

PROBLEMS

12.1 A computer manufacturer has the possibility of expanding the company's operations into New Zealand with the potential of greatly increased profits. The managing director feels that the move would lead to a 20% chance of making the company $1m in profits, but that there would be a 80% chance of losses of $200,000 if the move were unsuccessful. If the expansion is not attempted the company would continue to trade with an 80% chance of making a profit of $200,000 and a 20% chance of breaking even. Based on expected monetary considerations what would you suggest ?

12.2 A government department associated with the Cooperative Research Centre for Low Grade Coal has come up with a new process to pre-treat the raw coal and has called for tenders to implement this at a new factory. As Research and Development manager of a consulting firm you are familiar with the theory behind the process, but are also aware of new developments overseas that may mean the technology is out of date in the very near future. You must decide on one of the three following courses of action, on behalf of your employer:

(a) do not tender.
(b) prepare and submit a tender, based on the designated process.
(c) work on the new overseas development and prepare a tender based on this new method.

You estimate that, if you get the contract for the CRC process there is a 70% chance of making a $60,000 profit, a 15% chance of breaking even, and a 15% chance of making a loss of $300,000. The cost of preparing the tender is $20,000, and this cost has been taken into account in the above profit-and-loss and break-even estimates. The chances of winning the contract for the CRC process are good: about 70%.

Although the chances of winning the contract on the basis of the new overseas developments are much lower, about 40%, you stand to make a much larger profit. You estimate that, if you get the contract, you would have a 70% chance of making a profit of $1,500,000. There is a 30% chance of losing $300,000. The cost of preparing the tender in this case is $30,000, but there is the added cost of $200,000 for the work involved in bringing the overseas method to tendering stage. Both of these costs have been included in the profit and loss estimates given.

Make any calculations you consider appropriate, state any assumptions you decide to make and hence indicate your decision. What is the expected value of your decision?

12.3 Re-analyse Question 12.1 using the utility information given in Table 12.7 which you have obtained from the managing director about her feeling for the utility value of the companies profits and losses. Does this change the earlier decision? Is the managing director showing risk aversion or risk acceptance in this case?

Table 12.7 Utility information for the manager.

Profit/Loss	Utility
+$1,000,000	100
+$500,000	30
break even	0
-$500,000	-20
-$1,000,000	-40

12.4 Assuming there is a 1 in a 600,000* chance of winning $10,000 in a $1 lottery, derive the expected value of the return of buying a ticket. *This figure comes from a television advertisement running in South Australia during April 2005.

12.5 In the design of a pulp mill, annual losses due to interruption of the electricity supply may be one of the following values depending on the time that the interruption lasts ($0, $2m, $6m). The losses in any one year will depend on the severity of the interruptions which actually occur.

There are three alternative actions which are being investigated, which have annual costs associated with them:

A: do nothing $0
B: provide electricity from diesel power generators with manual operation for critical areas of the mill when power blackout occurs. $200,000
C: provide diesel power generators for all mill operations with automatic cut in control system. $650,000

The probabilities of the annual losses will depend on the action taken by the designing engineer. Estimated values of the probabilities are given in the following table.

Action	Annual Cost of Action($)	Probability of Incurring Annual Loss		
		$0	$2 m	$6 m
A	0	0.60	0.25	0.15
B	200,000	0.75	0.20	0.05
C	650,000	0.90	0.09	0.01

12.6 In the following situation, represented only by its payoff matrix, determine for each player: (a) strictly dominant strategies, (b) strictly dominated strategies, (c) weakly dominant strategies and (d) weakly dominated strategies. Find also all Nash equilibrium points.

		A		
		short	long	mid
B	short	(2,-2)	(2,-2)	(2.5,-2.5)
	long	(1,-1)	(3,-3)	(2.5,-2.5)
	mid	(2,-2)	(2,-2)	(2.5,-2.5)

12.7 Assume it was not possible to estimate the probability of the various outcomes occurring in Question 12.1. What decision would be made if the Laplace Criterion was used? What decision would be made if the Pessimistic Criterion was used to make a decision? What decision would be made if it was wished to minimise regret? Base any calculations on the monetary values.

REFERENCES

Bazerman, M., 1998, *Judgement in Managerial Decision-making*. (John Wiley & Sons), 200pp.

Bennett, D., 2001, Development of Sustainability Concepts in Australia. In: *Towards Sustainability. Emerging Systems for Informing Sustainable Development* (Eds. Venning, J. and Higgins, J.) (University of New South Wales Press), 22–47.

Bierman, H.S. and Fernandez, L., 1998, *Game Theory with Economic Applications*, 2nd Edition, (Addison-Wesley), 452pp.

Chan, F.T.S., Chan, M.H. and Tang, N.K.H., 2000, Evaluation Methodologies for Technology Selection. *Journal of Materials Processing Technology*, 107, 330–337.

Dörner, D., 1989, *The Logic of Failure*. (Basic Books), 222pp.

Dreizin, Y., 2006, Ashkelon Seawater Desalination Project – Off-Taker's Self Costs, Supplied Water Costs, Total Costs and Benefits. *Desalination*, 190, 104–116.

Feynman, R., 1999, *The Best Works of Richard P. Feynman. The Pleasure of Finding Things Out*. (Ed. J. Robbins), (Penguin), 270pp.

Greening, L.A. and Bernow, S., 2004, Design of Coordinated Energy and Environmental Policies: Use of Multi-Criteria Decision-making. *Energy Policy*, 32, 721–735.

Hobbs, B.F., 1980, A Comparison of Weighting Methods in Power Plant Siting. *Decision Sciences*, 11, 725–737.

Holland, J.H., 1998, *Emergence from Chaos to Order*. (Oxford University Press), 258pp.

Kmietowicz, A.W. and Pearman, A.D., 1981, *Decision Theory and Incomplete Knowledge*, (Gower), 121pp.

Knutson, B. and Peterson, R., 2005, Neurally Reconstructing Expected Utility. *Games and Economic Behaviour*, 52(2), 305–315.

Manhart, K., 2005, Lust for Danger. *Scientific American Mind*, 16(3), 24–31.

Mankuta, D.D., Leshno, M.M., Menasche, M.M. and Brezis, M.M., 2003, Vaginal Birth after Cesarean Section: Trial of Labor or Repeat Cesarean Section? A Decision Analysis. *American Journal of Obstetrics and Gynaecology*, 189(3), 714–719.

Mochtar, K. and Arditi, D. (2001) Role of Marketing Intelligence in Making Pricing Policy in Construction. *Journal of Management in Engineering*, 17(3), 140–148.

Paulos, J.A., 1988, *Innumeracy*. (Penguin), 135pp.

Peterson, I., 2002, Tricky Dice Revisited. *Science News Online*, 161(15) http://www.sciencenews.org/articles/20020413/mathtrek.asp (downloaded 21st June, 2006)

Renn, O., 2003, Social Assessment of Waste Energy Utilization Scenarios. *Energy*, 28, 1345–1357.

ReVelle, C.S., Whitlatch, E.E. Jnr. and Wright, F.R., 2004, *Civil and Environmental Systems Engineering*, (Pearson Prentice Hall), 552pp.

Rorie, A.E. and Newsome, W.T., 2005, A General Mechanism for Decision-making in the Human Brain? *Trends in Cognitive Science*, 9(2), 41–43.

Sarvary, M. and Parker, P.M., 1997, Marketing information: A competitive analysis. *Marketing Science*, 16(1), 24–38.

Schefter, J., 2000, *The Race. The Definitive Story of America's Battle to Beat Russia to the Moon*, (Arrow Books), 303pp.

Schwartz, J.M. and Begley, S., 2002, *The Mind and The Brain*. (Regan Books), 420pp.

Shiv, B., Loewenstein, G. and Bechara, A., 2005a, The Dark Side of Emotion in Decision-making: When Individuals with Decreased Emotional Reactions Make More Advantageous Decisions. *Cognitive Brain Research*, 23, 85–92.

Shiv, B., Loewenstein, G., Bechara, A., Damasio, H. and Damasio, A.R., 2005b, Investment Behaviour and the Negative Side of Emotion. *Psychological Science*, 16(6), 435–439.

Squyres, S., 2005, *Roving Mars*. (Hyperion), 422pp.

Tabatabaei, J. and Zoppou, C., 2000, Remedial Works on Cotter Dam and the Risk to These from Floods. *ANCOLD Bulletin*, 114, 97–111.

Winman, A., Hansson, P. and Juslin, P., 2004, Subjective Probability Intervals: How to Reduce Overconfidence by Interval Evaluation. *Journal of Experimental Psychology: Learning, Memory and Cognition*, 30(6), 1167–1175.

CHAPTER THIRTEEN

Optimisation

This chapter introduces general concepts of optimisation. Optimisation is a process aimed at identifying the "best" solution to a planning or design problem. Optimisation may be carried out using judgement, by comparing a large number of options or by using analytical techniques. Analytical optimisation techniques may be applied to problems that can be represented by one (or more) objective functions and a number of constraints. Specific optimisation techniques including linear programming, differential calculus, separable programming, dynamic programming and genetic algorithms are considered in detail in this chapter.

13.1 INTRODUCTION

As outlined in Chapter 3, engineering problems are often ill-defined and open-ended. As such, there is usually no single "correct" answer to an engineering problem. The challenge for the engineer is to find the "best" solution to a particular problem.

The engineering methodology outlined in Chapter 3 emphasizes the identification of objectives to be achieved by a particular plan or design and measures of effectiveness for ranking alternative solutions. Furthermore, most engineering problems have a number of constraints that limit the available solutions to the problem.

Optimisation is a process of identifying the "best" solution to a problem i.e. that solution that achieves the maximum (or minimum) level of the objectives while satisfying all of the constraints. Usually there are several objectives to be achieved. The choice of a final solution may involve a compromise between these. For example, a reservoir for a town water supply is to be designed to achieve the following objectives:

1. maximise reliability of the water supply;
2. minimise cost; and
3. minimise environmental impact.

Clearly, no single size of reservoir will satisfy all three of these objectives. A large reservoir would be needed to maximise reliability. This is unlikely to have the minimum cost or the minimum environmental impact. The reservoir that has minimum cost is small in size (or, in fact, no reservoir at all). This will provide a water supply with low reliability, although the environmental impact could be low.

Ultimately, the selection of a final design involves compromise between the various objectives which may involve the input of community values and a judgement to be made about what is in the best interest of the community as a whole.

Optimisation involves the selection of values for a number of **decision variables**. In the above example, the primary decision variable is the capacity of the reservoir. In general, the decision variables may be **continuous** or **discrete**. The capacity of a reservoir can take on any positive value within a defined range and so is a continuous decision variable. The selection of the site for a new airport is an example of a discrete decision variable, as there will be a finite set of sites available and only one of them will be chosen. The following is an example of simple optimisation problem with continuous variables that can be solved using calculus.

Example 13.1: A tank of volume 200 m^3 is to be made from stainless steel. It will have a cylindrical body and hemispherical ends as shown in Figure 13.1. If the material for the ends costs twice as much per square metre as that for the body, find the dimensions of the tank that minimise its cost.

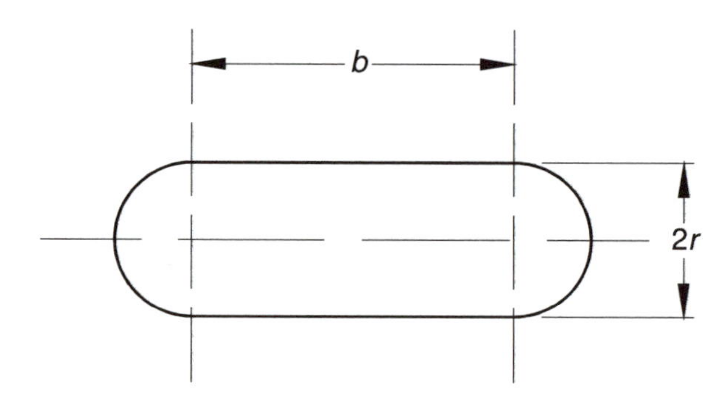

Figure 13.1 A stainless steel tank.

Solution : In this example, the decision variables are the length of the body of the tank and its radius (both in metres). These will be denoted by b and r (respectively). The objective is to minimise the cost of the design. If the cost per unit area of the body is C \$/m^2, then the total cost of the tank is given by:

$$Y = 2C\pi rb + 2C(4\pi r^2)$$

$$Y = 2C\pi rb + 8C\pi r^2 \qquad (13.1)$$

Note that the first term in Equation (13.1) represent the cost of the tank walls and the second term represent the cost of the ends. As C is constant, it can be ignored in the optimisation and the problem becomes:

Minimise $Z = 2\pi rb + 8\pi r^2$ (13.2)

Having identified the objective function, we need to consider what constraints exist. In this case, we require the volume of the tank to be at least 200 m^3. This can be expressed by the following inequality:

$$\pi r^2 b + 4\pi r^3/3 \geq 200 \quad (13.3)$$

The first term on the left hand side represents the volume of the cylindrical body of the tank and the second term represents the volume of its hemispherical ends. Further constraints are required to represent the physical reality that both dimensions of the tank must be non-negative, i.e.

$$b \geq 0 \quad (13.4)$$
$$r \geq 0 \quad (13.5)$$

Expressions (13.2) through (13.5) represent a simple optimisation model that can be solved to find the dimensions of the tank that will minimise cost while satisfying the relevant constraints. The solution of this particular problem is discussed in Sections 13.11 and 13.12 of this chapter.

The above example involved only two decision variables and one objective function. Real engineering optimisation problems may have hundreds or even thousands of decision variables, several objectives and hundreds of constraints. A large number of techniques have been developed for the solution of engineering optimisation problems. Several of these techniques will be discussed in this chapter including linear programming, separable programming, dynamic programming and genetic algorithms. The optimisation techniques considered in this chapter deal only with problems that have a single objective function. Interested readers are referred to Goicoechea *et al.* (1982) for a discussion of optimisation techniques for multi-objectives.

13.2 APPROACHES TO OPTIMISATION

In general the following three approaches to optimisation, which are explained in more detail below, can be identified (Meredith et al., 1985):

1. subjective optimisation;
2. combinatorial optimisation; and
3. analytical optimisation.

Subjective Optimisation

The simplest type of optimisation is based on engineering experience and judgement. This often takes the form of comparing a small number of alternatives and selecting the best. In these circumstances, it may be claimed that the design has been "optimised". In other cases, "rules of thumb" based on experience are used to identify an efficient design. For example, for the design of water supply pipelines a number of water authorities try to choose pipe sizes such that the pressure loss per kilometre of length falls within defined bounds. A typical value is 3 m head loss per kilometre length of pipe.

Subjective optimisation is used in cases where the objective(s) and/or the constraints cannot be formulated in mathematical terms. In other cases, there may be so many constraints that the principal task is to identify a design that satisfies all of them (i.e. a feasible design). In any case, the application of judgement by experienced engineers is an essential part of the design process and should not be undervalued.

Combinatorial Optimisation

A number of engineering planning and design problems involve the comparison of a finite number of discrete alternatives according to a set of objectives (or selection criteria). For problems of this type, combinatorial optimisation techniques are appropriate. The simplest combinatorial technique is termed complete enumeration and involves the evaluation of all of the alternatives according to the criteria and the selection of one alternative using appropriate multiobjective techniques (as discussed in Section 9.5). This may be used where the number of combinations is relatively small.

An example of where complete enumeration of alternatives has been applied successfully is Emerson Crossing in Adelaide, Australia. Emerson Crossing is an intersection of two roads in the City of Adelaide that has the unusual feature of having a dual railway line crossing diagonally through it (Figure 13.2).

This came about from a desire by the early transportation engineers to reduce the number of level crossings where cars and trains were in potential conflict. It made excellent sense when traffic volumes were quite low, but by the 1960s the traffic volumes on Cross Road and South Road had built up to such levels that the delays at this intersection at peak hour were extremely long. This was aggravated by the fact that the intersection was closed to road traffic in both directions for 15 minutes of each peak hour due to the passage of trains. A study of ways to alleviate the congestion was carried out by the South Australian Highways Department (the responsible authority at the time). A large number of options were considered to reduce the traffic delays. One class of these options involved the use of grade separation at the intersection to provide a spatial separation of road and railway traffic. Other options considered included diverting traffic away from the intersection, the use of roundabouts and moving the railway line so that it had separate intersections with South Road and Cross Road. For various reasons, the grade separation options were considered to be better than the other options. In fact, a number of grade separation options are available. These include the following:

1. South Road on an overpass or underpass;
2. Cross Road on an overpass or underpass; and
3. the railway on an overpass or underpass.

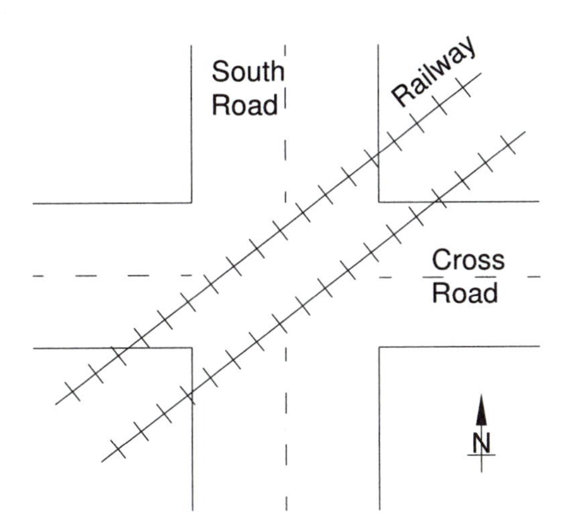

Figure 13.2: Layout of Emerson Crossing before the construction of the overpass (Not to scale)

These six basic solutions can be combined with each other (e.g. South Road on an underpass and the railway on an overpass) to give an additional 6 combinations. To these can be added the "do nothing" alternative to give a total of 13 alternatives. This is a small enough number for the application of complete enumeration to be carried out.

All 13 alternatives were considered in a feasibility study (Section 3.5) in order to eliminate the less competitive options. Using appropriate mathematical models the options were compared in terms of construction and operational costs, reduction in delays to vehicular traffic, likely reduction in accidents and their environmental and social impacts. Eventually, the option of putting South Road on an overpass was identified as the best option. This was implemented in 1982 and has proven to be quite successful.

In many real engineering problems, the number of options can be extremely large. This is particularly the case in systems where discrete sizes need to be chosen for a large number of elements. For example, the design of engineering networks such as road networks, water supply networks, electricity networks and communication networks require sizes to be chosen for a large number of elements such as roads, pipes, wires or communication channels. As each element is usually only available in discrete sizes e.g. number of lanes in a road, pipe sizes and wire sizes the problem becomes a large combinatorial optimisation problem. Techniques for optimisation of large combinatorial problems are discussed in Section 13.15.

Analytical Optimisation

Analytical optimisation can be applied when the objective function and constraints can be expressed in mathematical form. It involves the application of calculus or advanced mathematical optimisation techniques, some of which are discussed in detail in Sections 13.3–13.15 of this chapter.

13. 3 LINEAR PROGRAMMING

The simplest case of the general mathematical optimisation problem consists of a single **linear** objective function and one or more **linear** constraints. This is called the **Linear programming** (LP) problem. It may be written as follows:

$$\text{Max (or Min) } Z = \sum_{j=1}^{n} c_j x_j \tag{13.6}$$

subject to

$$\sum_{j=1}^{n} a_{ij} x_j \{\leq, =, \text{ or } \geq\} b_i \quad (i = 1, \ldots, m) \tag{13.7}$$

$$x_j \geq 0 \qquad (j = 1, \ldots, n) \tag{13.8}$$

where x_j are the decision variables, Z is the objective function, c_j are the cost coefficients, a_{ij} are the constraint coefficients and b_i are the right hand sides. c_j, a_{ij} and b_i are assumed to be constant and known. Note that the non-negativity conditions (13.8) are assumed in all LP problems.

Linear programming has been used to solve many large scale problems in engineering, economics, and commerce. Examples include optimising the mix of products from a refinery, optimising the assignment of crews to vehicles for a public transport system, and optimising the allocation of water among conflicting uses.

Example 13.2 The engineer for the Bullbar City Council must decide how to allocate the Council's workforce and machinery among various activities. For simplicity it will be assumed that only two activities are possible, constructing roads and constructing drains. The engineer has estimated that each kilometre of new road brings a net benefit of $30 000 per year to the community, whereas each kilometre of drain brings a net benefit of $20 000 per year. A kilometre of road requires 250 person-days of labour and 640 machine-hours to construct. On the other hand, a kilometre of drain requires 500 person-days of labour and 320 machine-hours to construct. The Council has a workforce of 50 people and 10 machines. Assuming 200 effective working days of 8 hours each per year, write an LP problem to determine the mix of roads and drains which the Council should undertake each year in order to maximise net benefits.

Solution Let x_1 be the number of km of roads to be constructed in a year, and x_2 be the number of km of drains to be constructed in a year. The objective function to be maximised is the total net benefits per year (in thousands of dollars), i.e.

$$\text{Max } Z = 30x_1 + 20x_2$$

Constraints are imposed by the available workforce and machinery. Considering the workforce first, the available resource is 50 people for 200 days, a total of 10,000 person-days. The required resource is $(250x_1 + 500x_2)$. Therefore:

$$250x_1 + 500x_2 \leq 10\,000$$

Similarly for machinery, the available resource is 10 machines x 200 days x 8 hours/day, i.e. 16 000 machine-hours. Therefore:

$$640x_1 + 320x_2 \leq 16\,000$$

In addition, the lengths of roads and drains constructed cannot be negative, and so

$$x_1 \geq 0 \qquad x_2 \geq 0$$

Therefore the problem formulation becomes:

$$\text{Max } Z = 30x_1 + 20x_2 \qquad (13.9)$$

subject to:

$$250x_1 + 500x_2 \leq 10\,000 \qquad (13.10)$$
$$640x_1 + 320x_2 \leq 16\,000 \qquad (13.11)$$
$$x_1 \geq 0 \qquad x_2 \geq 0 \qquad (13.12)$$

13.4 GRAPHICAL SOLUTION OF LP PROBLEMS

Before considering the solution of LP problems, some definitions are required. Any set of decision variables that satisfies all of the constraints is called a **feasible** solution. For a maximisation problem, any feasible solution that produces a value of the objective function not less than that produced by any other feasible solution is called an **optimum** solution.

A graphical solution technique will be illustrated for Example 13.2 given above. Firstly, consider constraint (13.10). If this were an equality it could be plotted as a straight line in (x_1, x_2) space as shown in Figure 13.3. All combinations of x_1 and x_2 which satisfy this constraint lie below it. Clearly any solution above this line is infeasible. This is indicated by shading on the infeasible side of the constraint as shown.

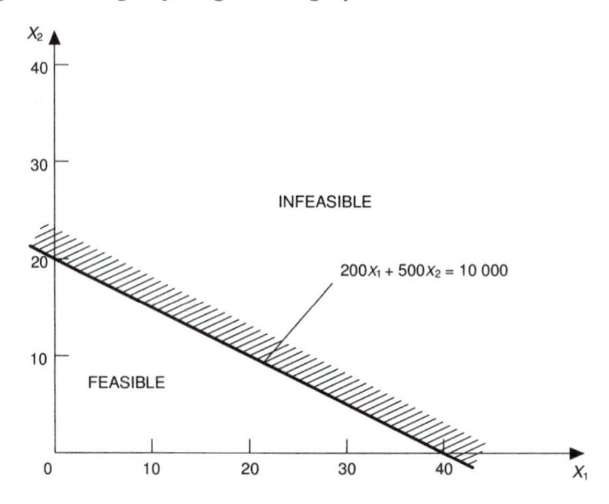

Figure 13.3 Workforce constraint for the Bullbar City Council problem.

In a similar fashion by plotting the other constraints (including the non-negativity constraints) the feasible solution space (or feasible region) can be identified. This is the region ABCD in Figure 13.4. The **feasible solution space** is the set of all combinations of the decision variables that satisfy all of the constraints of the problem.

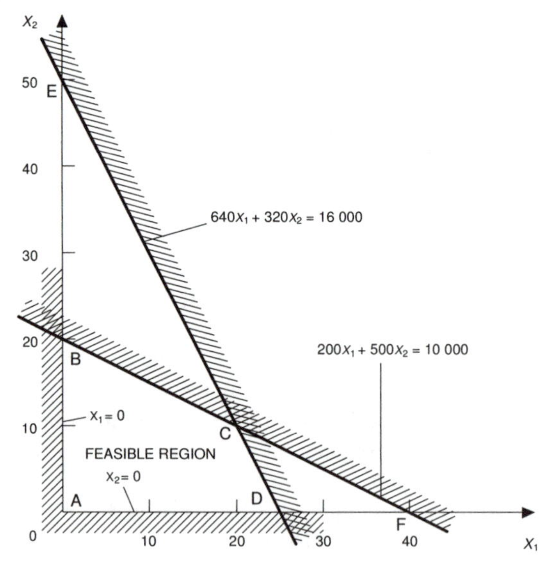

Figure 13.4 Feasible region for the Bullbar City Council problem.

The optimum solution may be determined by plotting contours of the objective function. As shown in Figure 13.5 these are parallel straight lines for any two-variable LP problem. It should be apparent that the optimum solution for this problem occurs at point C, i.e. at the intersection of the constraints (13.10) and (13.11). This corresponds to the highest value of the objective function Z that can be achieved within the feasible region.

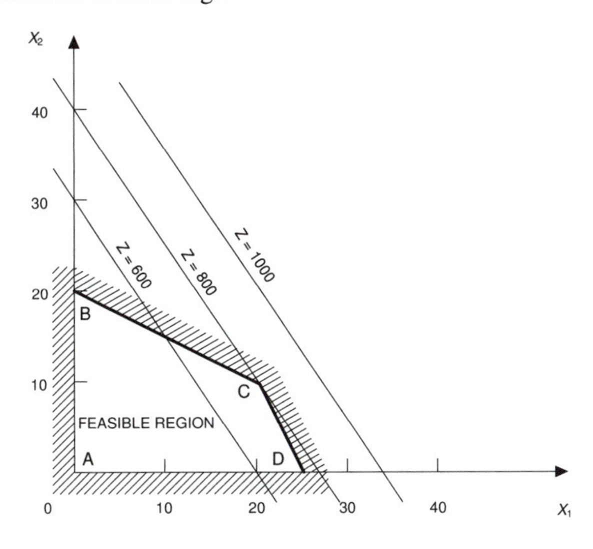

Figure 13.5 Optimum solution to the Bullbar City Council problem.

The optimum solution is:

$x_1 = 20$ km, $x_2 = 10$ km, $Z = \$800\,000$ per year

If the net benefits of drains relative to roads were changed by a sufficient amount, point B or D would become the optimum solution.

It should be noted that for this example, and indeed all LP problems, the feasible solution space is a **convex region.** A convex region is one that satisfies the property that a straight line joining any two points in the region only contains points in the region. Any region that does not satisfy this property is called a **non-convex region.**

Examples of convex and non-convex regions in two dimensions are illustrated in Figure 13.6. It should also be apparent from Figure 13.5 that the optimal solution(s) to any two-dimensional LP problem always occurs at one (or two) **corner points.** A corner point is a feasible solution defined by the intersection of two constraints. It can be seen that if the contours of Z were parallel to BC, both B and C and the points in between would be optimum solutions.

In m dimensions, an **extreme point** is a feasible solution that occurs at the intersection of m constraints. It can be shown that the optimum solution to a general LP problem always occurs at one (or more) extreme points.

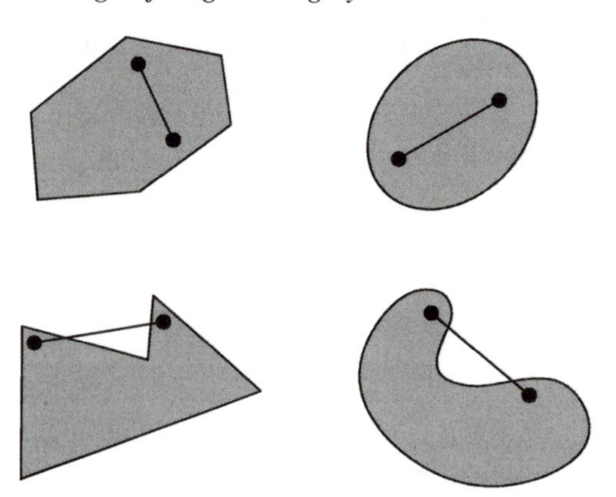

Figure 13.6 Examples of convex (top two) and non-convex (bottom two) regions.

13.5 CANONICAL AND STANDARD FORM OF AN LP PROBLEM

It is often useful in the solution of LP problems to reformulate them into either canonical form or standard form. The **canonical form** is

$$\text{Max } Z = \sum_{j=1}^{n} c_j x_j \tag{13.13}$$

subject to:

$$\sum_{j=1}^{n} a_{ij} x_j \leq b_i \qquad (i = 1, \ldots, m) \tag{13.14}$$

$$x_j \geq 0 \qquad (j = 1, \ldots, n) \tag{13.15}$$

It has the following features:

- the objective function is of the *maximisation* type;
- all constraints are of the 'less than or equal to' type;
- the right-hand side constants, b_i, may be positive or negative; and
- all decision variables are non-negative.

The canonical form of LP problems is used in Section 13.9 of this Chapter. Any LP problem may be rearranged into canonical form using one or more of the following operations:

- An objective function of the minimisation type can be made into the maximisation type by multiplying it by - 1, e.g. minimising $Z = 2x_1 - 3x_2$ is the same as maximising $Y = -2x_1 + 3x_2$.

- 'Greater than or equal to' constraints can be changed to 'less than or equal to' constraints by multiplying both sides by -1, e.g.
  $5x_1 - 4x_2 \geq 12$ is equivalent to $-5x_1 + 4x_2 \leq -12$

- Equality constraints can be converted into two inequalities, e.g.
  $4x_1 + 2x_2 = 12$
  is equivalent to
  $4x_1 + 2x_2 \leq 12$ and $4x_1 + 2x_2 \geq 12$
  i.e.
  $4x_1 + 2x_2 \leq 12$ and $-4x_1 - 2x_2 \leq -12$

- Variables which are unrestricted in sign may be replaced by two nonnegative variables; for example, consider the problem

  Max $Z = 3y_1 + 2y_2$

  subject to:
  $y_1 + y_2 \leq 6$
  $y_1 \leq 8$
  $y_1 \geq 0$
  y_2 unrestricted in sign

  This may be reformulated by using the substitution $y_2 = (y_2^+ - y_2^-)$ where $y_2^+ =$ the positive part of y_2 and $y_2^- =$ the negative part of y_2. Thus, the problem becomes:

  Max $Z = 3y_1 + 2y_2^+ - 2y_2^-$

  subject to:
  $y_1 + y_2^+ - y_2^- \leq 6$
  $y_1 \leq 8$
  $y_1, y_2^+, y_2^- \geq 0$

  It can be shown that if y_2^+ is positive in the final solution, y_2^- will be zero and vice versa.

  The **standard form** may be written as follows:

$$\text{Max (or Min) } Z = \sum_{j=1}^{n} c_j x_j \qquad (13.16)$$

subject to:

$$\sum_{j=1}^{n} a_{ij} x_j = b_i \qquad (i = 1, \ldots, m) \qquad (13.17)$$

$$x_j \geq 0 \qquad (j = 1, \ldots, n) \qquad (13.18)$$

The standard form has the following features:

- the objective function may be of the maximisation or minimisation type;
- all constraints are **equations,** with the exception of the non-negativity conditions which remain as inequalities of the 'greater than or equal to' type;
- the right-hand side of each constraint must be non-negative; and
- all variables are non-negative.

The standard form LP problems is used in Section 13.6 of this Chapter. Any LP problem can be transformed into standard form by using one or more of the following operations (in addition to those listed above):

- A negative right-hand side can be made positive by multiplying through by - 1 (and reversing the sign of the inequality).

- Inequality constraints may be converted to equalities by the inclusion of **slack or surplus** variables. For example, the 'less than or equal to' constraint,

  $$2x_1 + 4x_2 \leq 15 \qquad (13.19)$$

  may be written as

  $$2x_1 + 4x_2 + S_1 = 15$$

  where S_1 is a **slack variable** which must be non-negative in order to satisfy the original constraint.

- Similarly, the 'greater than or equal to' constraint

  $$3x_1 + 5x_2 \geq 18 \qquad (13.20)$$

  may be written as

  $$3x_1 + 5x_2 - S_2 = 18$$

  where S_2 is a **surplus variable which** must be non-negative.

The term slack variable is used because the right-hand side of Equation (13.19) often represents the availability of a particular resource (such as machine-hours in the case of the Bullbar City Council). If S_1 is positive then not all of the available resource is being utilized and there is 'slack' in the solution.

Constraints of the 'greater than or equal to' type often represent minimum requirements. For example, the right-hand side of Equation (13.20) may represent the minimum required output of steel from a particular process. If S_2 is positive the minimum output is being exceeded and there is a 'surplus' output.

13.6 BASIC FEASIBLE SOLUTIONS

By examining the standard form of an LP problem it can be seen that the constraints expressed by Equations (13.17) are a set of m linear equations in n unknowns (including slack and surplus variables).

Clearly we would expect n to be greater than m. If n is equal to m there is a unique solution to the constraint set and optimisation is not required. If n *is* less than m, the problem is over-constrained and there is, in general, no solution that satisfies all of the constraint equations.

A **basic** solution to an LP problem is one in which $(n - m)$ of the variables are equal to zero. A **basic feasible** solution to an LP problem is a basic solution in which all variables have non-negative values. Thus a basic feasible solution satisfies the constraints (13.17) and the non-negativity conditions (13.18).

Let us examine Example 13.2 further. In order to identify the basic solutions, we need to express the problem in standard form. This is achieved by introducing slack variables S_1 and S_2 into constraints (13.10) and (13.11) so that the problem becomes:

$$\text{Max } Z = 30x_1 + 20x_2 \qquad (13.21)$$

subject to:

$$250x_1 + 500x_2 + S_1 = 10\,000 \qquad (13.22)$$

$$640x_1 + 320x_2 + S_2 = 16\,000 \qquad (13.23)$$

$$x_1, x_2, S_1, S_2 \geq 0 \qquad (13.24)$$

Thus, excluding the non-negativity conditions, there are two constraint equations in four variables (including the slack variables S_1 and S_2). A basic solution may be found by setting any two of the variables equal to zero, and solving Equations (13.22) and (13.23) for the other two variables. In this manner the six basic solutions shown in Table 13.1 can be identified. Each of these basic solutions is identified by a letter in Figure 13.4. Clearly only the first four of these solutions are basic feasible solutions as the other two do not satisfy the non-negativity conditions (13.21).

Table 13.1 Basic solutions to the Bullbar City Council problem.

Solution	x_1	x_2	S_1	S_2
A	0	0	10 000	16 000
B	0	20	0	9600
C	20	10	0	0
D	25	0	3750	0
E	0	50	-15 000	0
F	40	0	0	-9600

It should be apparent from Figure 13.4 that each basic feasible solution is an **extreme point** of the feasible region. As it has already been shown that the optimum solution to an LP problem is an extreme point it follows that *the optimum solution to an LP problem is always a basic feasible solution.*

Now consider the basic solutions to this problem a little more carefully. A basic solution occurs when two of the variables x_1, x_2, S_1 and S_2 equal zero. If we set S_1 equal to zero in Equation (13.22) we obtain

$$250x_1 + 500x_2 = 10\,000 \qquad \text{(i.e. the line BCF in Figure 13.4)}$$

Similarly, by setting S_2 equal to zero in Equation (13.23) we obtain

$$640x_1 + 320x_2 = 16\,000 \qquad \text{(i.e. the line ECD in Figure 13.4)}$$

The line where $x_1 = 0$ is the x_2 axis, and vice versa. Thus each constraint line (including the x_1 and x_2 axes) in Figure 13.4 corresponds to one variable being equal to zero. The **basic solutions** (where two variables equal zero) occur at the intersections of two constraint lines. Furthermore each **basic feasible solution** corresponds to the intersection of two constraints within the feasible region. Therefore, the basic feasible solutions must be the extreme points of the feasible region.

13.7 THE SIMPLEX METHOD

The simplex method is a numerical procedure commonly used for solving LP problems. It is based on the property that the optimal solution to an LP problem is always a basic feasible solution. The simplex method uses the following steps:

Step 1 Identify an initial basic feasible solution (i.e. an extreme point)

Step 2 Consider movement from this extreme point to all adjacent extreme points and determine whether it is possible to improve the value of the objective function.

Step 3 If improvement is possible choose the next extreme point such that the maximum rate of change of objective function is achieved and move to this point. Return to Step 2.

Step 4 If no improvement is possible, an optimum solution has been obtained.

The interested reader is referred to Wagner (1975) or Taha (2003) for the details of the simplex method. Many computer packages are now available for solving very large LP problems using the simplex method.

13.8 SOME DIFFICULTIES IN SOLVING LP PROBLEMS

Occasionally difficulties may occur with LP problems caused by poor problem formulation or for other reasons. These problems and possible remedies are discussed in this Section.

Degenerate solutions

A degenerate solution occurs when one or more of the basic variables is equal to zero. One degenerate solution may be followed by another degenerate solution in which no improvement in the objective function occurs. This does not cause a problem with the simplex method and it is quite possible that several degenerate solutions may be followed by non-degenerate solutions in which an improvement in objective function is obtained. On some occasions, cycling between several degenerate solutions may occur. This rarely happens in practice and most commercial computer packages can identify it.

Unbounded solutions

It is sometimes possible to increase the value of the objective function without limit in an LP problem. This indicates a poorly formulated problem. For example:

$$\text{Max } Z = x_1 + 2x_2$$

subject to:
$2x_1 + x_2 \geq 12$
$x_1 + x_2 \geq 8$
$x_1, x_2 \geq 0$

The feasible region for this problem is shown in Figure 13.7. Clearly the objective function may be increased without limit in this feasible region. If the objective function were of the minimisation rather than maximisation type, an optimum solution could be found (assuming positive coefficients in the objective function).

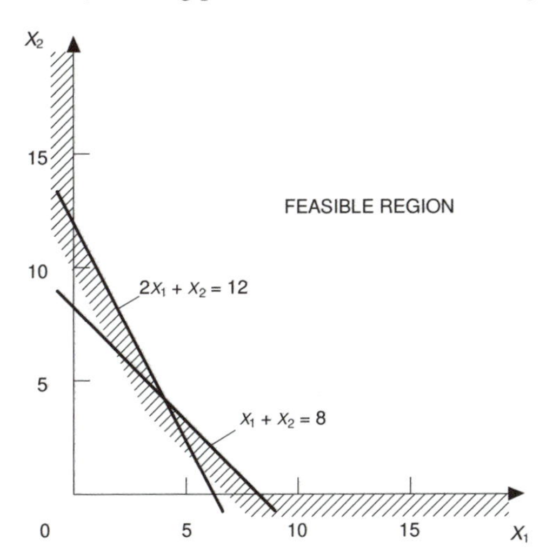

Figure 13.7 Feasible region for an unbounded problem.

No feasible solution

Another problem which may arise due to poor problem formulation is that there may be no solution which satisfies all of the constraints. For example:

$$\text{Max } Z = 2x_1 + x_2$$

subject to:

$$2x_1 + 3x_2 \geq 30$$
$$2x_1 + x_2 \leq 8$$
$$x_1 + 3x_2 \leq 9$$
$$x_1, x_2 \geq 0$$

The constraints to this problem are shown in Figure 13.8. Clearly there is no feasible region and hence no feasible solution to this problem.

Alternative optima

Occasionally an LP problem has more than one optimum solution. This occurs when the objective function is parallel to a boundary of the feasible region. For example:

$$\text{Max } Z = 2x_1 + x_2$$

subject to:

$$x_1 + x_2 \leq 8$$
$$2x_1 + x_2 \leq 12$$
$$x_1, x_2 \geq 0$$

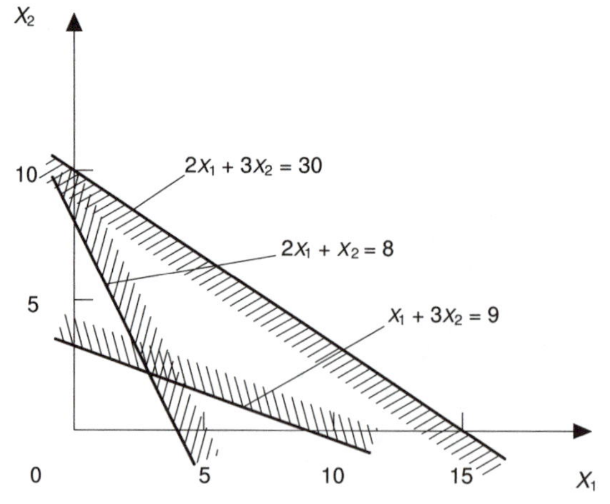

Figure 13.8 No feasible solution.

The feasible solution space for this problem is shown in Figure 13.9.

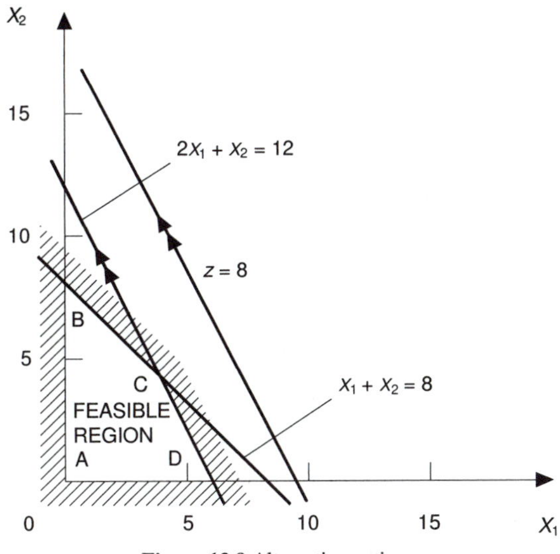

Figure 13.9 Alternative optima.

Also shown is the contour for $Z = 18$. Because contours of the objective function are parallel to the second constraint for this problem, it should be clear that the value of the objective function will be the same at points C and D. So both C and D and all points on the boundary between them are alternative optimum solutions to this problem.

The simplex method will identify either C or D as the optimum solution and then stop. The existence of alternative optima may be indicated as part of the solution technique.

Redundant constraints

In some LP problems not all of the constraints are boundaries of the feasible region. Constraints that are not boundaries are called **redundant constraints** because they have no effect on the solution. For example:

Max $Z = 2x_1 + x_2$

subject to:

$x_1 + x_2 \leq 8$

$x_1 + 2x_2 \leq 12$

$6x_1 + 5x_2 \leq 60$

$x_1, x_2 \geq 0$

The constraints and feasible region for this problem are shown in Figure 13.10. Clearly the constraint $6x_1 + 5x_2 \leq 60$ is a redundant constraint as it has no effect on the feasible region.

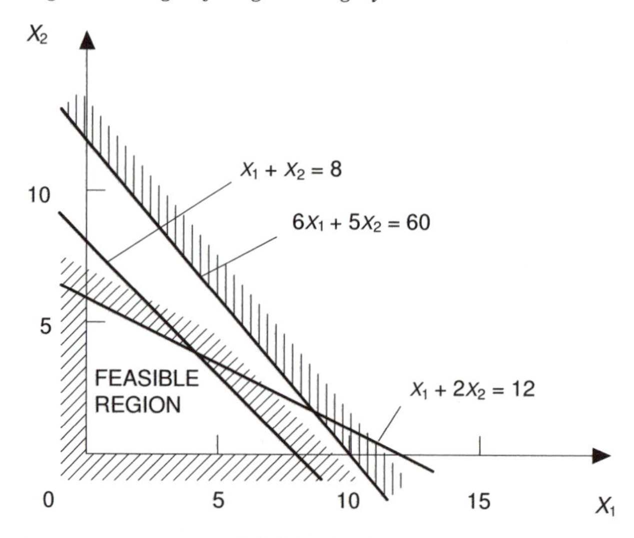

Figure 13.10 Redundant constraint.

The presence of redundant constraints does not in any way inhibit the simplex method from finding an optimum solution. However, because their presence affects the size of the problem and hence the computer time for its solution it is preferable to remove redundant constraints from the formulation if they can be readily identified.

13.9 EXAMPLES OF LP PROBLEMS

LP problems occur in many engineering design and planning problems. Some examples are outlined below.

Reservoir design

A common problem in water supply engineering is to determine the required size of a reservoir to provide a specified set of releases to a town or irrigation area. Consider the proposed reservoir shown schematically in Figure 13.11. The design variable is the capacity of the proposed reservoir, K. Flow data have been collected at the proposed site for the last T years. Q_t ($t = 1, \ldots, T$) is the known flow in year t. The known annual water demand in the city is R. E_t ($t = 1, \ldots, T$) represents the annual evaporation and other losses from the reservoir in year t and is assumed to be known. S_t ($t = 0, \ldots, T$) represents the volume of storage in the reservoir at the end of year t. O_t ($t = 1, \ldots, T$) is the spill from the reservoir in year t. S_t ($t = 1, \ldots, T$) and O_t ($t = 1, \ldots, T$) are derived variables whose values will be determined in the course of the optimisation.

The reservoir size will be determined on the assumption that the historical series of flows will be repeated. The annual demand of the city must be met in every year without the reservoir becoming empty. The objective is therefore to minimise the required capacity, K, subject to the constraints imposed by:

- continuity; and
- the reservoir capacity not being exceeded.

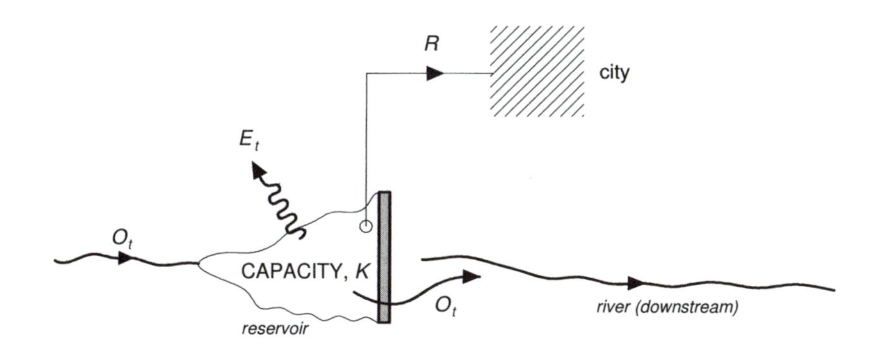

Figure 13.11 Reservoir design problem.

The continuity constraint for year t is as follows:

$$S_t - S_{t\text{-}1} = Q_t - E_t - R - O_t$$

i.e. the change in storage in the reservoir in year t equals the total inflow minus the total outflow. The constraint on capacity at the end of year t is simply

$$S_t \leq K$$

It is usual to assume that the reservoir has at least the same storage level at the end of the T years as at the beginning. This allows future demand to be met if the historical sequence repeats itself indefinitely, i.e.

$$S_T \geq S_0$$

Therefore, the LP problem is:

$$\text{Min } Z = K \tag{13.25}$$

subject to:

$$S_t - S_{t\text{-}1} + O_t = Q_t - E_t - R \quad (t = 1, \ldots, T) \tag{13.26}$$

$$S_t - K \leq 0 \qquad (t = 0, \ldots, T) \tag{13.27}$$

$$S_T - S_o \geq 0 \tag{13.28}$$

$$S_t, O_t, K \geq 0 \tag{13.29}$$

where decision variables have been placed on the left-hand side of constraints and known values on the right-hand side.

Road network analysis (adapted from Stark and Nicholls 1972, Example 4-5)

A common problem in transportation planning is the allocation of traffic volumes to a road network. A simple road network is shown in Figure 13.12.

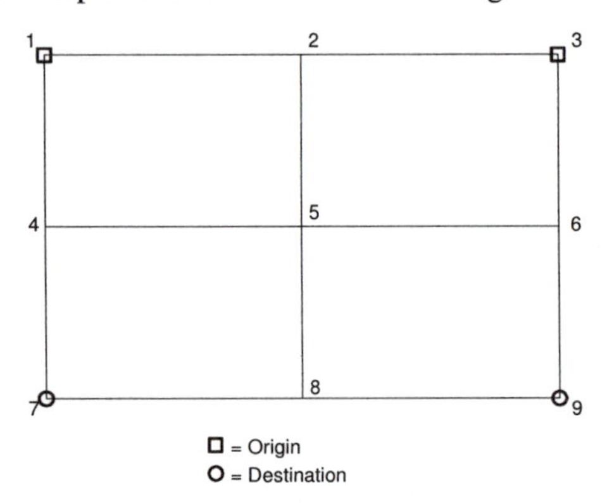

Figure 13.12 Portion of a road network.

The usual demand information for the network will be expressed in terms of an origin-destination matrix, i.e. the volume of traffic wishing to travel from each origin to each designation in a defined time period. Nodes 1 and 3 in Figure 13.12 are origins and nodes 7 and 9 destinations. A typical origin-destination matrix is given in Table 13.2. Other information available is the capacity (vehicles/h) and the travel time in each direction (minutes) for each road. Typical values are given in Table 13.3. In this example the capacity and travel time is the same in both directions for each road, although this is not necessarily always the case. The objective is to minimise the total time of travel for all traffic.

Table 13.2 Origin-Destination matrix (vehicle/h)

Origin node	Destination	Node
	7	9
1	4000	3000
3	3000	2000

Let $x_{ij}^{(k)}$ be the volume of traffic (vehicles/h) which originates from node k and travels on road ij, and let K be the set of all origin nodes. For all traffic originating at node k the total travel time is given by

$$Z^{(k)} = \sum_{i=1}^{n}\sum_{j=1}^{n}\delta_{ij}t_{ij}x_{ij}^{(k)} \qquad (13.30)$$

where

$$\delta_{ij} = \begin{cases} 1 \text{ if there is a road from node } i \text{ to node } j \\ 0 \qquad \text{otherwise} \end{cases}$$

and t_{ij} is the travel time (minutes) on road ij and n is the total number of nodes.

Table 13.3 Capacities and travel times for road network problem (these are the same in both directions).

Road	Capacity (vehicles/h)	Travel time (minutes)
1-2	4000	10
2-3	3000	12
3-4	4000	15
4-5	6000	13
5-6	6000	16
6-7	2500	8
7-8	4000	6
8-9	3500	12
9-10	5000	10
10-11	6000	5
11-12	6000	8
12-13	2500	7

The constraints represent continuity at each node, i.e.

$$\sum_{i=1}^{n} \delta_{ip} x_{ip}^{(k)} - \sum_{j=1}^{n} \delta_{pj} x_{pj}^{(k)} = D_p^{(k)} \; (p = 1, \ldots, n) \tag{13.31}$$

where

$$D_p^{(k)} = \begin{cases} \text{the volume of traffic which has origin } k \text{ and} \\ \quad \text{destination } p \text{ (for } p \neq k) \\ \\ \text{minus the volume of traffic which originates at node } k \\ \quad \text{(for } p = k) \end{cases}$$

When considering all origin nodes the objective function becomes

$$\text{Min } Z = \sum_{k \in K} \sum_{i=1}^{n} \sum_{j=1}^{n} \delta_{ij} t_{ij} x_{ij}^{(k)} \tag{13.32}$$

where the first summation is over all origin nodes. The continuity constraints (13.31) become

$$\sum_{i=1}^{n} \delta_{ip} x_{ip}^{(k)} - \sum_{j=1}^{n} \delta_{pj} x_{pj}^{(k)} = D_p^{(k)} \quad \begin{matrix} (p = 1,\ldots,n) \\ (k \in K) \end{matrix} \tag{13.33}$$

In addition, capacity constraints for each road must be considered. These are

$$\sum_{k \in K} x_{ij}^{(k)} \le C_{ij} \quad \text{(for all } i, j \text{ which are connected by a road)} \tag{13.34}$$

where C_{ij} the capacity of road ij. In addition,

$$x_{ij}^{(k)} \ge 0 \quad (\text{all } i, j, k) \tag{13.35}$$

The LP problem to be solved is defined by the objective function (13.32) and the constraints (13.33), (13.34), and (13.35).

In a more realistic case the travel time on each road depends on the volume of traffic on it. In addition, each driver tries to minimise his or her own travel time instead of the system being at a global minimum. In these cases, the objective function becomes non-linear but the constraints have the same form.

13.10 DUALITY

Duality is an important concept in mathematical analysis that often gives valuable insights into the physical problem being solved. In LP, every problem has another problem associated with it called the **dual.** By solving the original or **primal** problem using the simplex method, the solution to the dual is found automatically. The dual variables have an important interpretation that adds considerable information to an LP solution.

Writing the dual problem

In Section 13.5 the canonical form of an LP problem was given as

$$\text{Max } Z = \sum_{j=1}^{n} c_j x_j \tag{13.13}$$

subject to:

$$\sum_{j=1}^{n} a_{ij} x_j \le b_i \quad (i = 1,\ldots,m) \tag{13.14}$$

$$x_j \ge 0 \qquad (j = 1, \ldots, n) \tag{13.15}$$

For an LP problem in canonical form the dual problem may be written using the following rules:

1. There is a dual variable corresponding to each primal constraint and a primal variable corresponding to each dual constraint.
2. If the primal problem is of the maximisation type, the dual problem is of the minimisation type and vice versa.
3. All of the constraints in the maximisation problem are of the 'less than or equal to' type. All of the constraints in the minimisation problem are of the 'greater than or equal to' type.
4. The coefficients of the objective function in the primal problem are the right-hand sides in the dual problem and vice versa.
5. The variables in both problems are non-negative.

To write the dual to the problem given by (13.13)-(13.15), we need to define m dual variables $y_i (i = 1, \ldots, m)$, one corresponding to each primal constraint. The dual problem is:

$$\text{Min } W = \sum_{i=1}^{m} b_i y_i \qquad (13.36)$$

subject to:

$$\sum_{i=1}^{m} a_{ij} y_i \geq c_j \quad (j = 1, \ldots, n) \qquad (13.37)$$

$$y_i \geq 0 \qquad (i = 1, \ldots, m) \qquad (13.38)$$

Note that if these rules are applied to the problem defined by Conditions (13.36) - (13.38), the 'primal' problem is regained (i.e. Conditions (13.13) - (13.15)). Clearly, symmetry exists between the two problems.

It can be shown that, for any feasible solutions of the primal and dual, $Z \leq W$. That is, the objective function value for a feasible solution of the maximisation problem is a lower bound on the objective function value of the minimisation problem. Furthermore, if Z^* is the maximum value of Z and W^* is the minimum value of W then

$$Z^* = W^* \qquad (13.39)$$

i.e. at optimality, both problems have the same objective function value. For example, the primal problem of Example 13.2 is given by (13.9) - (13.12):

$$\text{Max } Z = 30x_1 + 20x_2 \qquad (13.9)$$

subject to:

$$250x_1 + 500x_2 \leq 10\,000 \qquad (13.10)$$

$$640x_1 + 320x_2 \leq 16\,000 \qquad (13.11)$$

$$x_1, x_2 \geq 0 \qquad (13.12)$$

As this is already in canonical form, the dual can readily be written as

$$\text{Min } W = 10\,000y_1 + 16\,000y_2 \qquad (13.40)$$

subject to:

$$250y_1 + 640y_2 \leq 30 \qquad (13.41)$$
$$500y_1 + 320y_2 \leq 20 \qquad (13.42)$$
$$y_1, y_2 \geq 0 \qquad (13.43)$$

Writing each problem in standard form we obtain for the primal problem:

$$\text{Max } Z = 30x_1 + 20x_2 \qquad (13.44)$$

subject to:

$$250x_1 + 500x_2 + S_1 = 10\,000 \qquad (13.45)$$
$$640x_1 + 320x_2 + S_2 = 16\,000 \qquad (13.46)$$
$$x_1, x_2, S_1, S_2 \geq 0 \qquad (13.47)$$

and, for the dual problem:

$$\text{Min } W = 10\,000y_1 + 16\,000y_2 \qquad (13.48)$$

subject to:

$$250y_1 + 640y_2 - S_3 = 30 \qquad (13.49)$$
$$500y_1 + 320y_2 - S_4 = 20 \qquad (13.50)$$
$$y_1, y_2, S_3, S_4 \geq 0 \qquad (13.51)$$

The solution to the primal problem was found previously (Sections 13.4 and 13.7) to be

$$Z = 800, x_1 = 20, x_2 = 10, S_1 = S_2 = 0$$

The dual problem defined by (13.40) to (13.43) can be solved graphically or by using an LP computer package to give:

$$W = 800, y_1 = 1/75, y_2 = 1/24, S_3 = S_4 = 0$$

thus verifying Equation (13.39). Furthermore it can be shown that the simplex solution to a primal problem contains the solution to the dual (and vice versa).

Most LP computer packages (including Excel Solver) will give the optimal values of the dual variables when they solve the primal problem.

Interpretation of the dual variables

Refer back to the primal problem defined by (13.13) - (13.15). In many problems, the right-hand side element of the *i*th primal constraint, b_i, may be interpreted as the availability of the *i*th resource. Combining Equations (13.36) and (13.39), we have at optimality:

$$Z^* = W^* = \sum_{i=1}^{m} b_i y_i^* \tag{13.52}$$

where y_i^* is the value of y_i at optimality ($i = 1, \ldots, m$).

Therefore, the optimum value of the *i*th dual variable, y_i^*, may be interpreted as the additional output Z which could be obtained by a unit increase in the *i*th resource. Thus, if b_i were increased by one unit, we would expect Z^* to increase by y_i^* units.

In the above problem, the dual variable y_1 represents the additional net benefits (in thousands of dollars) if the right-hand side of constraint (13.10) were increased by one unit. In this case this represents an additional person-day of labour. Similarly y_2 represents the additional net benefits (in thousands of dollars) if the right-hand side of constraint (13.10) were increased by one unit, i.e. by having one extra machine-hour available. Therefore

y_1 = 1/75 thousand dollars/person-day

y_2 = 1/24 thousand dollars/machine-hour

y_1 = \$13.33/person-day

y_2 = \$41.67/machine-hour

Expressed in this form the dual variables are **shadow prices** for the corresponding resources. The shadow price of a resource is the economic value of having an additional unit of the resource available.

If additional labour can be hired at less than the shadow price of \$13.33 per day the Council would make a net gain by doing so. Similarly if additional machine time can be obtained at less than its shadow price of \$41.67 per hour a net gain can be made. Of course, this gain by employing additional labour or machines will only apply for increases that are not so large as to cause a change in the optimum solution.

13.11 NON-LINEAR OPTIMISATION

In Sections (13.3)–(13.10) a special case of optimisation called linear programming (LP) is examined. This applies to problems with a single linear objective function and a set of linear constraints. Many general concepts of optimisation can be demonstrated using LP. The assumption of linearity is often not appropriate to real engineering systems. In general, the objective function and/or constraints may be non-linear, if indeed they can be expressed in mathematical form.

There are many techniques available for the solution of non-linear optimisation problems. In the following sections, the techniques of separable programming and dynamic programming are considered. Many practical engineering problems do not lend themselves to the use of formal optimisation techniques. In some cases it is possible to develop a mathematical model of the system but not to optimise it. Computer simulation (see Section 3.6) can then be used in a trial-and-error search for the optimum solution.

In many other design situations, formal optimisation is not possible but an approximation to the optimum design is found by evaluating a range of alternative solutions.

The general single-objective, non-linear optimisation problem may be written as:

$$\text{Max } Z = f(x_1, x_2, \ldots, x_n) \qquad (13.53)$$

subject to:

$$h_i(x_1,\ldots,x_n) \begin{Bmatrix} \leq \\ = \\ \text{or} \geq \end{Bmatrix} b_i (i = 1,\ldots,m) \qquad (13.54)$$

This problem differs from the general LP problem (Conditions 13.6 - 13.8) in three major respects:

1. The feasible solution space for a non-linear problem may be either convex or non-convex. (This follows from the fact that the boundaries of the feasible region may, in general, be generated by non-linear functions. Figure 13.6 illustrates some non-convex feasible regions.)
2. An optimum solution to a non-linear problem may occur at any point in the feasible region (not necessarily at an extreme point).
3. It is possible for more than one 'optimum' to exist. The solution corresponding to the absolute maximum (or minimum) of the objective function is called the **global optimum;** all other optima are called **local optima.**

Two examples are used to illustrate the second point.

Example 13.3 Consider the following problem:

$$\text{Min } Z = (x_1 - 2)^2 + (x_2 - 3)^2$$

subject to:

$2x_1 + x_2 \leq 14$

$x_1 + 3x_2 \leq 15$

$x_1, x_2 \geq 0$

The feasible solution space and contours of the objective function for this problem are shown in Figure 13.13. The minimum value of Z equals zero and occurs at $x_1 = 2$, $x_2 = 3$. This is in the interior of the feasible region.

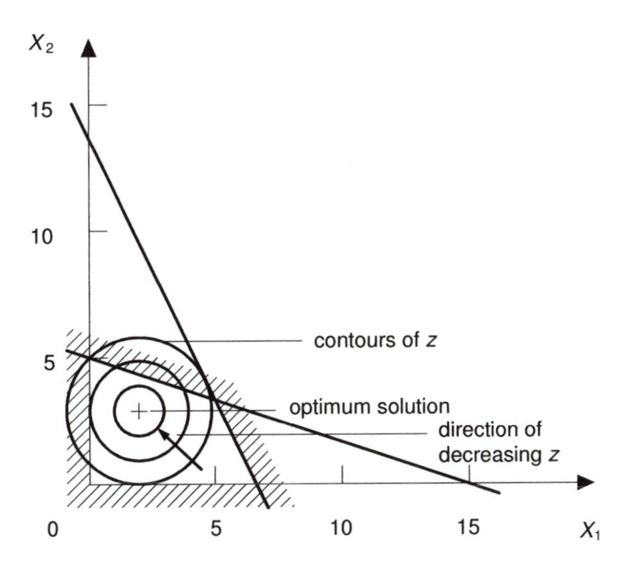

Figure 13.13 Optimum solution at interior of the feasible region.

Example 13.4 Consider the following problem:

$$\text{Max } Z = x_1 x_2$$

subject to

$$x_2 + 0.1x_1^2 \leq 10$$
$$x_1 \leq 7$$
$$x_1, x_2 \geq 0$$

The feasible region and contours of Z are shown in Figure 13.14. In this case the optimum solution ($Z = 38.49$ at $x_1 = 5.77$, $x_2 = 6.67$) lies on the boundary of the feasible region but not at an extreme point.

In relation to the third point above, it is useful to think of the objective function surface as a range of mountains. In a maximisation problem every peak is a local optimum, and the highest peak is the global optimum. Of course, in many problems, the objective function has only one peak and there are no local optima. Depending on the constraints this may or may not correspond to the optimum solution.

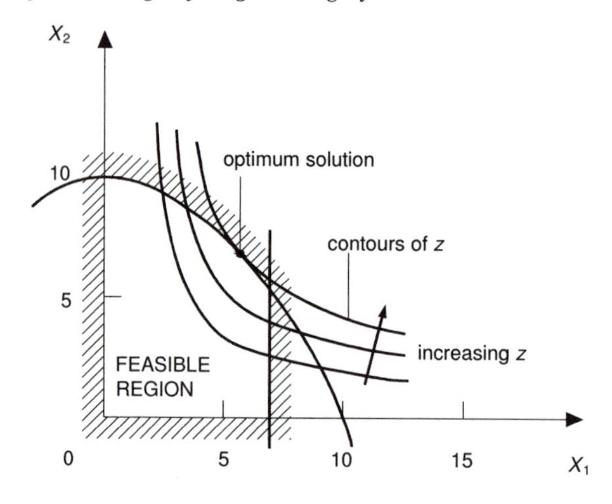

Figure 13.14 Optimum solution at boundary of the feasible region.

In linear programming, the feasible solution space is always convex, the optimum solution always occurs at an extreme point and local optima do not occur. The three differences between linear and non-linear optimisation listed above make non-linear optimisation considerably more difficult than linear optimisation. In fact, there is no single solution technique that works for all non-linear optimisation problems. Many specific techniques have been developed, each of which is applicable to a certain category of non-linear problems. The techniques may be grouped into the following categories:

- the graphical technique;
- classical optimisation techniques;
- numerical techniques.

The graphical technique may be used for problems with only one or two decision variables. It involves plotting the constraints and the objective function in terms of the decision variables in order to identify the optimum solution. The technique is illustrated by Examples 13.3 and 13.4 considered earlier. It is generally only suitable for simple functions.

13.12 UNCONSTRAINED PROBLEMS USING CALCULUS

The general unconstrained optimisation problem is:

$$\text{Max } Z = f(x_1, x_2, \ldots, x_n) \qquad (13.55)$$

where $f(x_1, x_2, \ldots, x_n)$ is any non-linear function of the n decision variables $x_1, \ldots, x_n$. This is perfectly general, as a minimisation problem can be converted into a maximisation problem by multiplying the objective function by -1. Firstly consider the single-variable case:

$$\text{Max } Z = f(x) \tag{13.56}$$

Classical differential calculus can be used to find the maxima of $f(x)$ provided that $f(x)$ and its first- and second-order derivatives exist and are continuous for all values of x.

A **necessary condition** for a particular solution x^o to be a maximum of $f(x)$ is:

$$\left.\frac{df(x)}{dx}\right|_{x=x^0} = 0 \tag{13.57}$$

The solutions to Equation (13.57) are called **stationary points** and may include maxima, minima, and points of inflection. A **point of inflection** is a point where $f(x)$ has zero slope but is neither a maximum nor a minimum. For example, the function $f(x) = x^3$ has a point of inflection at $x = 0$ as shown in Figure 13.15.

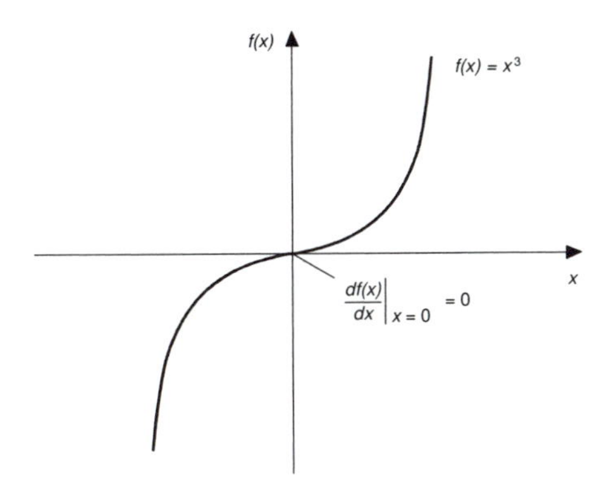

Figure 13.15 Point of inflection.

A **sufficient** condition for a stationary point x^o to be a maximum of $f(x)$ is:

$$\left.\frac{d^2 f(x)}{dx^2}\right|_{x=x^0} < 0 \tag{13.58}$$

A **sufficient** condition for a minimum at x^o is:

$$\left.\frac{d^2 f(x)}{dx^2}\right|_{x=x^0} > 0 \tag{13.59}$$

Example 13.1 (repeated): A tank of volume 200 m^3 is to be made from stainless steel. It will have a cylindrical body and hemispherical ends as shown in Figure 13.15. If the material for the ends costs twice as much per square metre as that for the body, find the dimensions of the tank that minimise its cost.
Solution : Let the length of the body of the tank be b m and its radius be r m. The formulation given in Section 13.1 is repeated below.

$$\text{Min } Z = 2\pi rb + 8\pi r^2 \tag{13.2}$$

subject to the volume constraint:

$$\pi r^2 b + 4\pi r^3/3 = 200 \tag{13.3}$$

This constraint can be written as an equality because the minimum cost tank will not have excess volume in this case. Using Equation (13.3) we can substitute for b in the objective function

$$b = \frac{[200 - 4\pi r^3/3]}{\pi r^2} \tag{13.60}$$

Thus (13.2) becomes:

$$\text{Min } Z = \frac{400}{r} + \frac{16}{3}\pi r^2 \tag{13.61}$$

which is an unconstrained single-variable optimisation. The necessary condition (13.61), is found by differentiating Z with respect to r,

$$\frac{\mathrm{d}Z}{\mathrm{d}r} = \frac{-400}{r^2} + \frac{32\pi r}{3} = 0$$

and then solving for r gives

$$r = 2.285 \text{ m}$$

and substituting into Equation (13.3),

$$b = 9.146 \text{ m}$$

Now checking the sufficiency condition (13.58):

$$\frac{\mathrm{d}^2 Z}{\mathrm{d}r^2} = \frac{+800}{r^3} + \frac{32\pi}{3}$$

which is positive for all positive values of r, therefore the above solution is a minimum. In practice the chosen dimensions would be rounded off, e.g. $b = 9$ m, $r = 2.3$ m.

The necessary and sufficient conditions for the unconstrained **multivariate** case are discussed in Appendix 13A. Classical optimisation techniques can be extended to solve **constrained** multivariate problems (particularly those with equality constraints). This is beyond the scope of this book and the interested reader is referred to Taha (2003) or Ossenbruggen (1984) for a treatment of this topic.

Numerical optimisation techniques are often preferred to calculus-based techniques because of the difficulty in applying the latter to problems of reasonable size. A large number of numerical techniques are available for solving non-linear optimisation problems.

The techniques considered in this chapter are separable programming and dynamic programming. Taha (2003), Beightler et al. (1979) and Wagner (1975) contain a more detailed treatment of various numerical techniques suitable for solving non-linear optimisation problems.

13.13 SEPARABLE PROGRAMMING

Separable programming is an extension of linear programming. It works by converting a non-linear optimisation problem into an LP problem that approximates the original non-linear one. Separable programming can only be applied to separable functions. A function $f(x_1, x_2, \ldots, x_n)$ is said to be separable if it can be expressed as the sum of n single-variable functions $f_l(x_l), \ldots, f_j(x_j), \ldots, f_n(x_n)$.

Although separable programming can be used to approximate any continuous non-linear function it is most effective when the following special conditions apply:

- the individual functions, $f_j(x_j)$, of the objective function are concave for a maximisation problem or convex for minimisation problems; and
- the constraints define a convex feasible region.

A convex function is one that has the property that a straight line joining any two points on the function always lies on or above the function. For a concave function a straight line joining any two points on the function always lies on or below the function. Examples of convex and concave functions are shown in Figure 13.16. A convex region has been defined in Section 13.4.

If the properties defined above do not apply, separable programming involves the use of integer variables. This is beyond the scope of this book. The interested reader is referred to Taha (2003).

Separable programming always yields a global optimum to the approximate linear problem. To illustrate the separable programming technique, consider the minimisation of the convex objective function shown in Figure 13.17. The non-linear function is approximated by a series of straight line segments between selected values of x, namely λ_1, λ_2, ..., λ_K which must be selected to span the full range of likely values of x.

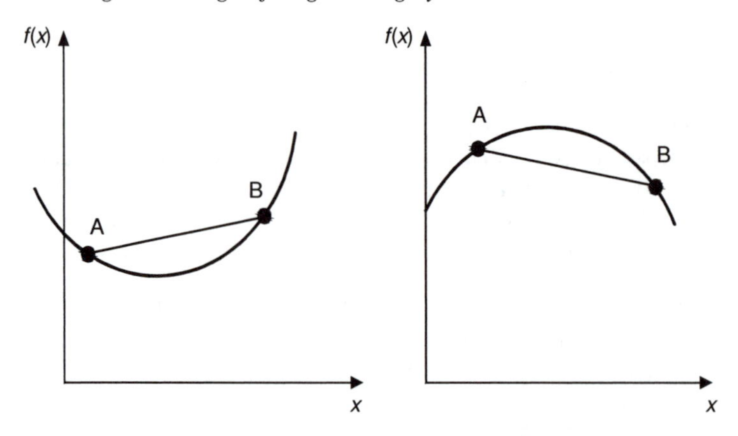

Figure 13.16 Convex and concave functions

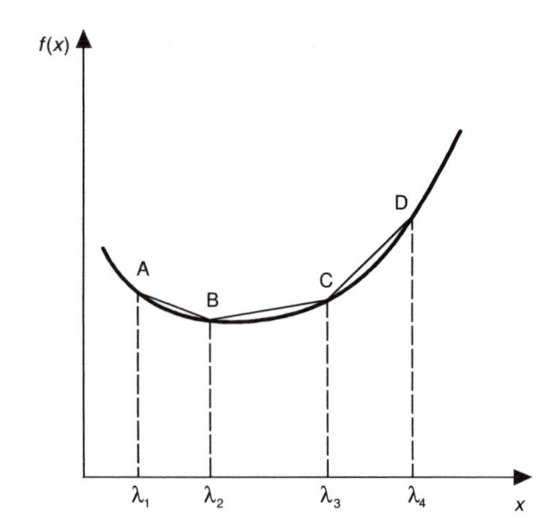

Figure 13.17 A convex objective function

We define a new set of variables $w_1, \ldots, w_K$ such that

$$x = \sum_{k=1}^{K} w_k \lambda_k \tag{13.62}$$

$$\sum_{k=1}^{K} w_k = 1 \tag{13.63}$$

and

$$w_k \geq 0 \qquad (k = 1, \ldots, K) \qquad (13.64)$$

In the final solution, x is a weighted average of the values $(\lambda_1, \lambda_2, \ldots, \lambda_K)$ where w_k $(k = 1, \ldots, K)$ are the relative weights. Provided at most two adjacent values of w_k are non-zero, $f(x)$ can be approximated by

$$f(x) = \sum_{k=1}^{K} w_k f(\lambda_k) \qquad (13.65)$$

For example, if $w_2 = w_3 = 0.5$ and all other values of w_k are zero, Equations (13.62) and (13.65) give:

$$x = 0.5\lambda_2 + 0.5\lambda_3$$
$$f(x) = 0.5f(\lambda_2) + 0.5f(\lambda_3)$$

which is a point on the straight line midway between points A and B in Figure 13.18.

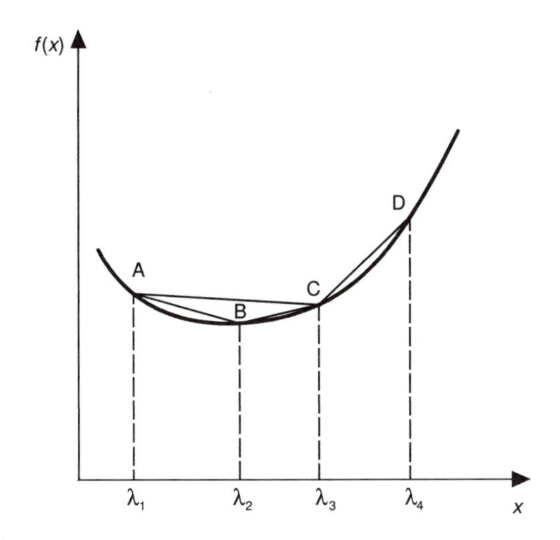

Figure 13.18 A convex objective function

It can be shown that, at most, two adjacent weights w_k will be non-zero when using LP and the special conditions above apply. This can be demonstrated intuitively for the case of minimising a convex function as shown in Figure 13.18. Suppose as a particular solution we obtain $w_1 = 0.6$ and $w_3 = 0.4$ with all other $w_k = 0$. Then

$$x = 0.6\lambda_1 + 0.4\lambda_3$$
$$f(x) = 0.6f(\lambda_1) + 0.4f(\lambda_3)$$

This is a point on the straight line segment between points A and C. Clearly the line segments AB and BC lie below this and an LP solution which is minimising $f(x)$ will achieve a solution on one of the lower line segments – i.e. with only two adjacent weights non-zero.

Example 13.5: A cylindrical tank is to be made from a 6 m square sheet of metal as shown in Figure 13.19. The tank will be h metres high and d metres in diameter. The height of the tank must not be more than twice its diameter. Use separable programming to find the dimensions of the tank that maximise its volume.

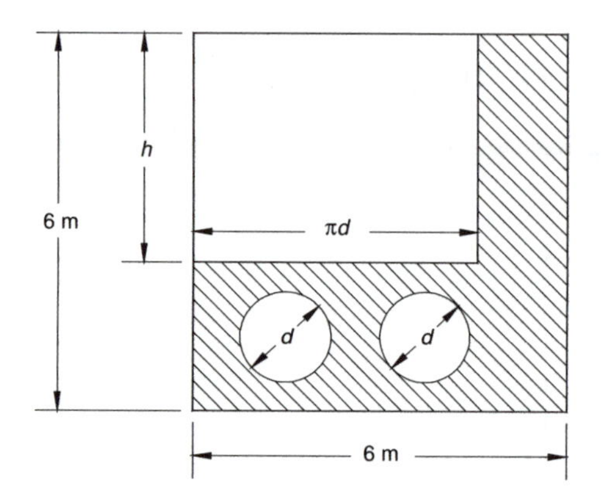

Figure 13.19 Plan for side and ends of a cylindrical tank

Solution: The optimisation problem is:

$$\text{Max V} = \pi \text{d}^2 h/4 \tag{13.66}$$

subject to

$$\pi d \leq 6$$

i.e. $d \leq 1.910$ (13.67)

$$h + d \leq 6 \tag{13.68}$$

$$h/d \leq 2$$

i.e. $h - 2d \leq 0$ (13.69)

$$h, d \geq 0 \tag{13.70}$$

The objective function can be made separable by taking its logarithm (to the base 10), i.e.

$$\log V = \log(\pi/4) + 2\log d + \log h \tag{13.71}$$

Clearly, maximising log *V* is equivalent to maximising *V*. As log *d* and log *h* are concave functions in a maximisation objective, we can use separable programming without integer variables.

It is now necessary to approximate log *d* and log *h* over a suitable range. First consider log *h*. From Equation (13.68) it is obvious that *h* must be less than 6 m. An approximation to log *h* for *h* between 0.5 m and 6 m is shown in Figure 13.20. Values for λ_k and $\log_{10}\lambda_k$ are given in Table 13.4. Applying Equations (13.62)–(13.65) to approximate the function log (*h*) gives

$$h = 0.5w_1 + w_2 + 2w_3 + 4w_4 + 6w_5 \tag{13.72}$$

$$w_1 + w_2 + w_3 + w_4 + w_5 = 1 \tag{13.73}$$

$$w_i \geq 0 \qquad (i = 1,\ldots,5) \tag{13.74}$$

$$\log h = -0.301w_1 + 0.w_2 + 0.301w_3 + 0.602w_4 + 0.778w_5 \tag{13.75}$$

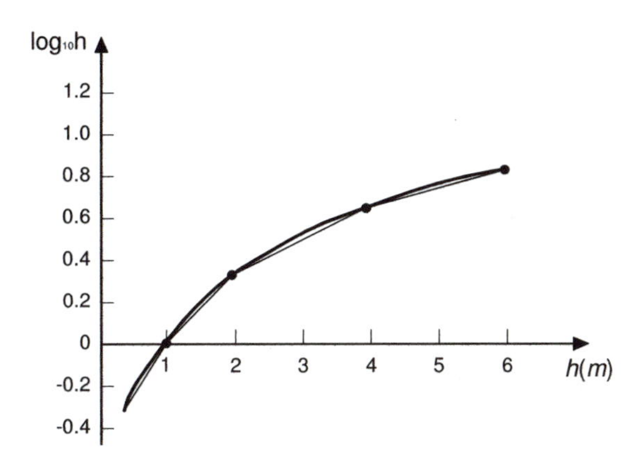

Figure 13.20 Approximation to $\log_{10} h$

As *d* must be less than or equal to 1.91 m, log *d* will be approximated over the range 0.5 m–2 m. Using Table 13.4, this gives

$$d = 0.5u_1 + u_2 + 2u_3 \tag{13.76}$$

$$u_1 + u_2 + u_3 = 1 \tag{13.77}$$

$$u_j \geq 0 \qquad (j = 1,\ldots,3) \tag{13.78}$$

$$\log d = -0.301u_1 + 0.u_2 + 0.301u_3 \tag{13.79}$$

Thus, combining Equations (13.71), (13.75), and (13.79), the objective function becomes

$$\text{Max } \log V = \log(\pi/4) - 0.602u_1 + 0.602u_3$$
$$-0.301w_1 + 0.301w_3 + 0.602w_4 + 0.778w_5$$

subject to the constraints (13.67) to (13.70), (13.72) to (13.74) and (13.76) to (13.78).

Solving this problem using LP gives the solution: h = 3.82 m, d = 1.91 m, and V = 10.95 m^3, which can be shown to be the global optimum.

Table 13.4 Values of λ_k and $\log_{10} \lambda_k$ for the cylindrical tank problem

k	λ_k	$\log_{10} \lambda_k$
1	0.5	-0.301
2	1.0	0.000
3	2.0	0.301
4	4.0	0.602
5	6.0	0.778

13.14 DYNAMIC PROGRAMMING

Dynamic programming is a non-linear optimisation technique that is applicable to a wide class of multi-stage decision problems. The term **dynamic** indicates the application of the technique to problems that involve movement from one point in time or space to another point by means of a number of intermediate stages (intervals of time or distance). At each stage a decision has to be taken, i.e. a choice from several alternatives has to be made, and the problem is to find the optimum sequence of decisions.

The method will be illustrated by the following example. Goods are to be transported by truck from city A to city K via the highway network shown schematically in Figure 13.21. The cost of transport on each length of highway in dollars is indicated on the figure. The problem is to find the minimum-cost route from city A to city K.

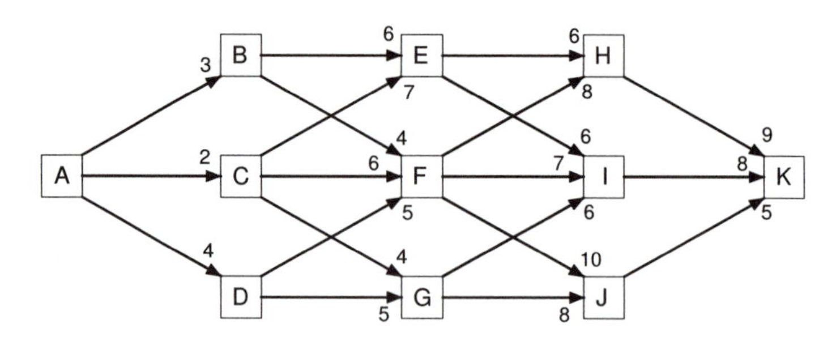

Figure 13.21 Highway routes and travel costs in dollars.

Inspection of Figure 13.21 indicates that, regardless of the route taken, the truck must pass through three other cities. Thus we may consider the problem to consist of four sequential decisions. We say that the problem consists of four stages. A **stage** is defined as the interval that separates consecutive decision points.

One method for determining the minimum-cost route would be to evaluate all alternatives. This is called **complete enumeration,** and in this case it would involve the costing of 17 feasible routes. For a larger problem containing, for example, 6 stages and 10 cities at each stage, there would be 10^5 alternative paths to evaluate. Clearly a more efficient approach is needed. Dynamic programming provides such an approach.

Dynamic programming is based on the **Principle of Optimality,** which may be stated as follows:

> *An optimal policy must have the property that regardless of the decisions leading to a particular state, the remaining decisions must constitute an optimal sequence for leaving that state.*

As discussed in Chapter 3, the 'state' of a system summarizes concisely its entire previous history and therefore allows the future behaviour to be determined for any future input. In this example the states are the cities passed through. When applied to the current problem, the principle of optimality means that if the optimum route passes through state F, it must conclude with the optimum route from F to K.

Thus the principle of optimality can be applied one stage at a time through the network in order to identify the minimum-cost path. Although it is possible to proceed either backwards or forwards in time, it is usually advantageous to work backwards. The present problem will be worked backwards.

With one stage to go, the truck could be in state H, I, or J. There is no choice of destination, which must be K. Nevertheless the calculations are set up in tabular form in order to be consistent with the rest of the solution. The information with one stage to go is summarized in Table 13.5(a). Regardless of the initial state, the decision must be to move to K. The optimum decisions and corresponding minimum costs for these trips are summarized in Table 13.5(a).

With two stages to go the truck could be in city E, F, or G and a decision must be made regarding the next state. The minimum cost for each decision is the sum of the cost for the first stage plus the minimum cost for the remaining stages. This information is summarized in Table 13.5(b).

If starting in city E there is no direct route to city J, so the choice is between cities H and I. The minimum cost of proceeding through H is the direct cost ($6) plus the minimum cost from H onwards which is obtained from Table 13.5(a) as $9. Similarly, the minimum cost of proceeding from E through I is the direct cost ($6) plus the minimum cost from I onwards ($8). The optimum decision from E corresponds to the smallest cost, namely $14 via city I. Similar calculations are made for the initial states F and G.

Table 13.5(b) indicates the optimum decisions from all cities with two stages to go. Thus if a truck driver should ever find himself in city G, the optimum decision is to go to city J, and so on.

Table 13.5 Decision tables for highway route problem.

(a) With one stage remaining

Initial state	Decision: *K*	Optimum decision	Minimum cost ($)
H	9	K	9
I	8	K	8
J	5	K	5

(b) With two stages remaining

Initial	Decision:			Optimum	Minimum
state	H	I	J	decision	cost ($)
E	6 + 9	6 + 8	—	I	14
F	8 + 9	7 + 8	10 + 5	I or J	15
G	—	6 + 8	8 + 5	I	13

(c) With three stages remaining

Initial	Decision:			Optimum	Minimum
state	E	F	G	decision	cost ($)
B	6 + 14	4 + 15	—	F	19
C	7 + 14	6 + 15	4 + 13	G	17
D	—	5 + 15	5 + 13	G	18

(d) With four stages remaining

Initial	Decision:			Optimum	Minimum
state	B	C	D	decision	cost ($)
A	3 + 19	2 + 17	4 + 18	C	19

At this point we have evaluated all possible paths over the last two stages. Now with three stages to go, dynamic programming enables us to evaluate only a subset of all possible paths.

Table 13.5(c) summarizes the calculations with three stages to go. If starting in city B, the choice is between proceeding through city E or F. The minimum cost of travelling through E is the sum of the direct cost ($6) plus the minimum cost from E onwards. The latter term is given in Table 13.5(b) as $14. The minimum cost of travelling through city F may be assessed in a similar way. Thus the optimum route from B is via city F at a cost of $19. The minimum cost route from states C and D are also indicated in Table 13.5(c).

Finally, with four stages to go the initial state is A and the choice is between cities B, C, and D. The minimum cost via B is the direct cost of \$3 plus the minimum cost from B onwards (\$19 from Table 13.5(c)). The minimum-cost route is via C at a cost of \$19.

The optimum route through the network can be determined by working forwards from the starting state A using Table 13.5. Table 13.5(d) indicates that from A we should proceed to C. Table 13.5(c) shows that G is the optimum choice from C. From G we go to J (Table 13.5(b)) and from J to K (Table 13.5(a)). Therefore the minimum cost route is A-C-G-J-K at a cost of \$19. Note also that the dynamic programming solution identifies the optimum solution extending from any non-optimal intermediate state. For example, should an incorrect initial decision lead to B instead of C, the optimal subsequent sequence is B-F-I-K or B-F-J-K.

The benefits of using dynamic programming to solve this problem should be apparent from Table 13.5. With three stages to go only seven comparisons had to be made, even though there are seventeen possible routes through the last three stages. For larger problems, the advantage of using dynamic programming is more marked. For example, a problem with six stages and ten states at each stage contains 10^5 paths, but dynamic programming involves only 410 comparisons.

It should be noted that dynamic programming involves the implicit evaluation of all possible paths. Therefore it always achieves the **global optimum.** It is clearly a very powerful technique for problems that can be organized into a suitable form.

The dynamic recursive equation

Although computer packages for solving general dynamic programming problems are available, it is often better for the analyst to write a computer program for the specific problem being solved. The tabular form of solution given above is suitable for hand solution of dynamic programming problems. Computer codes for solving dynamic programming problems rely on the development of a suitable recursive equation as follows:

Let $c_n(i,j)$ be the cost of proceeding from state i with n stages to go to state j with $(n$-1) stages to go.
Let $f_n(i)$ be the cost of the optimum policy to the end of the final stage when starting in state i with n stages to go.
Let N be the total number of stages in the problem.

Then the dynamic recursive relationship for a minimisation problem may be written as:

$$f_n(i) = \min_{j} \left[c_n(i, j) + f_{n-1}(j) \right] \quad (n = 1, \ldots N) \qquad (13.80)$$

Thus, the cost of the optimum policy when starting in state i with n stages to go is the minimum over all feasible states j of going from i to j plus the cost of the optimum policy when in state j with $n-1$ stages to go.

The use of the dynamic recursive equation will be demonstrated for a problem of reservoir operation.

Example 13.6 A single reservoir of capacity 6 units is situated on a river and is used to regulate flow releases to downstream irrigators. A maximum of 4 units can be released in one season. An optimum release policy is required over the next four seasons beginning with winter. The economic benefits to irrigators for various releases are shown in Table 13.6. The expected inflows to the reservoir over the next four seasons are 4, 3, 1, and 2 units, respectively. At the beginning of winter the reservoir starts with 3 units in storage and it must have at least 3 units in storage at the end of autumn. Dynamic programming should be used to find the optimum release policy.

Table 13.6 Seasonal benefits for release of water for irrigation.

Release	Benefits ($m) for release during:			
(units)	winter	spring	summer	autumn
1	0.2	1.5	1.3	0.5
2	0.4	2.5	2.3	1.0
3	0.5	3.3	3.0	1.2
4	0.6	3.8	3.3	1.3

Solution: The stages in this problem are the four seasons; the state variable is the storage in the reservoir at the end of each season as this summarizes the effects of all previous inputs and decisions. The objective function (to be maximised) is the total benefits of water released for irrigation.

Figure 13.22 shows the feasible states for this problem. Given the initial storage of 3 units and the winter inflow of 4 units the minimum storage at the end of winter is 3 units corresponding to a release of 4 units. The maximum storage is 6 units which corresponds to the release of 1 unit. (A release of zero units is not possible because a spill of 1 unit will occur in any case.) At the end of spring a minimum storage of 2 units can be achieved by releasing 4 units from the initial storage of 3 units.

At the end of summer, the minimum feasible storage is 1 unit which can be achieved via several paths. Although a storage of zero units can be achieved it is not possible to come up to the desired storage of 3 units at the end of autumn from this state. The storage at the end of autumn will be 3 units in most cases as there is no advantage in having more water in storage. However, if the storage at the end of summer is 6 units, the minimum achievable storage at the end of autumn is 4 units.

The dynamic recursive equation (13.80) will be applied backwards in time to this problem. In this case, let i_n be the reservoir storage with n stages remaining, x_n be the reservoir release during the current stage with n stages remaining, and q_n the inflow during the current stage with n stages remaining (known).

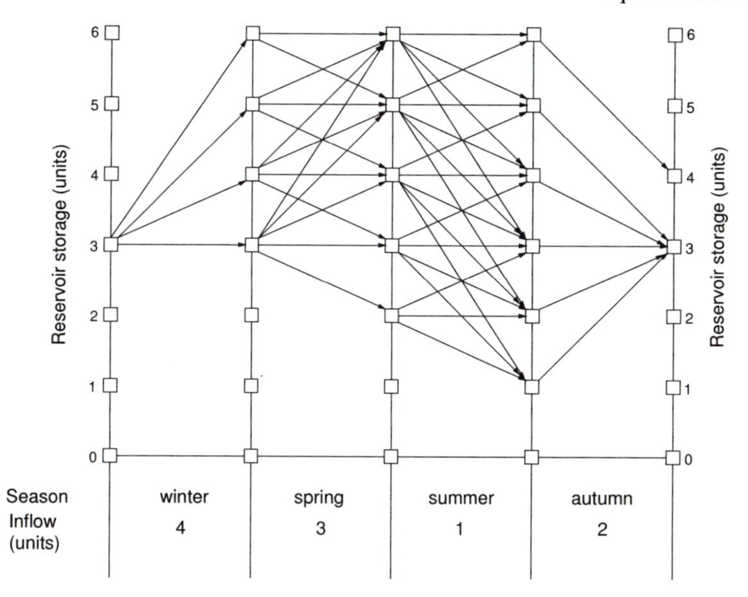

Figure 13.22 Feasible states for the reservoir release problem.

Clearly

$$i_{n-1} = i_n - x_n + q_n \tag{13.81}$$

i.e. the reservoir storage with (n-1) stages remaining is the storage with n stages remaining minus the release plus the inflow during the current stage. Therefore, the dynamic recursive relation for this problem becomes:

$$f_n(i_n) = \max_{0 \le x_n \le 4} \left[c_n(x_n) + f_{n-1}(i_{n-1}) \right] \quad \begin{matrix} (n = 0,1,...,4) \\ \text{all feasible } i_n \end{matrix} \tag{13.82}$$

where $f_n(i_n)$ is the maximum benefit when in state i with n stages remaining and $c_n(x_n)$ is the benefit of releasing x_n when n stages remain. Substituting Equation (13.82) into Equation (13.81) gives

$$f_n(i_n) = \max_{0 \le x_n \le 4} \left[c_n(x_n) + f_{n-1}(i_n - x_n + q_n) \right] \quad \begin{matrix} (n = 0,1,...,4) \\ \text{all feasible } i_n \end{matrix} \tag{13.83}$$

Now to apply Equation (13.83) backwards in time, we start by defining

$$f_o(i_o) = 0 \quad (i_o = 0,1,...,6) \tag{13.84}$$

(a) *With one stage remaining:*

$$f_1(i_1) = \max_{0 \le x_1 \le 4} \left[c_1(x_1) + f_o(i_1 - x_1 + 2) \right] \quad (\text{for } i_1 = 1,\ldots,6)$$

where values of $c_1(x_1)$ are given in the right hand column of Table 13.6. Clearly, there is no choice at this stage, so we simply determine:

$$f_1(1) = [c_1(0) + f_0(1-0+2)] = [c_1(0) + f_0(3)] = 0 + 0 = \$0 \quad (\text{with } x_1^*(1) = 0)$$

where $x_n{}^*(i_n)$ denotes the optimum release when in state i_n with n stages remaining.

$$f_1(2) = [c_1(1) + f_0(2-1+2)] = 0.5 + 0 = \$0.5\,\text{m (with } x_1{}^*(2)=1)$$
$$f_1(3) = [c_1(2) + f_0(3-2+2)] = 1.0 + 0 = \$1.0\,\text{m (with } x_1{}^*(3)=2)$$
$$f_1(4) = [c_1(3) + f_0(4-3+2)] = 1.2 + 0 = \$1.2\,\text{m (with } x_1{}^*(4)=3)$$
$$f_1(5) = [c_1(4) + f_0(5-4+2)] = 1.3 + 0 = \$1.3\,\text{m(with } x_1{}^*(5)=4)$$
$$f_1(6) = [c_1(4) + f_0(6-4+2)] = 1.3 + 0 = \$1.3\,\text{m (with } x_1{}^*(6)=4)$$

These calculations are summarized in Table 13.7(a).

(b) With two stages remaining:

Equation (13.82) becomes:

$$f_2(i_2) = \max_{0 \le x_2 \le 4} \left[c_2(x_2) + f_1(i_2 - x_2 + 1) \right] \quad (\text{for } i_2 = 2,3,\ldots,6)$$

with $i_2 = 2$ only releases of 0, 1, or 2 units are feasible. Comparing these,

$$f_2(2) = \max \begin{bmatrix} c_2(0) + f_1(2-0+1) \\ c_2(1) + f_1(2-1+1) \\ c_2(2) + f_1(2-2+1) \end{bmatrix} = \max \begin{bmatrix} 0 + 1.0 \\ 1.3 + 0.5 \\ 2.3 + 0 \end{bmatrix}$$

$= \$2.3\text{m}$ (with $x_2{}^*(2) = 2$)

Note that the values of $f_1(1)$ - $f_1(3)$ are given in part (a) above and in Table 13.7(a). The calculations with two stages remaining are summarized in Table 13.7(b).

(c) With three stages remaining:
Equation (13.82) becomes

$$f_3(i_3) = \max_{0 \le x_3 \le 4} \left[c_3(x_3) + f_2(i_3 - x_3 + 3) \right] (\text{for } i_3 = 3,4,5,6)$$

The calculations are summarized in Table 13.7(c).

Table 13.7 Decision tables for reservoir release problem.

(a) With one stage remaining

Initial storage i_1	Release (x_1) 0	1	2	3	4	Optimum release $x_1^*(i_1)$	Maximum benefit ($M) $f_1(i_1)$
1	0+0	—	—	—	—	0	0
2	—	0.5+0	—	—	—	1	0.5
3	—	—	1.0+0	—	—	2	1.0
4	—	—	—	1.2+0	—	3	1.2
5	—	—	—	—	1.3+0	1	1.3
6	—	—	—	—	1.3+0	1	1.3

(b) With two stages remaining

Initial storage i_2	Release (x_2) 0	1	2	3	4	Optimum release $x_2^*(i_2)$	Maximum benefit ($M) $f_2(i_2)$
2	0+1.0	1.3+0.5	2.3+0	—	—	2	2.3
3	0+1.2	1.3+1.0	2.3+0.5	3.0+0	—	3	3.0
4	0+1.3	1.3+1.2	2.3+1.0	3.0+0.5	3.3+0	3	3.5
5	0+1.3	1.3+1.3	2.3+1.2	3.0+1.0	3.3+0.5	3	4.0
6	0+1.3	1.3+1.3	2.3+1.3	3.0+1.2	3.3+1.0	4	4.3

(c) With three stages remaining

Initial storage i_3	Release (x_3) 0	1	2	3	4	Optimum release $x_3^*(i_3)$	Maximum benefit ($M) $f_3(i_3)$
3	0+4.3	1.5+4.0	2.5+3.5	3.3+3.0	3.8+2.3	3	6.3
4	0+4.3	1.5+4.3	2.5+4.0	3.3+3.5	3.8+3.0	3 or 4	6.8
5	0+4.3	1.5+4.3	2.5+4.3	3.3+4.0	3.8+3.5	3 or 4	7.3
6	0+4.3	1.5+4.3	2.5+4.3	3.3+4.3	3.8+4.0	4	7.8

(d) With four stages remaining

Initial storage i_4	Release (x_4) 0	1	2	3	4	Optimum release $x_4^*(i_4)$	Maximum benefit ($M) $f_4(i_4)$
3	0+7.8	0.2+7.8	0.4+7.3	0.5+6.8	0.6+6.3	1	8.0

(d) *With four stages remaining:*
The initial storage is 3 units and Equation (13.82) becomes

$$f_4(3) = \max_{0 \le x_3 \le 4} \left[c_4(x_4) + f_3(3 - x_3 + 4) \right]$$

The calculations are summarized in Table 13.7(d). To find the optimum release policy over the four seasons we work backwards through Table 13.7 as follows: release 1 unit in winter (leaving a final storage of 6), release 4 units in spring (final storage = 5), release 3 units in summer (final storage =3), and release 2 units in autumn (final storage = 3). The economic benefits are $8.0M for the year.

In actual reservoir operations the future inflows are usually not known with certainty, but only in a probabilistic sense. In this case an extension of dynamic programming called **stochastic dynamic programming** may be applied. This will not be discussed in this book.

Dimensionality in dynamic programming

The examples considered above both contained only one state variable. Many real problems contain more than one state variable. An example is the operation of a system of n interconnected reservoirs. Here the state of the system at any time is summarized by the storage in each of the reservoirs and there are n state variables.

An increase in the number of state variables significantly increases the amount of computational effort to solve dynamic programming problems. This has been referred to as the **curse of dimensionality.** It may be illustrated by the reservoir operating problem. Suppose it is desired to apply dynamic programming to a single reservoir. If we divide the volume of the reservoir into 10 discrete levels, there are 10 feasible states at each stage. Suppose we now consider a system of n reservoirs, with the storage in each one being able to take on any one of 10 discrete values. The total number of states at each stage of this problem is 10^n. For example, if n=4 there are 10 000 feasible states at each stage. In practical terms, dynamic programming problems are limited to around four state variables because of this difficulty with dimensionality.

13.15 HEURISTIC OPTIMISATION METHODS

In recent years quite a deal of research has been carried out into a new class of optimisation techniques that might loosely be called "heuristic optimisation algorithms". These include techniques such as genetic algorithms, the evolutionary algorithm, differential evolution, ant colony optimisation, particle swarm optimisation and shuffled complex evolution (Michalewicz and Fogel, 2004). All of these methods have the following features in common:

1. They deal with a population of solutions rather than a single solution at any one time.

2. They are effectively "guided search methods" that interact with a simulation model of the system under study. The model can include any non-linear or logical constraints or variables. Thus any system that can be modelled can be optimised using one of these techniques.
3. They usually involve some random processes, so that they may be considered to be "stochastic" optimisation methods.
4. They usually do not reach a single optimum solution but progressively improve the population of solutions that they are working with over time.
5. One cannot prove that the true optimal solution has been obtained. Effectively the methods identify "near-optimal" solutions.

In this text only the genetic algorithm (GA) technique will be considered in detail. This is discussed in the next section.

13.15.1 Genetic Algorithm Optimisation

Holland (1975) was one of the early advocates of genetic algorithms, although they could be viewed as descended from the work of Box (1957) who used evolutionary techniques for optimisation. The publication of the book by Goldberg (1989) has inspired a large number of engineering applications of GA optimisation. The applications include structural optimisation (Goldberg and Samtani, 1986), optimising the operation of pumps in gas pipelines (Goldberg and Kuo, 1987), control system optimisation for aerospace applications (Krishnakumar and Goldberg, 1990) and the optimum design of water supply networks (Simpson et al., 1994).

In essence GAs are a set of guided search routines that work in conjunction with a computer simulation model of a system in order to optimise aspects of the system's design or operation. Some distinguishing features of GAs compared to the traditional optimisation techniques such as linear programming and dynamic programming considered earlier in this chapter are as follows (Simpson et al., 1994):

1. GAs work directly with a population of solutions rather than a single solution. This population is spread throughout the solution space, so the chance of reaching the global optimum is increased significantly;
2. GAs deal with the actual discrete sizes available so that roundoff of continuous variables is not required;
3. Because GAs work with a population of solutions, they identify a number of near-optimal solutions. These solutions could correspond to quite different configurations that may have advantages in terms of non-quantified objectives such as environmental or social objectives; and
4. GAs only use information about the objective or fitness function and do not require the existence or continuity of derivatives of the objectives.

GAs work by analogy to population genetics and involve operators such as selection, crossover and mutation. Unlike techniques such as linear programming, they do not necessarily converge to the global optimal solution. However, because they work in conjunction with a simulation model they can handle any non-linear, discontinuous or logical set of objective functions or constraints. In essence, any system that can be simulated on a computer can be optimised using GAs. The "theory of schemata" suggests that increasingly fit solutions will be present in exponentially increasing numbers in succeeding generations in a GA run.

13.15.2 Application of Genetic Algorithms

Combinatorial optimisation was introduced in Section 13.2. It was noted that some engineering problems involve extremely large combinations of decision variables. The number of combinations can be so great that it is not feasible to simulate all alternatives in a reasonable amount of computer time. Genetic algorithms can be applied to these large combinatorial optimisation problems. An example of such problems is the New York Tunnels problem that was first studied by Schaake and Lai (1969). The basic layout of the system is shown in Figure 13.23.

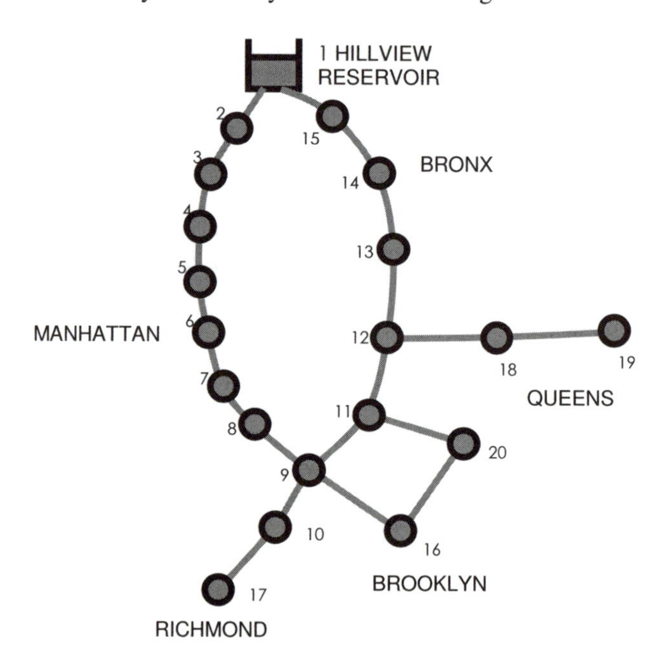

Figure 13.23 Layout of the New York Water Supply Tunnels

Water is supplied from Hillview Reservoir through 21 major tunnels ranging in size from 1.524 m to 5.182 m in diameter. In Figure 13.23 the lines represent the various tunnels and the nodes represent junctions in the tunnels where water is distributed to many smaller pipes in the network. These smaller pipes are not shown. At the time of the study, it was anticipated that the system would be unable to meet forecast increases in demand while supplying adequate pressures. Hence the problem was to identify which tunnels should be duplicated and with what diameter tunnels in order to meet the forecast increased demand.

The tunnels can only be drilled in discrete sizes of which 16 options (ranging between no duplication to duplication with a tunnel 5.182 m in diameter cover the practical sizes available for this problem. In practice this is a small network as the distribution system for a large city could contain more than 100,000 pipes. Nonetheless the number of options is not insignificant and equals 16^{21} or 1.934 x 10^{25}. Such networks are usually designed using a trial-and-error procedure whereby the engineer chooses a trial design and then uses a hydraulic analysis computer model to estimate the pressures throughout the network under the design demand conditions. If the pressures are inadequate, some pipe (or tunnel) sizes are increased in the model and it is rerun. If the pressures are higher than the minimum requirements, some pipe sizes are reduced in the model and it is rerun. Eventually, the engineer decides that the design is adequate i.e. provides the desired pressures at a reasonable cost.

Although this is a relatively small network, the identification of the minimum cost solution is a non-trivial task. For example, the use of complete enumeration would require 1.934 x 10^{25} computer runs. If each takes 1 millisecond, the total run time would be 6.13 x 10^{14} years. This can be compared with the estimated life of the universe of 15 x 10^{9} years! In practice of course, engineers can eliminate many of the infeasible solutions using experience. Suppose that we can eliminate 99.99% of the potential solutions and then enumerate the remaining 0.01%. Unfortunately the computer run time will still be 6.13 x 10^{10} years or 4 times the life of the universe. Computer speeds are doubling every 18 months or so. If that trend continues, the computer run time to completely enumerate the New York tunnels problem will be down to 1 day in about 60 years time. Until that time, we need to rely on other techniques to identify optimal or near-optimal solutions to problems like the New York Tunnels problem.

It should be noted that the New York water supply system has changed considerably since the study by Schaake and Lai in 1969. However, it has been used by a number of researchers over the last 37 years as a benchmark problem to test their pipe network optimisation routines (Dandy et al., 1996). The best solution found so far is that by Maier et al. (2003) using the technique of Ant Colony Optimisation.

Representation of solutions

A large number of variants of GAs have been developed over the last 15 years or so. In this Section only a simple GA will be considered. A GA operates on a string of numbers that represents the decision variables. By analogy to population genetics, this string is sometimes called a chromosome while the individual numbers in the string are called genes.

For the New York tunnels problem the string would consist of 21 integers each one being between 0 and 15 as shown in Figure 13.24.

[3|12|0|5|0|1|0|0|9|3|0|7|5|11|0|2|2|0|7|1|0|0]

Figure 13.24 A GA String for the New York Tunnels problem

The first integer in the string represents the size of the duplicate tunnel for tunnel # 1, the second integer the duplicate for tunnel #2, and so on. The integers are decoded into actual pipe sizes using Table 13.8. Note that the original problem has been converted into SI units from the original USA customary units of feet and inches.

Table 13.8 Coding of New Tunnels and their costs for the New York Tunnels problem (adapted from Dandy et al. 1996)

GA code	Tunnel Diameter (m)	Cost ($/m)
0	0	0
1	0.914	306.8
2	1.219	439.6
3	1.524	577.4
4	1.829	725.1
5	2.134	876.0
6	2.438	1036.7
7	2.743	1197.5
8	3.048	1368.1
9	3.353	1538.7
10	3.658	1712.6
11	3.962	1893.0
12	4.267	2073.5
13	4.572	2260.5
14	4.877	2447.5
15	5.182	2637.8

The solution given by the string in Figure 13.24 therefore represents duplicating Tunnel #1 with a tunnel of 1.524 m diameter, duplicating Tunnel #2 with a tunnel of diameter 4.267 m, not duplicating Tunnel #3, and so on.

Fitness function

The objective function in GA optimisation is called the fitness function. GA optimisation is usually aimed at maximising the fitness function, so if the problem involves minimisation, the fitness function is written as minus the value of the objective function or as the inverse of the objective function. The later transformation can only be used if the objective function is strictly positive.

Steps in the GA

The steps involved in the operation of a simple GA are given below:

1. generation of the initial population;
2. computation of the objective function for each solution in the population;
3. evaluation of the performance of each solution in the population relative to the constraints;
4. computation of penalty cost for not meeting the constraints;
5. computation of the fitness function for each solution;
6. check for convergence of the population. If convergence has occurred, then stop; otherwise continue.
7. selection of a set of parent strings for the next generation;
8. crossover of pairs of parents; and
9. mutation of selected strings. Return to step 2.

Each of these steps will be explained in more detail in relation to the New York Tunnels problem. Firstly, some additional data are provided for the problem. Table 13.9 contains information on the design demands and elevations of each of the nodes.

Table 13.9 Nodal Data for the New York City Water Supply Tunnels (adapted from Dandy et al., 1996)

Node	Demand (m^3/s)	Minimum total head (m)
1	0	91.44 (actual)
2	2.710	77.72
3	2.710	77.72
4	2.587	77.72
5	2.587	77.72
6	2.587	77.72
7	2.587	77.72
8	2.587	77.72
9	4.986	77.72
10	0.029	77.72
11	4.986	77.72
12	3.435	77.72
13	3.435	77.72
14	2.710	77.72
15	2.710	77.72
16	4.986	79.25
17	1.686	83.15
18	3.435	77.72
19	3.435	77.72
20	4.986	77.72

The design demands are for a future target year, in this case 1994 (25 years on from the time of the study). Table 13.10 contains information about the lengths and diameters of the existing tunnels in the system (as they were in 1969).

The New York Tunnels system needs to be designed to meet the design demands at minimum cost while ensuring that the minimum total heads at all nodes (given in Table 13.9) are satisfied. The difference between the total head and the elevation at a node is the pressure head at that node. A minimum level of pressure head is required to ensure reasonable levels of service for consumers. The total heads at each node for any particular design can be computed using a combination of the continuity and head loss equations for the system. The details of computing the total heads are given in Appendix 13B. The Hazen Williams coefficient for all tunnels is 100.

Table13.10: Data for the existing New York Tunnels (adapted from Dandy et al., 1996)

Tunnel	Start Node	End Node	Length (m)	Existing Diameter (m)
[1]	1	2	3535.7	4.572
[2]	2	3	6035.0	4.572
[3]	3	4	2225.0	4.572
[4]	4	5	2529.8	4.572
[5]	5	6	2621.3	4.572
[6]	6	7	5821.7	4.572
[7]	7	8	2926.1	3.353
[8]	8	9	3810.0	3.353
[9]	9	10	2926.1	4.572
[10]	11	9	3413.8	5.182
[11]	12	11	4419.6	5.182
[12]	13	12	3718.6	5.182
[13]	14	13	7345.7	5.182
[14]	15	14	6431.3	5.182
[15]	1	15	4724.4	5.182
[16]	10	17	8046.7	1.829
[17]	12	18	9509.8	1.829
[18]	18	19	7315.2	1.524
[19]	11	20	4389.1	1.524
[20]	20	16	11704.3	1.524
[21]	9	16	8046.7	1.829

As this is a minimisation problem and the cost is strictly positive, the fitness function used will be the inverse of the total cost. The total cost will be the sum of the cost of constructing duplicate tunnels plus the penalty cost (if any) of not satisfying the minimum head constraints at nodes.

The steps involved in applying GA optimisation to the New York Tunnels problem are described below.

1. Generation of the initial population.
As stated previously, a GA works with a population of solutions at any one time. The population size is one of the parameters that must be chosen for a GA. It is typically in the range 20–1000. The initial population is usually generated randomly, so for the New York Tunnels each string would consist of a set of 21 randomly generated integers in the range 0–15. Each integer would have a specified probability of being selected. These probabilities are usually equal, but they could be selected to have a bias towards zero in the case of the New York Tunnels.

2. Computation of the objective function for each solution in the population.
The objective function can be now computed for each solution in the initial population. In the case of the New York Tunnels, this equals the cost of the solution and can be computed by knowing the diameters of the duplicate tunnels (from the values in the string), their lengths (given in Table 13.10) and the costs per unit length (given in Table 13.9).

3. Evaluation of the performance of each solution in the population relative to the constraints.
The hydraulic performance of each design can be evaluated by calculating the total head at each node in the network using a hydraulic computer model or by solving the non-linear equations given in Appendix 13B.

4. Computation of penalty cost for not meeting the constraints.
If the total head at any node is below the minimum value (given in Table 13.9) a penalty cost is added, for example $16 million per metre of total head below the minimum required value. This is an arbitrary high figure chosen to penalise solutions that do not satisfy the minimum pressure constraints. If several nodes have total heads below the minimum values, the penalty will be based on the node with the largest deficit.

5. Computation of the fitness function for each solution.
The objective function value (i.e. the cost of the solution computed in step 2) is added to the penalty cost (step 4) to give the total cost for each design. The fitness function is the inverse of the total cost so that fitness function is to be minimised.

6. Check for convergence of the population. If convergence has occurred, then stop; otherwise continue.
Unlike a number of other optimisation techniques discussed earlier in this Chapter, genetic algorithms do not have an easy way of determining if the global optimum solution has been achieved. In practice the algorithm is run for a specified number of generations or until little improvement is achieved in the best solution over a number of generations. If one of these criteria is met, the algorithm is stopped, otherwise go on to step (7).

7. Selection of a set of parent strings for the next generation.
By analogy to natural selection, not all solutions will be parents (i.e. produce offspring) in the next generation. Using the principle of "survival of the fittest", the GA will select solutions with the highest fitness values to be parents for the next generation. A common way to carry out this process is using tournament selection. In tournament selection, two solutions are selected at random from the current population, they are compared in terms of their fitness and the string with the higher fitness becomes a parent for the next generation. This process is repeated (without replacing the first two strings) to give another parent. The process continues until all strings have been involved in tournaments. At this stage $N/2$ parents have been selected, where N is the population size. The full process is carried out a second time starting with the full population from the current generation, so that a total of N parents have been identified.

8. Crossover of pairs of parents.
Now parents are "mated" to produce the next generation. This is carried out by selecting a pair of parents at random from the N available. Now crossover of the genetic material of the parents will occur with a probability p_c. A uniformly distributed random number between zero and one is selected. If this number is less than or equal to p_c, crossover occurs and some exchange occurs in the parental genetic material passed on to the offspring. This is simulated by determining a point to cut the two strings as shown in Figure 13.25(a).

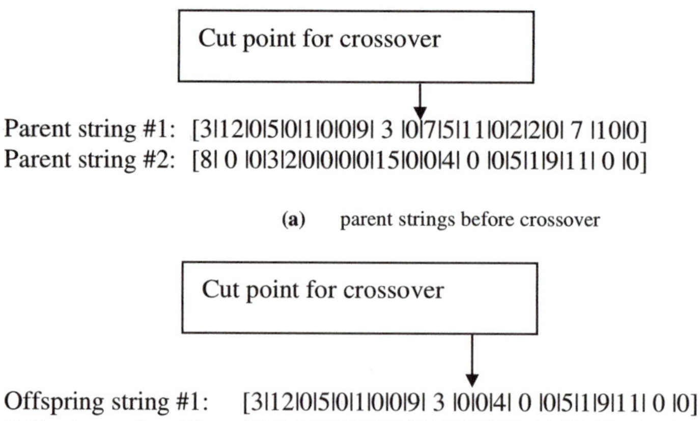

(a) parent strings before crossover

(b) offspring strings after crossover

Figure 13.25 Crossover of two strings

The point at which the strings will be cut is determined randomly. With 21 numbers (genes) in the string, there are 20 possible places to cut the strings. An integer between 1 and 20 is chosen randomly and is used to identify where the strings will be cut. The right hand tails are then switched between the two strings. Suppose the random number chosen is 11. The offspring strings are shown in Figure 13.25(b).

If crossover does not occur, the two offspring strings are identical to their parents, so the two parent strings are passed into the next generation. This process is repeated for pairs of strings that are selected (without replacement) from the parent population, so that a population of offspring strings is formed.

9. Mutation of selected strings.

In nature, there is some chance that mutation will occur and the offspring will have some genetic differences from their parents. This serves to keep some diversity in the population and allows all species to adapt to a changing environment. The GA process allows some small percentage of strings to mutate every generation. This percentage is determined by the GA parameter called the probability of mutation, p_m. After the crossover process, each string is tested to see if mutation occurs. A uniformly distributed random number between zero and one is chosen. If it is less than or equal to p_m mutation occurs. One of the numbers in the string will change to a randomly chosen value. The number that mutates is chosen at random by selecting an integer random number between one and l (where l = the length of the string). The new value is randomly chosen over the range of possible values for this location in the string. In the case of the New York Tunnels, the new value will be randomly chosen between 0 and 15. Now return to step 2.

Convergence of the GA is illustrated in Figure 13.26 which shows the minimum cost solution in each generation. It can be seen from this figure that the GA converges rapidly for a start, flattens out and then only improves as a result of occasional mutations that identify areas of better solutions.

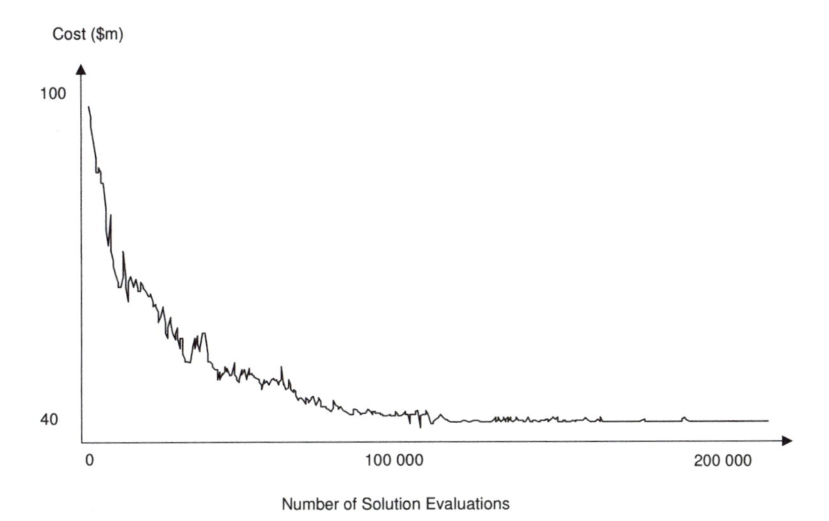

Figure 13.26 Convergence of the GA for the New York tunnels problem (adapted from Dandy et al. 1994)

The solution identified by Dandy et al. (1996) is given in Table 13.11 and has a total cost of $38.80 m compared to the original design of Schaake and Lai (1969) which had a cost of $78.09 m. Only the tunnels listed in the table are to be duplicated. Schaake and Lai (1969) used linear programming to solve this problem but required assumptions to be made about the total head at each node. This method is unlikely to lead to the global optimal solution and the Schaake and Lai (1969) solution had much more duplication of pipes than necessary. Over the years a number of other researchers have applied various optimisation techniques to solve this problem. The best solutions have been found using evolutionary algorithms such as genetic algorithms or ant colony optimisation.

Table 13.11: Solution to the New York Tunnels problem by Dandy et al. (1996)

Duplicate Tunnel Number	Duplicate Diameter
[15]	3.048
[16]	2.134
[17]	2.438
[18]	2.134
[19]	1.829
[21]	1.829

13.15.3 Choice of GA Parameters

The following parameters need to be specified for a GA: population, probability of crossover, probability of mutation, tournament size (this can be more than two), and maximum number of generations to run. These values need to be chosen based on experience in the particular problem regime. Indicative ranges for each parameter are given in Table 13.12.

In addition, penalty costs need to be chosen for each set of constraints. It is difficult to give general advice for these values because they depend on the units of the objective function and constraints. In general the penalties need to be large enough to ensure that the constraint are satisfied, but small enough to ensure that the optimum solution can be approached from both the infeasible as well as the feasible region. In practice, the GA is usually run a number of times using different combinations of parameters and a range of starting seeds for the sequence of random numbers and best overall solution is identified.

Table 13.12: Indicative ranges for values of the GA parameters

GA parameter	Symbol	Indicative range
Population size	N	50–1000
Probability of crossover	p_c	0.7–1.0
Probability of mutation	p_m	0.001–0.02
Tournament size	T	2–4
Maximum number of generations	G	500–2000

13.16 SUMMARY

Optimisation is a process of attempting to find the "best" solution for an engineering planning or design problem. The "best" solution is the one that best achieves the defined objectives for the particular system while satisfying all of the constraints. There are a number of approaches to optimisation including subjective, combinatorial and analytical optimisation. Subjective optimisation is based on engineering judgement and experience and is used in practice when the problem is too complicated to allow combinatorial or analytical methods to be used.

Combinatorial optimisation involves evaluating and comparing all (or a large number) of the possible combinations of solutions for a particular problem. Complete enumeration of all alternatives can be used when there are a small number of possible combinations, otherwise heuristic methods (such as genetic algorithms) can be applied to identify near-optimal designs.

Analytical optimisation techniques can be used where the objective(s) and constraints for the system can be written in mathematical form. There are a large number of techniques that can be used to solve analytical optimisation problems. In this text the following techniques have been described in detail: linear programming, differential calculus, separable programming and dynamic programming.

The choice of the most appropriate optimisation technique to use for a particular problem is a skill based on experience and a good knowledge of the problem and the available techniques.

PROBLEMS

13.1 Solve the following LP problems graphically:

(a) Max $Z = 3x_1 + 5x_2$

subject to $2x_1 + x_2 \leq 12$

$4x_1 + 5x_2 \leq 40$

$x_1, x_2 \geq 0$

(b) Min $Z = 10x_1 + 8x_2$

subject to $x_1 \geq 4$

$x_1 + 3x_2 \geq 12$

$x_1 + x_2 \geq 8$

$x_1, x_2 \geq 0$

(c) Max $Z = 4x_1 - 3x_2$

subject to $x_1 \leq 10$

$2x_1 + 3x_2 \geq 12$

x_1, x_2 unrestricted in sign.

13.2 Formulate the following LP problems into canonical and standard form:

(a) Min $Z = 4x_1 + 2x_2$

subject to $2x_1 + 3x_2 \geq 9$

$$3x_1 + x_2 \geq 6$$
$$x_1, x_2 \geq 0$$

(b) Max Z = $x_1 + 2x_2 + 4x_3$

subject to

$$x_1 + x_3 \leq 5$$
$$2x_1 - x_2 \geq 4$$
$$x_1, x_3 \geq 0 \; x_2 \text{ unrestricted in sign.}$$

13.3 Portion of a road network is shown in Figure 13.27. The travel times and capacities in both directions for all roads are given in Table 13.13. During peak conditions, 5000 vehicles per hour are travelling from node 1 to node 6 and 4000 vehicles per hour are travelling from node 4 to node 2.

(a) Formulate an LP problem which could be used to find the volume of traffic in each direction on each road during peak conditions.
(b) Solve the LP problem using a standard computer package.

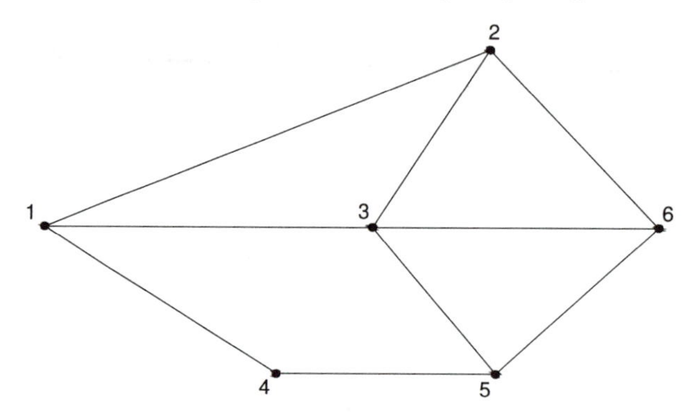

Figure 13.27 Portion of a road network.

Table 13.13 Travel times and capacities for a road network.

Road	Travel time (minutes)	Capacity (vehicles/h)
1–2	15	3000
1–3	10	4000
1–4	8	1000
2–6	6	3000
3–2	7	2000
3–5	6	3000
3–6	5	2000
4–5	9	3000
5–6	10	4000

13.4 A reservoir and pipe network is shown in Figure 13.28. The hydraulic grade line at locations C and D must be at or above RL 20 m and 0 m respectively. The elevation of the hydraulic grade line at B is to be determined. Three sizes of pipe are available according to Table 13.14.

Table 13.14 Pipe sizes and costs.

Pipe diameter (mm)	Cost ($/m length)
600	180
750	320
900	500

Each of the sections AB, BC, and BD may be made up of lengths of any or all pipe sizes. However, only 1000 m of 750 mm diameter pipe is available in time for the job.

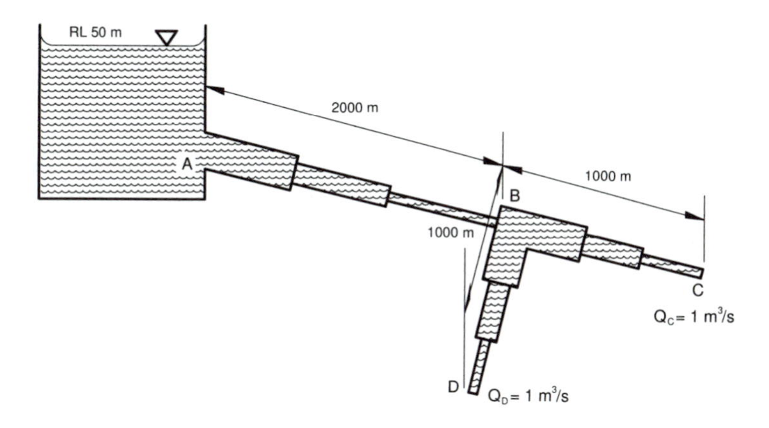

Figure 13.28 A reservoir and pipe network.

The head loss in any length of pipe is given by:

$$h_l = 0.001\frac{V^2 L}{D}$$

where h_l *is* the head loss (m), V is the average velocity in pipe (m/s), L is the pipe length (m), and D is the pipe diameter (m). Ignore the velocity head and head losses at entrance, exit, contraction, expansion, and bends.

(a) Set up an LP model which could be used to determine the lengths of pipe of each diameter to be used in each of the sections AB, BC, and BD so as to minimise the total cost.

(b) Solve the model using a suitable computer package.

13.5 A river receives organic waste from the two factories shown in Figure 13.29 The organic waste load is 1000 kg/day from the first factory and 1400 kg/day from the second. The streamflows (including waste input) at locations 1 and 2 are 3 m^3/s and 4 m^3/s (respectively).

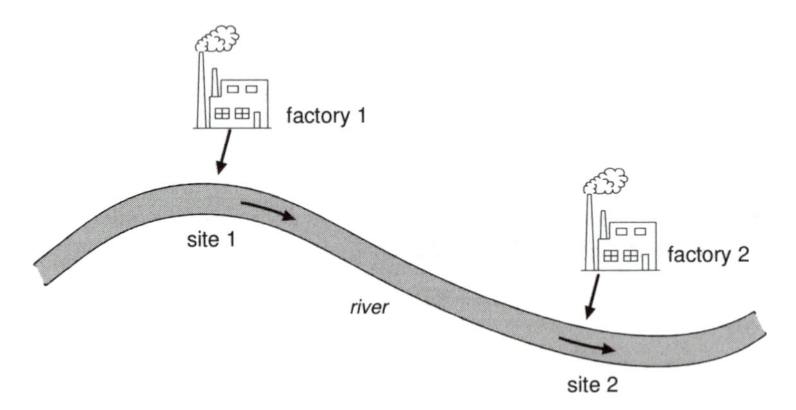

Figure 13.29 A river receiving organic waste from two factories.

It is proposed to install waste treatment plants at each factory so as to reduce the maximum waste concentration in the river to 0.002 kg/m^3 at all locations. The cost of waste removal is $0.30/kg at factory 1 and $0.20/kg at factory 2. A maximum of 90% of the waste from either factory can be removed. Waste decomposition in the river may be ignored.

Formulate an LP model which could be used to find the minimum cost combination of waste treatment plants at factories 1 and 2 such that the maximum waste concentration is not exceeded. Solve the model graphically.

13.6 Write the duals to Problems 13.1(a)-(c) and 13.2(a)-(b).

13.7 Carry out the following:

(a) Write the dual to the LP model formulated in Problem 13.5.

(b) From the graphical solution of the primal, determine the optimum values of the dual variables. (Hint: the dual variables will be zero for those primal constraints which have some slack or surplus at the optimum solution. Why?)

(c) Verify the optimum values of the dual variables found in part (b) by solving the dual problem using the simplex method.

(d) Use the value of the dual variables to estimate the reduction in treatment cost if the maximum allowable waste concentration at site 2 is increased to 0. 0022 kg/m^3.

13.8 An optimal policy is required to maximise the net daily benefits due to the operation of a reservoir from which water may be extracted for both power generation and irrigation purposes.

Each 1000 m^3 of water released through the power station results in the generation of 150 kWh of electrical energy, with all of the water passing from the turbines into the river downstream of the dam. The net benefit to the community of power generation at this site is 10 c/kWh for the first 3000 kWh, 6 c/kWh for the next 3000 kWh and 4 c/kWh thereafter (up to a maximum of 15 000 kWh).

Each 1000 m^3 of water released for irrigation purposes produces a net benefit to the community of \$10. In addition, 30% of the water released for irrigation is returned to the river downstream of the dam.

The total volume of water available for power generation and irrigation purposes is 100 000 m^3/day. In order to maintain water quality standards in the river downstream of the dam, a minimum flow of 10,000 m^3/day must be maintained.

Formulate the problem of finding the daily volumes of water to be used for power generation and irrigation as an LP problem using the separable programming technique.

13.9 Rework Problem 13.5 assuming that the cost of waste removal is not constant at each site but is given in Table 13.15. Linear interpolation may be used for intermediate values of plant capacity. Why is it possible to use separable programming for this problem?

Table 13.15 Costs of waste removal

Capacity of plant (pollutant removal kg/day)	Cost (\$x10^3) of installing a plant at	
	Industry 1	Industry 2
0	0	0
500	100	120
1000	250	300
1400	—	500

13.10 A trucking company wants to transport goods from city A to city B at minimum cost. The cost of travel on each link (in dollars) is shown in Figure 13.30. Find the minimum cost route from A to B. (Note that it may be necessary to introduce dummy states in order to apply dynamic programming to this problem.)

13.11 A quarry produces crushed rock for road-making. It can produce up to 5000 m^3 per month. The quarry's orders and cost of production per 1000 m^3, for the next four months are given in Table 13.16

The quarry may store up to 4000 m^3 of crushed rock from one month to the next at a cost of \$50/1000 m^3.

Use dynamic programming to find the quarry's minimum cost production schedule over the four months assuming it starts and finishes with no crushed rock in storage.

Table 13.16 Orders and production costs.

Month	1	2	3	4
Orders (x 1000 m^3)	2	4	4	3
Production cost for the first 1000 m^3 ($)	500	550	600	550
Production cost for each additional 1000 m^3 ($)	300	250	350	300

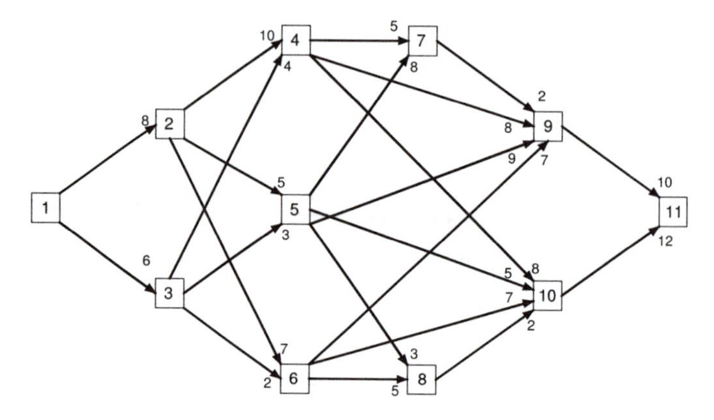

Figure 13.30 A highway network.

13.12 Water is pumped from a river into a reservoir and is used to supply water to a nearby city. The maximum volume of water which can be pumped into the reservoir in any month is 30 000 ML.

It is desired to determine the minimum-cost pumping schedule over the next four months so as to meet the known demands of the city. The demands and pumping costs are given in Table 13.17.

Table 13.17 Demands and pumping costs.

Month	Demand (ML)	Cost ($M) of pumping a volume of:			
		0	10 000 ML	20 000 ML	30 000 ML
1	20 000	0.0	1.0	2.5	3.0
2	40 000	0.0	1.5	2.7	3.3
3	30 000	0.0	1.2	2.0	3.0
4	10 000	0.0	0.9	2.6	3.5

The reservoir contains a volume of 30 000 ML, at the start of the first month and must contain at least 20 000 ML, at the end of the fourth month. Evaporation and seepage losses are negligible.

Using dynamic programming and discrete volumetric units of 10 000 ML, find the minimum-cost set of monthly volumes to be pumped to the reservoir over the period.

13.13 The expansion of a wastewater collection system for a major city is to be designed. It consists of 20 major pipelines (each with 8 possible sizes), 50 smaller pipelines (6 possible sizes each), 12 pumps (12 possible sizes each) and 6 detention storages (5 possible sizes each).

(a) What is the number of possible combinations for this design?

(b) How does this compare with the estimated number of atoms in the Universe (10^{75})?

(c) Based on the comparison in part (b), comment on the difficulty of finding the true optimum solution to this problem.

(d) If it takes 0.001 seconds to simulate the performance of each solution to this problem using an appropriate simulation package, how long would it take to simulate all of the possible combinations?

13.14 In order to demonstrate the operation of a simple genetic algorithm, consider the simple problem of minimising the following function:

$$Z = 0.5x_1 + x_2 + 2x_3 + 3x_4 + 5x_5$$

Where all x_i *(i = 1., …5)* equal either 0 or 1. Each solution to this problem can be represented by a binary string of length five. The optimum solution to the problem should be obvious.

(a) Using a random number generator or table of random numbers, generate a starting population of 20 solutions and determine the value of the objective function of each one.

(b) Using tournament selection, a crossover probability of 0.7 and mutation probability of 0.02 per bit, develop a further population of 20 solutions from the initial population.

(c) Evaluate the objective function for each new solution and compare the average value and the lowest value of the objective function for each generation. Comment on the results.

(d) Repeat this process for another four generations and verify that the best solution is improving over the generations.

(e) How would you modify the genetic algorithm solution process if the following constraint needed to be included: $x_1 + x_2 + x_3 + x_4 + x_5 \geq 3$?

13.15 The westbound portion of a proposed new freeway in a city in the USA is shown in Figure 13.31.

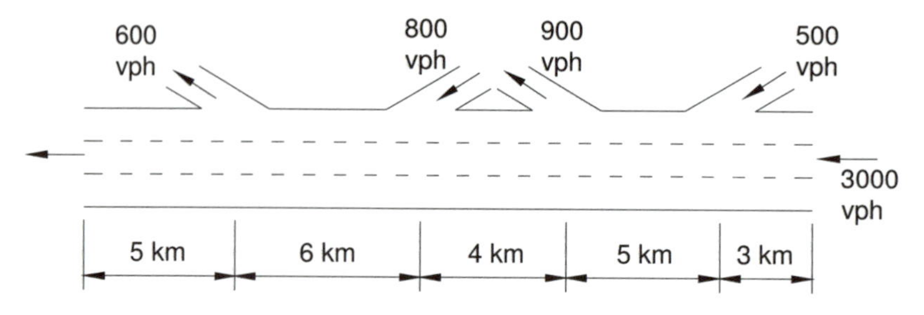

Figure 13.31: Portion of a proposed new freeway (not to scale)

The average travel time on each section of the freeway is given by the following equation:

$$t_i = t_{i0}[1 + (V_i / C_i)^4]$$

where t_i = the travel time on section i (minutes); t_{0i} = the "free speed" travel time on section i (minutes) based on a speed of 100 km /hour; V_i = volume of traffic on section i (vehicles per hour); and C_i = the nominal "capacity" of section i (vehicles per hour).

The nominal capacity per lane is 1200 vehicles per hour and the cost to build 1 km of freeway is \$1m per lane. There is a minimum of 2 lanes and a maximum of 6 lanes per section of freeway.

(a) Write the objective function and constraints for an optimisation model that can be used to find the optimal number of lanes for each section of the freeway so as to minimise the total travel time by all vehicles on the freeway subject to a budget of \$70m.

(b) Outline how this problem could be solved using the genetic algorithm (GA) technique. Describe the structure of a typical string and outline the steps involved in running the GA.

REFERENCES

Beightler, C., Phillips, D. & Wilde, D., 1979, *Foundations of Optimisation,* 2nd ed. (Englewood Cliffs, NJ: Prentice-Hall).

Box, G.E.P., 1957, Evolutionary operation: a method for increasing industrial productivity, *Applied Statistics*, 6(2), 81–101.

Dandy, G.C., Simpson, A.R. and Murphy, L.J., 1993, A review of pipe network optimisation techniques. *Proceedings of Watercomp 93*, (Melbourne, Institution of Engineers, Australia) National Conference Publication No 93/2, 373–383.

Dandy, G.C., Simpson, A.R. and Murphy, L.J., 1996, An improved genetic algorithm for pipe network optimisation, *Water Resources Research*, 32(2), 449–458.

deNeufville, R. & Stafford, J.H., 1971, *Systems Analysis for Engineers and Managers* (New York: McGraw-Hill).

Goicoechea, A., Hansen, D.R. & Duckstein L., 1982, *Multiobjective Decision Analysis with Engineering and Business Applications* (New York: Wiley).

Goldberg, D. E., 1989, *Genetic Algorithms in Search, Optimisation, and Machine Learning* (Addison-Wesley Publishing Company, Inc.)

Goldberg, D.E. and Samtani, M.P., 1986, Engineering optimization via genetic algorithms. *Proceedings 9th Conference on Electronic Computation*, ASCE, New York, NY, 471–482.

Goldberg, D.E. and Kuo, C.H., 1987, Genetic algorithms in pipeline optimization, *Journal of Computing in Civil Engineering*, 1(2), 128–141.

Holland, J.H., 1975, Adaptation in natural and artificial systems, (University of Michigan Press, Ann Arbor).

Krishnakumar, K. and Goldberg, D.E., 1990, Control system optimization using genetic algorithms. *Proceedings AIAA Guidance, Navigation and Control Conference*, American Institute of Aeronautics and Astronautics (AIAA).

Maier H.R., Simpson, A.R., Zecchin, A.R., Foong, W.K., Phang, K.Y., Seah, H.Y. and Tan, C.L., 2003, Ant colony optimisation for design of water distribution systems, *Journal of Water Resources Planning and Management, ASCE,* 129(3), 200–209.

Meredith, D.D., Wong, K.W., Woodhead, R.W. & Wortman R.H., 1985, *Design and Planning of Engineering Systems,* 2nd ed. (Englewood Cliffs, NJ: Prentice-Hall).

Michalewicz, Z. and Fogel, D.B., 2004, *How to Solve It: Modern Heuristics*, 2nd ed. (Berlin: Springer).

Ormsbee, L.E. and Wood, D.J., 1986, Hydraulic design algorithms for pipe networks, *Journal of Hydraulic Engineering*, ASCE, 112(12), 1195–1207.

Ossenbruggen, P.J. 1984. *Systems Analysis for Civil Engineers* (New York: Wiley).

Schaake, J., and Lai, D. 1969. Linear Programming and Dynamic Programming Applications to Water Distribution Network Design. *Report 116*, Department of Civil Engineering, Massachusetts Institute of Technology, Cambridge, Massachusetts.

Shamir, U. and Howard, C.D.D., 1968, Water distribution systems analysis. *Journal of Hydraulics Division*, ASCE, 94(HY1), 219–234.

Simpson, A.R., Dandy, G.C. and Murphy, L.J., 1994 Genetic algorithms compared to other techniques for pipe optimisation. *Journal of Water Resources Planning and Management, ASCE*, 120 (4), 423–443.

Smith, A.A., Hinton, E. and Lewis R.W., 1983, *Civil Engineering Systems. Analysis and Design.* (New York: Wiley).

Stark, R.M. & Nicholls, R.L., 1972, *Mathematical Foundations for Design. Civil Engineering Systems.* (New York: McGraw–Hill).

Streeter, V.L. and Wylie, E.B., 1979, Fluid mechanics. (McGraw–Hill, New York).

Taha, H.A., 2003, *Operations Research. An Introduction,* 7th ed. (Upper Saddle River , NJ: Pearson Education).

Wagner, H.M., 1975, *Principles of Operations Research with Applications to Managerial Decisions,* 2nd ed. (Englewood Cliffs, NJ: Prentice-Hall).

APPENDIX 13A: NECESSARY AND SUFFICIENT CONDITIONS FOR THE SOLUTION OF UNCONSTRAINED MULTIVARIATE OPTIMISATION PROBLEMS

The necessary and sufficient conditions for an unconstrained multivariate optimisation problem are given below. For convenience define $\underline{\mathbf{X}} = (x_1, x_2, \ldots, x_n)$, then condition 13.55 becomes:

$$\text{Max } Z = f(\underline{\mathbf{X}}) \tag{13.85}$$

As in the single-variable problem, differential calculus can be used to solve this problem provided $f(\underline{\mathbf{X}})$ and its first- and second-order partial derivatives exist and are continuous for all values of $\underline{\mathbf{X}}$.

A **necessary** condition for a particular solution $\underline{\mathbf{X}}^o$ to be a maximum of $f(\underline{\mathbf{X}})$ is:

$$\left.\frac{\partial f(\underline{\mathrm{X}})}{\partial x_j}\right|_{\underline{\mathrm{X}} = \underline{\mathrm{X}}^o} = 0 \quad (j = 1, \ldots, n) \tag{13.86}$$

The solutions to Equation (13.86) are called stationary points and include maxima, minima, points of inflection, and saddlepoints. In order to determine which of the stationary points are maxima (or minima) it is necessary to examine the second-order partial derivatives of $f(\underline{\mathbf{X}})$. These are contained in the Hessian matrix, $\underline{\mathbf{H}}$, which is defined as follows:

$$\underline{\mathbf{H}} = \begin{bmatrix} \frac{\partial^2 f}{\partial x_1^2} & \frac{\partial^2 f}{\partial x_1 \partial x_2} & \cdots & \frac{\partial^2 f}{\partial x_1 \partial x_n} \\ \frac{\partial^2 f}{\partial x_2 \partial x_1} & \frac{\partial^2 f}{\partial x_2^2} & \cdots & \frac{\partial^2 f}{\partial x_2 \partial x_n} \\ \cdot & \cdot & \cdot & \cdot \\ \cdot & \cdot & \cdot & \cdot \\ \cdot & \cdot & \cdot & \cdot \\ \frac{\partial^2 f}{\partial x_n \partial x_1} & \frac{\partial^2 f}{\partial x_n \partial x_2} & \cdots & \frac{\partial^2 f}{\partial x_n^2} \end{bmatrix}$$

A **sufficient** condition for a stationary point, $\underline{\mathbf{X}}^o$ to be a minimum is that the Hessian matrix evaluated at $\underline{\mathbf{X}}^o$ is **positive definite.** For $\underline{\mathbf{X}}^o$ to be a maximum, the Hessian matrix evaluated at $\underline{\mathbf{X}}^o$ should be **negative definite.**

A matrix is positive definite if all of its **principal minor determinants** λ_k ($k = 1, \ldots, n$) are positive.

The *k*th principal minor determinant of $\underline{\mathbf{H}}$ is defined as:

$$\lambda_k = \begin{vmatrix} h_{11} & h_{12} & \cdots & h_{1k} \\ h_{21} & h_{22} & \cdots & h_{2k} \\ . & . & & . \\ . & . & & . \\ h_{k1} & h_{k2} & \cdots & h_{kk} \end{vmatrix}$$

A matrix is negative definite if its *k*th principal minor determinant has the sign $(-1)^k$ for $k = 1, 2, \ldots, n$.

APPENDIX 13B: HYDRAULIC ANALYSIS OF PIPE NETWORKS

This appendix explains how flows and pressure heads are determined in a pipe network. It is based on Dandy et al (1993).

Consider a general pipe network with *NP* pipes, *NJ* junction nodes (excluding fixed grade nodes or reservoirs) and *NF* fixed grade reservoirs. The following are assumed to be known:

1. the demands at all nodes ($q_1, q_2, \ldots, q_{NJ}$);
2. the diameters of all pipes ($D_1, D_2, \ldots, D_{NP}$); and
3. the total head at one or more nodes (e.g. at fixed grade reservoirs).

The total head at each node is the sum of the elevation head and the pressure head at the node. It is assumed that velocity heads, minor losses and losses at junctions are negligible. The analysis described in this appendix will be a steady state analysis assuming that all demands are constant.

The basic unknowns are the discharge in each pipe, ($Q_1, Q_2, \ldots, Q_{NP}$) and the total head at each node ($H_1, H_2, \ldots, H_{NJ}$). The pressure head above ground level at each node can be determined by subtracting the elevation of ground level at the node from the total head at that node.

The basic equations are:

(i) the continuity equations:

$$\sum_{j=1}^{NPJ} Q_j + q_i = 0 \quad \text{for all nodes } i = 1, \ldots, NJ \qquad (13.87)$$

where the summation is made across the *NPJ* pipes connected to node *i* (flow away from the node is taken as positive).

(ii) the head loss equations:

for pipe j between nodes i and k, the head loss is related to the discharge in the pipe as well as its diameter. One head loss equation commonly used in water supply engineering is the Hazen-Williams equation given here in SI units (Streeter and Wylie, 1979).

$$H_i - H_k = h_j = \frac{10.675 L_j Q_j \left|Q_j\right|^{0.852}}{C_j^{1.852} D_j^{4.8704}} \quad \text{for all pipes } j = 1, \ldots, NP \qquad (13.88)$$

where, h_j = head loss in pipe j

L_j = length of pipe j

C_j = Hazen-Williams coefficient for pipe j

Equation sets (13.87) and (13.88) provide *(NJ + NP)* equations in the same number of unknowns (pipe discharges and nodal heads).

The number of equations can be reduced by summing the head losses around loops and equating the sum to zero or by transposing Equation sets (13.88) to get expressions for Q_j $(j = 1,2, \ldots NP)$ in terms of H_i and H_k and then substituting into Equation sets (13.87) to eliminate the pipe discharges (Ormsbee and Wood, 1986).

All solution techniques involve the solution of non-linear equations and may require the use of techniques such as the Newton-Raphson method (Shamir and Howard, 1968).

CHAPTER FOURTEEN

Epilogue

The main premise of this book is that planning and design are the key activities which, together with management, allow any engineering project to be taken from the initial concept stage through to successful implementation. Although modern engineering encompasses an incredibly broad range of specialised fields, from traditional civilian and military engineering, through relatively new fields like biomedical engineering and mechatronics, to newly emerging areas such as nanotechnology, each engineering project relies for its success on the application of the basic processes of planning, design and management. Our prime aim here has been to show how the processes of planning and design are carried out.

In the early chapters of the book we have emphasised that planning and design are very similar in nature: they are both, in essence, problem-solving activities. Engineering problems are characterised on the one hand by their complexity and, on the other hand, by a characteristic that we have called open-endedness. Open ended problems do not have a single "correct" solution, but rather a whole range of alternative possible solutions. A number of basic systems concepts were introduced in Chapter 2 to deal with complexity. In Chapter 3 a methodology for solving complex, open-ended problems was presented, which provides an effective basis for dealing with the processes of planning and design.

In order to undertake engineering planning and design, it is necessary to be able to think creatively and laterally, as well as rationally and logically. These contrasting ways of thinking are sometimes characterised as left-brain and right-brain thinking. Because of early education, training and background, many engineering students are potentially well suited to tasks that involve analytic, rational thinking. It can therefore come as a surprise and a hurdle to young engineers to have to be able to move away from the left-brain thinking processes they are comfortable with and engage in creative and lateral thinking. Chapter 4 discusses various aspects of creative thinking, as well as some techniques that can be used to develop right-brain activities. Following on from the information presented in Chapter 3 on engineering methodology, Chapter 5 looks at a range of practical techniques that are used in the planning of engineering projects.

Engineers need to develop a range of human skills if they are to succeed in planning and design work in particular, and engineering work generally. Of particular importance are the skills of personal and personnel management, and the ability to work with others in teams. These skills are discussed in Chapter 6.

Engineers must also be able to communicate effectively and unambiguously, not only with their colleagues but also with their clients and with the community. At present we are seeing the strong participation of clients and other interested parties (the "stakeholders") in the decision making processes of engineering work.

This places added demands on the communication skills of engineers and in particular on their ability to explain technical problems to non-experts. The topic of communication is dealt with in Chapter 7.

In addition to having a range of human skills, engineers must be reasonably familiar with various other areas of knowledge, ranging from economics, ethics, law and sociology through to environmental matters including sustainability. We have provided an introduction to the most important of these areas in Chapters 8, 9 and 10 of this book. Sooner or later, of course, engineers have to face special problems where they lack the necessary background knowledge and expertise. In such situations they have to be able to acquire and absorb new information and knowledge rapidly, and deal with experts in the appropriate fields.

In all engineering work there are inherent risks that that have to be evaluated and managed. In planning and design the goal is to create systems and processes that are reliable. Reliability and safety are prime objectives in modern engineering work. Chapter 11 discusses various aspects of the over-riding concepts of engineering risk and reliability.

As a result of the open-ended nature of engineering work, decision making is an important aspect of planning and design. In each phase of an engineering project a large number of decisions (or choices) have to be made. We have provided an introduction to mathematical decision making in Chapter 12, not in the expectation that the idealised decision models presented there will always be directly applicable to real situations, but rather with the intention of providing a conceptual framework for dealing with real decision problems.

Chapter 13 provides an introduction to the wide range of optimisation techniques that are available to engineers, including those that use traditional mathematical techniques as well as the modern search approaches that can be used in association with non-linear and discontinuous modelling and computer simulation. Optimisation techniques frequently find application in the detailed phases of planning and design.

Various other techniques are used in the detailed phase of engineering design that we have not mentioned in this book. The reason for this is that design techniques tend to be very field-specific. In structural engineering design, for example, much of the analytical work is based on solid mechanics and materials technology. These disciplines provide the theoretical basis for the design calculations that have to be made to ensure that the strength and serviceability of specific structural systems are adequate. In water engineering, however, fluid mechanics and hydrology play the essential roles. Because of such field-specific reliance on specialised knowledge and theory, we have only attempted to discuss what we might call the "shell" within which engineering design is conducted. Specific methods of analysis and design form a significant part of the curriculum of the later years of engineering undergraduate courses. Our aim here has been to provide a context for further studies in the specialised design areas.

Planning and design, in broad terms, are by no means unique to engineering. Planning activities form the backbone of professional work in fields as diverse as architecture, financial planning and agricultural economics. For example, architects are deeply engaged in both planning and design work when they create, together with engineers and building contractors, shopping centres and retirement villages. Looking more broadly at society, we find that many professionals and experts, apart from engineers, engage in design. We see this in the diversity of design products available to us, ranging from jewellery, fine art and furniture, through to zoos, fun parks, industrial equipment and consumer products.

What is unique to engineering planning and design is not the techniques and procedures used, but rather the nature of the problems that are dealt with. In Chapter 1 we saw how engineering work is aimed at maintaining, improving and extending the physical infrastructure that society relies on in order to function effectively. We also saw that there is a close two-way relationship between society and its supporting engineering infrastructure: on the one hand, engineering work is dictated by the needs of society; on the other hand, the very nature of society is shaped by the solutions provided by engineers to infrastructure problems. The inter-relationship between engineering work and human society is as strong today as it was in previous centuries, and indeed in the earliest civilisations.

In early civilisations, engineers had the task of developing the basic infrastructure that was needed by the society of the time: buildings and fortifications to house and protect the population; water for drinking and washing and for growing crops; roads for transport and communication. Tools and machines were engineered to allow routine work to be undertaken more efficiently. The engineered infrastructure of the early civilisations, which was developed progressively over many generations, is impressive even by modern standards. Whenever new archaeological work is commenced, the remnants of the buildings, walls and irrigation channels usually provide the first indication (and, often, striking evidence) of the level of sophistication of the civilisation.

The basic needs of the society we live in today are not substantially different from those of earlier civilisations: shelter, protection and defence, clean water, systems for handling waste and sewerage, and efficient systems for transport and for communications. In addition, however, we rely today on enormous amounts of energy to heat and cool our workplaces and homes and to run the numerous labour saving devices we use, such as computers, printers, data storage devices and office equipment at work and washing machines, cleaning devices and microwave ovens at home. In effect, a broad array of cheap manufactured goods has replaced the animals, slaves and servants that were exploited by the affluent classes in previous societies to deal with the routine work of everyday life. Our engineered devices, together with the cheap energy they rely on, today allow almost everybody to be affluent in our society. However, this inevitably raises basic questions concerning equity and sustainability when we cease to think locally and look at the problems that engineers will have to face world wide in the immediate future.

While the infrastructure needs of society do not alter substantially over time, the means by which they are satisfied change, usually from one generation to the next. One reason for this has been our steadily increasing exploitation of energy sources and our increasing use of energy. When the ancient Egyptians constructed their buildings, temples and pyramids, the energy they used was all stored in human muscle. It was derived from the sun and the soil, via the bread, beer and meat that the workers lived on. Today, massive amounts of energy are used on our construction sites but it does not come from beer, bread and meat. It is derived largely from petroleum and from coal-generated electricity.

New solutions to engineering problems often come from scientific and mathematical knowledge via progressive improvements in technology. Modern engineering work is also reliant on the new knowledge that is obtained from ongoing engineering and scientific research. While there is a strong continuing need for empirical investigation in engineering work, the emphasis over the past century has tended to be away from the use of empirical *ad hoc* solutions, towards the use of fundamental principles and general concepts. The ever-widening scope of engineering analysis is in large part a reflection of ongoing research. This is the critical factor that helps engineers in each generation to bring new, better and more efficient solutions to the problems facing their society.

Consider, for example, the solutions that have been developed to satisfy the need for distance communication. The earliest means of rapid communication over long distances were visual: bonfires were lit successively on adjacent hills and high vantage points in order to send simple but important messages from one part of a country to another, for example to announce important matters of state. Complex and detailed information had to be transported physically and it could take months, for example, for important government documents to be distributed by road throughout the Roman Empire. Although communication was slow, the administration of the business of empire seems to have been remarkably efficient. However, it was not until the discoveries and inventions of people such as Hertz, Marconi, Morse and Edison that the rapid transmission of detailed and complex information over large distances became possible. Today, after generations of research and development, we can speak directly with others located almost anywhere on earth and even in space. It is also possible to transmit enormous amounts of information, in the form of complex images, documents and data files, almost anywhere and almost instantaneously using a variety of channels of telecommunication.

We have emphasised in this book that engineering is a profession in which both lateral thinking and analytic thinking are needed to develop solutions to our problems. Creativity, organisational ability, effective oral communication and an ability to write clearly and concisely are as important as a good knowledge of science and mathematics for success in the profession of engineering.

As already mentioned, this is not appreciated by many engineering students when they commence their studies. The choice of engineering as a career is usually made in the final year of high school and seems as often as not to be based on a preference for analytical school subjects like science and mathematics. Furthermore, when they commence their studies, engineering students often have little idea of the nature of engineering, or of the kind of work that they will be required to undertake in their chosen profession. The belated realisation that engineering requires creativity, organisational ability, effective oral communication and an ability to write clearly and concisely can come as a surprise.

One of the underlying goals in preparing this book has been to explain to students commencing their engineering studies the kind of work they will later engage in, and the skills that they will have to acquire, in order to pursue a successful career in engineering.

At the end of a text book on the nature of engineering and the way engineers undertake their work, it is natural for both reader and writer to reflect on the future of engineering. While we inevitably face change in an uncertain future that will be influenced by unpredictable world events, it is comforting to realise that there are a few things in engineering that are unlikely to change radically, and there are other things that will change in a more or less predictable fashion.

The earth we inhabit is not in stasis: it is not a system that is, or ever was, or ever will be, in a state of static and stable equilibrium. Rather, it is continually evolving and changing. Generally speaking though, change to date has tended to occur relatively slowly, at least when judged in terms of the human life span. This can give the impression that we live in an unchanging world which can be moulded to our wishes. Unfortunately, there is now ample evidence that our planet is currently undergoing accelerating changes that are, in part, caused by our own actions. This leaves no room for complacency; it leads rather to feelings of disquiet and unease.

Nevertheless, we can expect to see engineers in fifty or a hundred years time still providing for basic human needs, such as fresh water and shelter in a sustainable environment, together with effective systems for communication and transport, but within a somewhat more difficult and less certain environment. It may well be that some engineers will be trying to provide for such basic societal needs in remote locations, for example on the moon.

Some imminent changes can be predicted from the current state of society and of the engineering profession. For example, resources that are now globally scarce, such as cheap fresh water and cheap energy, are inevitably going to become much scarcer as the demand for them increases globally. Other resources that are relatively abundant at present, such as clean air, are likely to become less so. The need for alternative sources of fresh water and energy in particular is going to intensify markedly in the coming decades. Much engineering effort and ingenuity will be required to solve the problems of water and energy. On the other hand, some resources on earth are going to remain plentiful and cheap, such as sunshine, sand, rock, and sea water.

Of course, projections into the future become irrelevant in the event of global catastrophes: we need only think of the effects of a large meteor strike or global nuclear war. Such extreme events could easily bring society back to another, more primitive, starting point. In such a situation, engineers will have to begin again to create the infrastructure that humans need for survival and comfort.

Although catastrophic events do occur and are impossible to predict, the process of scenario planning, briefly alluded to in Chapter 3, can be used to examine and, to some extent, plan for them. It is not the place here, at the conclusion of this book, to go into the details of yet another planning technique. Nevertheless, important topics for scenario planning studies come readily to mind. For example, climate change is going to impact on society and on the nature of engineering work in different ways in different parts of the world. The future of the nuclear energy industry, nuclear proliferation, and the possible employment of nuclear devices in armed conflict (whether by nations or small dissident groups) are questions of great concern that will have a direct and specific bearing on our future and hence on the nature of engineering work. The devastating effects of nuclear accident, biological warfare, biological terrorism, as well as natural catastrophes, all deserve scenario planning studies in order to look intelligently at their implications.

Taking a minimalist but essentially optimistic view, we can be reasonably confident that human society will survive the coming decades and century, and will continue to demand a physical, engineered infrastructure that makes life not only possible but reasonably enjoyable. On the other hand, we face apparently intractable political and social problems, including global inequity, that are not as susceptible to solution as the associated engineering problems. We can only hope that human society will continue to progress, despite ups and downs, with an ever broadening demand for improvements and extensions of the engineered infrastructure, on earth and possibly in neighbouring parts of the solar system.

On an optimistic note, we can look forward with some confidence to the benefits that will come from developments in new and expanding engineering areas such as medical and biomedical engineering and nanotechnology. We might even expect to see some success in the search for clean, energy-efficient and water-efficient engineering innovations with direct applications in industry and manufacturing, in construction and agriculture, and in everyday life. Even so, it is inevitable that we will see curbs being placed on the use of energy and water and other scarce resources, rather than attempts to meet ever increasing demands. Reduced demand must occur, together with progressive improvements in efficiencies, if we are to achieve a world-wide balance between the demand and supply of energy, water and other scarce resources, and hence provide a sustainable basis for the work of engineers. Such success will require a great deal of innovative engineering planning and design.

Index

A library at your fingertips!

eBooks are electronic versions of printed books. You can store them on your PC/laptop or browse them online.

They have advantages for anyone needing rapid access to a wide variety of published, copyright information.

eBooks can help your research by enabling you to bookmark chapters, annotate text and use instant searches to find specific words or phrases. Several eBook files would fit on even a small laptop or PDA.

NEW: Save money by eSubscribing: cheap, online access to any eBook for as long as you need it.

Annual subscription packages

We now offer special low-cost bulk subscriptions to packages of eBooks in certain subject areas. These are available to libraries or to individuals.

For more information please contact webmaster.ebooks@tandf.co.uk

We're continually developing the eBook concept, so keep up to date by visiting the website.

www.eBookstore.tandf.co.uk

RECEIVED
JAN 14 2008
ENGINEERING LIBRARY